T0231139

*Douglas D. Buhler, PhD*
*Editor*

# Expanding the Context of Weed Management

*Expanding the Context of Weed Management* has been co-published simultaneously as *Journal of Crop Production*, Volume 2, Number 1 (#3) 1999.

*Pre-publication*
*REVIEWS,*
*COMMENTARIES,*
*EVALUATIONS . . .*

"*Expanding the Context of Weed Management* will recast future weed science discussions and future weed science education in language and examples that are solidly rooted in weed biology and weed ecology. Future weed managers will see this as a major collection of essays on ecologically based weed management principles. Buhler has helped interject significant new science and rational into the debate of how global weed management will change in the 21st century."

**Philip Westra**
*Professor of Weed Science*
*112 Weed Lab*
*Colorado State University*
*Ft. Collins, CO 80523*

*More pre-publication*
*REVIEWS, COMMENTARIES, EVALUATIONS . . .*

"Weed science and weed management are evolving faster than any weed. This book brings together some of the best thinkers in the discipline and addresses the future of weed research and improved management systems. The editor and the authors of most of the 12 chapters represent a new generation of weed scientists and they bring new insights and approaches for anyone interested in the future of weed management. The topics covered can be broadly categorized as weed biology, dynamics, management and integration. The complexities of interactions in the biological, managerial and economic arenas are well illustrated and most authors give suggestions on how to advance the discipline beyond our present level of understanding of these complexities.

. . . this book is a most valuable resource for those who seek to chart new courses for the discipline of weed science. The final chapter is one of the best as it brings together many of the concepts presented in earlier chapters. It is an excellently developed framework that captures the multifaceted impacts of and approaches to managing weeds from the field, farm, landscape, regional and global perspectives.

In addition to weed science faculty, staff and students at universities, this book will broaden the perspectives and possibilities of weed management for other agricultural professionals in academia and the private sector, extension educators, and crop consultants as well as producers seeking to move from weed control to weed and crop management. All readers will appreciate the fresh approach to one of agriculture's oldest problems–weeds."

**Jerry Doll, PhD**
*Professor of Weed Science*
*University of Wisconsin*

# Expanding the Context of Weed Management

*Expanding the Context of Weed Management* has been co-published simultaneously as *Journal of Crop Production*, Volume 2, Number 1 (#3) 1999.

# Expanding the Context of Weed Management

Douglas D. Buhler, PhD
Editor

*Expanding the Context of Weed Management* has been co-published simultaneously as *Journal of Crop Production*, Volume 2, Number 1 (#3) 1999.

CRC Press
Taylor & Francis Group
Boca Raton   London   New York

CRC Press is an imprint of the
Taylor & Francis Group, an informa business

*Expanding the Context of Weed Management* has been co-published simultaneously as *Journal of Crop Production*, Volume 2, Number 1 (#3) 1999.

© 1999 by The Haworth Press, Inc. All rights reserved. No part of this work may be reproduced or utilized in any form or by any means, electronic or mechanical, including photocopying, microfilm and recording, or by any information storage and retrieval system, without permission in writing from the publisher. Printed in the United States of America.

Reprinted 2009 by CRC Press

The development, preparation, and publication of this work has been undertaken with great care. However, the publisher, employees, editors, and agents of The Haworth Press and all imprints of The Haworth Press, Inc., including The Haworth Medical Press® and Pharmaceutical Products Press®, are not responsible for any errors contained herein or for consequences that may ensue from use of materials or information contained in this work. Opinions expressed by the author(s) are not necessarily those of The Haworth Press, Inc.

Cover design by Thomas J. Mayshock Jr.

**Library of Congress Cataloging-in-Publication Data**

Expanding the context of weed management / Douglas D. Buhler, editor.
    p. cm.
    Co-published simultaneously as Journal of crop production, volume 2, number 1 (#3) 1999.
    Includes bibliographical references and index.
    ISBN 1-56022-063-5 (alk. paper)
    1. Weeds–Control. I. Buhler, D. D. (Douglas D.) II. Journal of crop production.
SB611.E96 1999
632'.5–dc21
                    98-53514
                      CIP

# INDEXING & ABSTRACTING

Contributions to this publication are selectively indexed or abstracted in print, electronic, online, or CD-ROM version(s) of the reference tools and information services listed below. This list is current as of the copyright date of this publication. See the end of this section for additional notes.

- *AGRICOLA Database*

- *Chemical Abstracts*

- *CNPIEC Reference Guide: Chinese National Directory of Foreign Periodicals*

- *Crop Physiology Abstracts*

- *Derwent Crop Production File*

- *Environment Abstracts*

- *Field Crop Abstracts*

- *Foods Adlibra*

- *Food Science and Technology Abstracts (FSTA)*

- *Grasslands & Forage Abstracts*

- *PASCAL International Bibliography T205*

- *Plant Breeding Abstracts*

- *Referativnyi Zhurnal (Abstracts Journal of the All-Russian Institute of Scientific and Technical Information)*

- *Seed Abstracts*

- *Soils & Fertilizers Abstracts*

- *Weed Abstracts*

(continued)

*Special Bibliographic Notes related to special journal issues
(separates) and indexing/abstracting:*

- indexing/abstracting services in this list will also cover material in any "separate" that is co-published simultaneously with Haworth's special thematic journal issue or DocuSerial. Indexing/abstracting usually covers material at the article/chapter level.

- monographic co-editions are intended for either non-subscribers or libraries which intend to purchase a second copy for their circulating collections.

- monographic co-editions are reported to all jobbers/wholesalers/approval plans. The source journal is listed as the "series" to assist the prevention of duplicate purchasing in the same manner utilized for books-in-series.

- to facilitate user/access services all indexing/abstracting services are encouraged to utilize the co-indexing entry note indicated at the bottom of the first page of each article/chapter/contribution.

- this is intended to assist a library user of any reference tool (whether print, electronic, online, or CD-ROM) to locate the monographic version if the library has purchased this version but not a subscription to the source journal.

- individual articles/chapters in any Haworth publication are also available through the Haworth Document Delivery Service (HDDS).

# Expanding the Context
# of Weed Management

## CONTENTS

Expanding the Context of Weed Management                 1
    *Douglas D. Buhler*

Weed Thresholds: Theory and Applicability               9
    *C. J. Swanton*
    *S. Weaver*
    *P. Cowan*
    *R. Van Acker*
    *W. Deen*
    *A. Shrestha*

Ecological Implications of Using Thresholds
    for Weed Management                                31
    *Robert F. Norris*

Increasing Crop Competitiveness to Weeds
    Through Crop Breeding                              59
    *Todd A. Pester*
    *Orvin C. Burnside*
    *James H. Orf*

Genetic Approach to the Development of Cover Crops
    for Weed Management                                77
    *Michael E. Foley*

Improving Soil Quality: Implications
    for Weed Management                                95
    *Eric R. Gallandt*
    *Matt Liebman*
    *David R. Huggins*

Soil Microorganisms for Weed Management 123
    *A. C. Kennedy*

Soil Weed Seed Banks and Weed Management 139
    *Jack Dekker*

A Risk Management Perspective on Integrated
    Weed Management 167
    *J. L. Gunsolus*
    *D. D. Buhler*

Maximizing Efficacy and Economics of Mechanical Weed
    Control in Row Crops Through Forecasts
    of Weed Emergence 189
    *Caleb Oriade*
    *Frank Forcella*

Multi-Year Evaluation of Model-Based Weed Control
    Under Variable Crop and Tillage Conditions 207
    *Melinda L. Hoffman*
    *Douglas D. Buhler*
    *Micheal D. K. Owen*

Knowledge-Based Decision Support Strategies: Linking
    Spatial and Temporal Components Within Site-Specific
    Weed Management 225
    *Gregg A. Johnson*
    *David R. Huggins*

Development of Weed IPM: Levels of Integration
    for Weed Management 239
    *J. Cardina*
    *T. M. Webster*
    *C. P. Herms*
    *E. E. Regnier*

Index 269

# ABOUT THE EDITOR

**Douglas D. Buhler, PhD,** is Research Agronomist at the USDA-Agricultural Research Service National Soil Tilth Laboratory in Ames, Iowa and Professor (Collaborator) of Agronomy at Iowa State University. His research centers on the effects of crop production practices on weed biology and ecology and the development of alternative weed management systems. He is author or coauthor of more than 250 scientific papers and abstracts, has served as an associate editor for the journals *Weed Science* and *Weed Technology,* and is a consulting editor for the *Journal of Crop Production* (The Haworth Press, Inc.). Dr. Buhler is a Fellow of the American Society of Agronomy and has received the Distinguished Young Scientist Award from the North Central Weed Science Society, the Outstanding Young Scientist and Outstanding Paper Awards from the Weed Science Society of America. He has also received the Midwest Area Early Career Scientist of the Year and Edminster Awards from USDA-ARS.

# Expanding the Context
# of Weed Management

Douglas D. Buhler

**SUMMARY.** Concerns about current weed control practices have increased the consideration of new weed management strategies. In recent times, weed control practices for major crops have been influenced greatly by the availability of selective herbicides. Herbicides are critical tools, but weed science must integrate more components to create weed management systems. Changes in weed management can be attained within the framework of existing cropping systems. However, for the longer term, new methods and approaches to weed management are needed. Weed scientists need to play a central role in the development of new cropping systems to make weed management an integral component of the system. This volume contains a series of review articles and original research that presents innovative approaches to weeds and weed management. It is our hope that these papers will stimulate discussion on a broader view of weeds and weed management. *[Article copies available for a fee from The Haworth Document Delivery Service: 1-800-342-9678. E-mail address: getinfo@haworthpressinc.com]*

**KEYWORDS.** Crop production systems, sustainable agriculture, weed biology, weed control

---

Douglas D. Buhler, Research Agronomist, United States Department of Agriculture–Agricultural Research Service, National Soil Tilth Laboratory, 2150 Pammel Drive, Ames, IA 50011 USA (E-mail: buhler@nstl.gov).

The author thanks those who have written previously on the future of weed science. Their views have been invaluable in the development of this paper.

[Haworth co-indexing entry note]: "Expanding the Context of Weed Management." Buhler, Douglas D. Co-published simultaneously in *Journal of Crop Production* (Food Products Press, an imprint of The Haworth Press, Inc.) Vol. 2, No. 1 (#3), 1999, pp. 1-7; and: *Expanding the Context of Weed Management* (ed: Douglas D. Buhler) Food Products Press, an imprint of The Haworth Press, Inc., 1999, pp. 1-7. Single or multiple copies of this article are available for a fee from The Haworth Document Delivery Service [1-800-342-9678, 9:00 a.m. - 5:00 p.m. (EST). E-mail address: getinfo@haworthpressinc.com].

© 1999 by The Haworth Press, Inc. All rights reserved.

*1*

## INTRODUCTION

Weeds have a significant economic impact on agricultural production. In the United States alone, it was estimated weeds and weed control have an annual economic impact of more than $15 billion (Bridges, 1994). Concerns over the economic costs and environmental impacts of current practices have led many weed scientists and crop producers to seek alternative strategies for weed control (Gressel, 1992; Wyse, 1992). Herbicides are important tools for modern agriculture but have become too dominant in many production systems. A goal for weed science should be to develop systems that give producers more options. New options will give producers more flexibility and help preserve the effectiveness of herbicides.

Herbicides have been the major focus of weed control research for the past several decades (Burnside, 1993). Herbicides have been valuable tools and have provided benefits to the farm and urban communities. However, weed management should be viewed as an integrated science. Increased attention must be given to biological, cultural, mechanical, and preventive tools and techniques to manage weeds. Refocusing on weed management as opposed immediate control will require a greater understanding of biological systems than we currently possess (Holt, 1994; Navas, 1991).

Weed management implies a shift away from reliance on control of existing weed problems and places greater emphasis on prevention of propagule production, reduction of weed emergence in a crop, and minimizing weed competition with the crop (Buhler, 1996; Zimdahl, 1991). Weed management emphasizes integration of techniques to anticipate and manage problems rather than reacting to them after they are present. Weed management does not eliminate the need for control. The goal is to optimize crop production and grower profit by integrating preventive techniques, scientific knowledge, and management skills. While additional knowledge is needed in all areas of weed management, the most important task of weed science is to increase knowledge of weed biology and ecology, creating a better understanding of weediness. This knowledge will lead to the use of appropriate management techniques rather than prophylactic approaches that produce short-term results, but may create or worsen long-term problems.

It was proposed that weed science research can be separated into two major categories: weed control science and technology, and weed science principles (Wyse, 1992). Research on weed control science and technology involves herbicides, tillage, biological control, and other methods to remove unwanted vegetation from a disturbed system (Table 1). Research on weed science principles focuses on weed biology and ecology, examining why weeds are present and their impacts on the production system.

Weed science principles provide an understanding for the basis of weed problems. Integrated pest management is also based on knowledge of pest

TABLE 1. Examples of weed science research categorized as: (1) weed control science and technology, and (2) weed science principles (modified from Wyse, 1992).

| Weed control science and technology | Weed science principles |
| --- | --- |
| * Herbicide development | * Propagule dormancy mechanisms |
| * Biological control | * Seed production and longevity |
| * Cover and smother crops | * Population genetics |
| * Tillage | * Emergence dynamics |
| * Herbicide selection decision aids | * Weed/crop interference |
| * Competitive crops | * Population shifts |
| | * Modeling weed/crop systems |

biology and ecology (Smith, 1978). Weed science has lagged behind other pest management disciplines in developing integrated management systems, largely due to our poor understanding of weed science principles. In the short-term, improved understanding of weed science principles will not revolutionize weed management. However, the knowledge gained will provide the foundation for development of new strategies and more efficient techniques, resulting in more reliable weed management systems that are cost effective and pose less of a threat to the environment.

## ALTERNATIVE WEED MANAGEMENT

Alternative weed management, like most contemporary issues in agriculture, is not easily defined and does not have a universally accepted definition. Because of this lack of clear definition, formulating the goals for weed management research is difficult. As agricultural issues are debated by policy makers, researchers, and producers, the role of weed management in a broad range of issues should be considered. Issues for weed scientists might include:

1. Should environmental issues be the primary focus of alternative weed management?
2. Is the goal to eliminate herbicide use or to provide new options and management flexibility?

3. Will weed science be reactive to government regulation or provide leadership to government agencies?
4. How will economics and risks associated with alternative systems be addressed?
5. How do we balance the desire to reduce tillage and preserve surface residues while reducing herbicide use?
6. What will be the role of the agricultural chemical industry in the development and application of alternative weed management systems?
7. Should weed management research encompass global and societal issues or be focused on in-field management practices?

## APPROACHES TO DEVELOPING ALTERNATIVE WEED MANAGEMENT SYSTEMS

Environmental regulations, reduced tillage systems, increasing farm size, economic pressure, and product cancellations are restricting the weed control options available to crop producers. In addition, weed populations continue to adapt to production practices through herbicide resistance and population shifts. Limited crop choices are reducing crop rotation and intensifying the selection pressure on weed communities. Many current cropping systems are highly simplified, allowing the best adapted weed species to proliferate.

Fundamental changes in weed management will not occur instantaneously. We need both short- and long-term approaches to developing new weed management systems. Short-term approaches will apply existing knowledge, provide immediate benefits, and provide transition to new systems (Table 2). Long-term approaches require basic changes in crop management systems and provide unique methods to manage weeds. Long-term alternatives will involve major investments in research and development before application.

Short-term approaches to changing weed management centers around reducing herbicide use or improving weed control in current cropping systems. Many methods exist to 'fine-tune' weed control systems (Table 2). However, the ability to make these changes varies among growers. For example, growers with good management skills and fields with low weed pressure can reduce herbicide use and incur only a slight risk of weed control failure. For others, the savings achieved by reducing herbicide use may be small compared with the risk of control failures, crop yield losses, and the additional time needed for weed control. Many of these issues are addressed later in this volume.

Long-term approaches to changing weed management (Table 2) are more difficult to define than short-term herbicide use reduction or improving control of specific weed species. Reconsidering all phases of the science and technology is important for weed scientists if they wish to explore alternative

TABLE 2. Potential approaches for short- and long-term development of alternative weed management systems.

| Short-term strategies | Long-term strategies |
|---|---|
| Application of existing knowledge | Weed biology and ecology |
|   - scouting |   - population biology |
|   - increased management |   - weed/crop interactions |
| Refinements of current systems |   - emergence and growth predictors |
|   - integrated management | Site-specific management |
|   - decision aids |   - herbicide application |
|   - appropriate use of tillage |   - soil and crop management |
|   - application technology | Redesign cropping systems |
|   - reduced rates of herbicides |   - increase rotation and diversity |
| |   - improve resource utilization |
| |   - new tillage systems |
| | New control methods |
| |   - biological control |
| |   - allelopathy |
| |   - managed competition |
| |   - competitive cultivars |

weed management systems constructively. Currently, weed control practices are justified by the knowledge that weed populations are very difficult to prevent once established, uncontrolled plants contribute to future infestations, and cause direct economic losses when left uncontrolled. However, we seldom examine the causes of the perpetual presence of weeds. Consideration of the continual adaptation of weed communities is an essential element for the development of weed management systems.

## CONCLUSIONS

The basis of weed science has been control technology rather than understanding weedy species and their roles in the agroecosystem (Wyse, 1992).

Weeds have been present since the beginnings of civilization and are not likely to disappear in the near future. All forms of disturbance result in survival and selection of the best adapted weeds (Holt, 1994). Any cropping system that exerts a continuous, strong selection pressure will cause a build-up of the best adapted weed species and biotypes. Development of improved cropping systems will require an approach that concentrates on the processes and patterns that link fundamental scientific, economic, and sociological disciplines to agricultural systems.

Agriculture and the public view of agriculture are changing. Weed science must also change by taking a broader view of weeds as part of the crop ecosystem and by addressing the long-term questions surrounding production practices and technology. Agricultural systems are composed of interacting production, environmental, biological, economic, and social components. These interactions require the study of not only the parts, but also the entire system (Holling, 1978). Long-term improvements in weed management and agricultural production systems, will require a convergence of traditional agriculture, ecological theory, economics, and sociology (Levins, 1986; Radosevich and Ghersa, 1992).

Hypothesizing about how new weed management systems can be developed is easy, delivering a product is another matter. There are many complicating factors, but we must view these as challenges rather than excuses. The material presented in this volume is an attempt to address some of these challenges.

## REFERENCES

Bridges, D.C. (1994). Impacts of weeds on human endeavors. *Weed Technology* 8:392-395.

Buhler, D.D. (1996). Development of alternative weed management strategies. *Journal of Production Agriculture* 9:501-505.

Burnside, O.C. (1993). Weed science–The step child. *Weed Technology* 7:515-518.

Gressel, J. (1992). Addressing real weed science needs with innovations. *Weed Technology* 6:509-525.

Holling, C.S. (1978). The nature and behavior of ecological systems. In *Adaptive Environmental Assessment and Management.* New York, NY: John Wiley and Sons, pp. 25-37.

Holt, J.S. (1994). Impact of weed control on weeds: New problems and research needs. *Weed Technology* 8:400-402.

Levins, R. (1986). Perspectives in integrated pest management: From an industrial to ecological model of pest management. In *Ecological Theory and Pest Management.* New York, NY: John Wiley and Sons, pp. 1-18.

Navas, M.L. (1991). Using plant population biology in weed research: A strategy to improve weed management. *Weed Research* 31:171-179.

Radosevich, S.R. and C.M. Ghersa. (1992). Weeds, crops, and herbicides: A modern-day "neckriddle." *Weed Technology* 6:788-795.

Smith, R.F. (1978). History and complexity of integrated pest management. In *Pest Control Strategies,* ed. E.H. Smith and D. Pimentel. New York, NY: Academic Press, pp. 41-53.

Wyse, D.L. (1992). Future of weed science research. *Weed Technology* 6:162-165.

Zimdahl, R.L. (1991). *Weed Science–A Plea for Thought.* Washington, DC: United States Department of Agriculture, Cooperative State Research Service.

RECEIVED: 05/27/97
ACCEPTED: 04/10/98

# Weed Thresholds:
# Theory and Applicability

C. J. Swanton
S. Weaver
P. Cowan
R. Van Acker
W. Deen
A. Shreshta

**SUMMARY.** Weed thresholds are an integral component of an integrated weed management system (IWM). In this paper we review the

C. J. Swanton, Professor, Department of Crop Science, University of Guelph, Guelph, Ontario, Canada N1G 2W1.

S. Weaver, Research Scientist, Agricultural Canadian Research Station, Harrow, Ontario, Canada N0R 1G0.

P. Cowan, Graduate Research Assistant, Department of Crop Science, University of Guelph, Guelph, Ontario, Canada N1G 2W1.

R. Van Acker, Assistant Professor, Department of Plant Science, University of Manitoba, Manitoba, Canada R3T 2N2.

W. Deen and A. Shreshta, Graduate Research Assistant and Research Associate, respectively, Department of Crop Science, University of Guelph, Guelph, Ontario, Canada N1G 2W1.

Address correspondence to: C. J. Swanton, Department of Crop Science, University of Guelph, Guelph, Ontario, Canada N1G 2W1 (E-mail: cswanton@crop.uoguelph.ca).

The authors thank Dr. Douglas D. Buhler and the anonymous reviewers for their excellent comments and suggestions on this manuscript. They also thank the Ontario Corn Producers' Association, Ontario Soybean Growers' Marketing Board, Ontario Bean Producers' Marketing Board, National Sciences Engineering Research Council of Canada, Ontario Ministry of Agriculture, Food and Rural Affairs, Food Systems 2002, and all the members of the Weeds Lab, Department of Crop Science, University of Guelph, for their input and support.

[Haworth co-indexing entry note]: "Weed Thresholds: Theory and Applicability." Swanton, C. J. et al. Co-published simultaneously in *Journal of Crop Production* (Food Products Press, an imprint of The Haworth Press, Inc.) Vol. 2, No. 1 (#3), 1999, pp. 9-29; and: *Expanding the Context of Weed Management* (ed: Douglas D. Buhler) Food Products Press, an imprint of The Haworth Press, Inc., 1999, pp. 9-29. Single or multiple copies of this article are available for a fee from The Haworth Document Delivery Service [1-800-342-9678, 9:00 a.m. - 5:00 p.m. (EST). E-mail address: getinfo@haworthpressinc.com].

© 1999 by The Haworth Press, Inc. All rights reserved.

literature on weed thresholds and examine the theory and applicability of thresholds within the context of a systems approach to IWM. The development of empirical models describing single and multi-species thresholds are reviewed and discussed in terms of the magnitude of weed threshold values in various crops and the importance and limitations of the parameters used to calculate these values. Mechanistic weed threshold crop competition models are suggested as a means of overcoming some of the limitations of empirically based threshold models. A mechanistic approach to the development of weed threshold models is desirable since relative crop and weed responses to environmental factors, cultural practices and the dynamic nature of competition are considered. Guidelines for the application of weed thresholds within a cropping system are outlined. *[Article copies available for a fee from The Haworth Document Delivery Service: 1-800-342-9678. E-mail address: getinfo@ haworthpressinc.com]*

**KEYWORDS.** Integrated weed management, weed competition, weed models

## INTRODUCTION

Weed thresholds represent an integral component of an integrated weed management (IWM) system. Several papers have been published on the principles of IWM but few have stressed the importance of implementing a systems approach to IWM (Swanton and Weise, 1991). There has been a tendency to develop weed control options in isolation of the larger issues of agroecosystem health (Swanton and Murphy, 1996). A systems approach to farm weed management implies that each weed control decision be evaluated in terms of its impact on the performance of the farming system as a whole (Ikerd, 1993). Integrated weed management incorporates crop breeding, fertilization, rotation, chemical and mechanical weed control, competition, and successional and soil management into a method of reducing weed interference while maintaining acceptable crop yields (Swanton and Weise, 1991; Thill et al., 1991). Integrated weed management is designed to be economically, environmentally and socially acceptable (Swanton and Murphy, 1996).

Predicting the outcome of crop-weed interference is part of an IWM system. A great deal of research has focused on estimating crop yield losses as a function of weed pressure, in order to identify the required level of control. Weed management decision rules based on weed density thresholds and critical periods of interference can help agricultural producers determine the necessity and timing of herbicide application or other control measures (Cousens, Wilson and Cussans, 1985; Swanton and Weise, 1991). However, only a small percentage of producers are actually using thresholds to make weed management decisions (Czapar, Curry, and Gray, 1995).

From the weed scientist's perspective, the utility of weed thresholds is limited by their dependency on the accuracy of the weed-crop competition model upon which they are based, and the stability of parameter estimates contained within the model over time and space (Kropff and Lotz, 1992; Chikoye and Swanton, 1995; Dieleman et al., 1995; Knezevic, Weise, and Swanton, 1995). From the grower's perspective, the utility of weed thresholds is limited by implications of leaving weeds in the field which may affect harvestability of the crop, landlord-tenant relationships and the long-term impact of weed seed return (Swanton and Weise, 1991; O'Donovan, 1996). In this paper, we review the literature on weed thresholds and examine their applicability in the context of a systems approach to IWM.

## *THRESHOLDS FOR SINGLE WEED SPECIES*

Weed thresholds have been defined in a variety of ways (Cousens, 1987). The two most commonly cited thresholds are "competitive" and "economic." A competitive threshold has been defined as the weed density above which crop yield is reduced by an unacceptable amount (Oliver, 1988). This amount is arbitrary, but usually ranges from 5 to 20%. An economic threshold is defined as the weed density at which the benefit derived from herbicide application just equals the cost of control. Unlike competitive thresholds, economic thresholds vary with assumptions made about weed-free yield, commodity price, and cost and efficacy of control. Both types of thresholds are usually derived from empirical regressions models of the relationship between crop yield and the density of a single weed species that emerges at the same time as the crop.

The relationship between crop yield loss and weed density has been described by various functions, including linear (Bauer and Mortensen, 1992), quadratic (Vail and Oliver, 1993), sigmoidal (Zimdahl, 1980) and the rectangular hyperbola (Spitters, 1983; Cousens, 1985). Each of these functions vary in their ability to describe the outcome of weed and crop competition. For example, Swinton and King (1994) pointed out that the shape of the yield loss function influenced the threshold level as a sigmoidal function tended to place the control threshold at a higher weed density than hyperbolic or linear functions. Likewise, Cousens (1985) compared 17 functional forms and found that the rectangular hyperbolic model provided the best fit to a large number of data sets and was most consistent with the underlying biology of crop-weed interference. This model gained general acceptance for documenting crop yield response to varying densities of single weed species (Stoller et al., 1987; Wilson and Wright, 1990; Weaver, 1991; Coble and Mortensen,

1992; Norris, 1992; Sattin, Zanin, and Berti, 1992; Berti and Zanin, 1994; Lindquist et al., 1996).

## Magnitude of Single Species Thresholds in Various Crops

A number of generalizations can be drawn about the magnitude of weed threshold levels in different crops. For the purposes of this discussion, "thresholds" will refer to either explicitly calculated economic thresholds, or to weed densities which result in a 5-to-10% yield loss, which in many crops is roughly equivalent to the cost of weed control. The majority of threshold studies in North America have dealt with annual broad-leaved and grass weeds in corn (*Zea mays* L.) and soybeans (*Glycine max* Merr.). Threshold densities for annual broad-leaved weeds emerging with the crop have been determined to be generally less than 5 plants m$^{-2}$ in corn, and often less than 1 plant m$^{-2}$ in soybeans (Moolani, Knake, and Slife, 1964; Marra and Carlson, 1983; Shurtleff and Coble, 1985; Weaver, 1986, 1991; Oliver, 1988; Bauer and Mortensen, 1992; Sattin, Zanin, and Berti, 1992; Knezevic, Weise, and Swanton, 1994; Dieleman et al., 1995, 1996; Lindquist et al., 1996). Whereas, the threshold levels for annual grasses have often been reported to be between 10 and 40 plants m$^{-2}$ as they have been found to be less competitive at low densities than broad-leaved weeds in corn and soybeans (Sibuga and Bandeen, 1980; Beckett, Stoller, and Wax, 1988; Vail and Oliver, 1993; Bosnic and Swanton, 1997a,b). Similar threshold levels have been reported for annual weeds in cotton (*Gossypium hirsutum* L.) and sunflower (*Helianthus annus* L.), although there have been far fewer studies in these crops (Buchanan and Burns, 1971; Buchanan et al., 1980; Onofri and Tei, 1994). These threshold densities are lower than the weed densities typically found in growers' fields in the absence of weed control. Their use therefore, will be limited in most cases to decisions about the need for control of weed "escapes" from soil applied herbicides or tillage operations used to control the initial flush of weeds.

Much of the threshold research in Europe and Australia has involved cereal crops, and varying threshold levels have been suggested. Cereals such as barley (*Hordeum vulgare* L.) and wheat (*Triticum aestivum* L.) are strong competitors with weeds, and economic thresholds in these crops are often higher than those in row crops. Gerowitt and Heitefuss (1990) recommended fixed thresholds in winter cereals of 20 to 30 plants m$^{-2}$ for grass weeds and 40 to 50 plants m$^{-2}$ for most broad-leaved weeds, with the exception of a few very competitive broad-leaved species, such as catchweed bedstraw (*Galium aparine* L.), which had thresholds of 0.1 to 2.0 plants m$^{-2}$. Similarly, Streibig (1989) and Zanin, Bertini, and Toniolo (1993) calculated threshold levels for a number of weeds in wheat and barley to be greater than 20 plants m$^{-2}$, and in some cases as high as 100 plants m$^{-2}$. Thresholds can possibly play a primary role in decisions for weed control in cereals, at least for fields of low to moderate weed pressure.

In contrast, estimates of weed thresholds in crops that have low competitive ability and/or are grown in wide rows, are lower than those in corn and soybeans. Weed densities as low as 1 plant m$^{-2}$ can reduce yield of sugarbeets (*Beta vulgaris* L.) or potato (*Solanum tuberosum* L.) by more than 20% (Schweizer and Lauridson, 1985; Kropff, 1988; VanGessel and Renner, 1990; Norris, 1992). Weakly competitive crops of high economic value such as many horticultural crops, may have thresholds near zero for many weeds (Norris, 1992). In such crops, yield potential may already have declined by the time weeds are counted and a postemergence herbicide is applied (Dunan et al., 1995). Furthermore, herbicide choices are often limited in horticultural crops and the weed management issue in these systems is how to achieve weed control rather than determining whether control is necessary. Thus, the importance of thresholds as a management tool will vary from one crop to another.

## THRESHOLDS FOR MULTIPLE WEED SPECIES

Traditional one-weed one-crop studies however, do not reflect the norm in actual farm situations (Stoller et al., 1987; Oliver 1988; Hume, 1989; Berti and Zanin, 1994; Zanin, Bertini, and Toniolo, 1993; Swinton et al., 1994a). An inherent weakness underlying studies on single weed/crop competition is reflected by Combellack and Friesen's (1992) statement that, "the large numbers of studies on weed/crop competition have had very little influence on weed control practices, which probably reflects an emphasis on single weed/crop interactions whereas in practice a complex of weeds is present." Farmers are faced with a mixed weed species infestation in their fields (Hume, 1989; Swinton et al., 1994a; Toler, Guice, and Murdock, 1996; Van Acker, Lutman, and Froud-Williams, 1997) and weed/crop interference studies must address this reality.

Estimation of crop yield loss as a function of multiple weed species is complicated by interactions between weeds as well as between the crop and weed (Connolly, 1988; Roush et al., 1989; Dale et al. 1991; Clements, Weise, and Swanton, 1994). The difficulty involved in executing multiple weed species research is reflected by the limited number of studies on multiple weed species interference (Van Acker, 1996). The size of the experiment is often restrictive and analysis and interpretation of the results are complicated. These problems have been handled by limiting weed densities or by analyzing the influence of naturally occurring multiple weed species infestations on crop yield.

The objective of most studies of multiple-weed species interference has been to determine if the outcome is additive, i.e., if the impact of multiple weed species on crop yield is equal to the sum of the impact of the individual species (Blackshaw, Anderson, and Dekker, 1987; Milroy, Turner, and Goodchild, 1990; Sims and Oliver, 1990). Results generally suggest that additivity is dependent on the intensity of interference, occurring at low but not at high weed densities. The intensity of interference is a function of many factors including the weed species present, their densities, and the environmental conditions under which the experiment was conducted (Flint and Patterson, 1983; Liebman and Robichaux, 1990).

Other studies have shown that the relative impact of weed species on crop yield varies not only with density but with ratio (i.e., frequency of weed vs. crop). For example, Kroh and Stephenson (1980) found the relative competitive ability of four weed species changed depending on whether the species were grown in pairs or in 4-way mixtures. Hume (1989) used linear multiple regression to describe the effect of weed communities dominated by green foxtail *(Setaria viridis* L. Beauv.) on the yield of wheat in western Canada, and found that the competition coeficients for each weed species varied according to their ratio in mixture. Van Acker, Lutman, and Froud-Williams (1997) used response surface models to analyze the interference of chickweed [*Stellaria media* (L.) Vill.] and barley on flax (*Linum usitatissimum* L.) or faba bean (*Vicia faba* L.) yield, and found that the parameter values for chickweed were affected by the presence of barley. They concluded that multiple weed interference models cannot be parameterized using data from single weed species experiments.

The results of these experiments and others (Blackshaw and Schaalje, 1993; Toler, Guice, and Murdock, 1996) suggest that the assumption of additivity in multiple species interference will not always be valid and needs to be tested further. Despite these theoretical difficulties, additivity of weed interference has been assumed and incorporated into yield loss models (Spitters, 1983; Cousens, 1985; Kropff and Spitters, 1991).

### Weed Competitive Indices

As a practical method of estimating crop yield losses from multiple weed species, weeds are often ranked in terms of their competitiveness against a crop. These weed "rankings" have led to the formation of competitive indices. Most researchers have sought to develop a competitive index that transforms all weed species into a common unit of weed pressure (Aarts and De Visser, 1985; Black and Dyson, 1993; Coble, 1985; Coble and Mortensen, 1992; Dew 1972; Lybecker, Schweizer, and Westra, 1991).

One approach for constructing competitive indices has been to survey the expert opinion of weed scientists about the relative competitive ability of various weeds in a particular crop throughout a geographic region (Lybecker,

Schweizer, and Westra, 1991; Black and Dyson, 1993). The most competitive weed is chosen as the reference species and all remaining weed species are converted to units of the reference species. Another approach has been to base competitive indices on weed biomass at a certain point in time. Wilson (1986) proposed the use of "Crop Equivalents," defined as the ratio between weed dry weight per plant in a crop and crop dry weight per plant on weed-free plots. Weed densities of each species in a field are then multiplied by their crop equivalent ratio and the products summed to characterize the competitive effect of the mixed weed infestation. Håkansson (1988) proposed a "Unit Production Ratio," based on the biomass production of one weed species relative to another without reference to crop biomass. Similarly, Jensen (1991) determined weed size hierarchies in cereals based on weed weights as an indirect way of ranking the competitive ability of different weed species. Indices based on weed size or weight are static measures that do not take into account inter- or intra-specific competition or the phenology of the individual weed species.

The most common method of deriving competitive indices has been to use the coefficients from regression equations. Dew (1972) was one of the first to calculate an index of competition for estimating crop losses due to weeds. His index of competition was expressed as the ratio between the slope and the intercept of a linear regression equation describing the relationship between crop yield and weed density for each species. Dew used this index to rank the competitive ability of barley, wheat, and flax against wild oats (*Avena fatua* L.). Coble (1985) expanded Dew's index by scaling it from 0 to 10, according to the most competitive species. The density of each weed species was then multiplied by its competitive index and the products summed to give a Total Competitive Load (TCL) and used to estimate soybean or corn yield loss (Coble, 1985; Coble and Mortensen, 1992). Aarts and De Visser (1985) introduced a system of "Standard Weed Units" to calculate yield loss in cereals from multiple weed species. The Standard Weed Units were calculated as the product of observed density and a scaled threshold value for each species summed over the species. Wilson and Wright (1990) ranked the competitive ability of weeds according to the initial slope ('I' value) of the rectangular hyperbolic yield loss function. A potential flaw of this approach is the use of low density (in the case of linear models), single weed species data to develop competitive indices for multiple weed species environments.

Swinton et al. (1994a) modified the rectangular hyperbolic yield loss equation (Cousens, 1985) to account for competitive indices developed from multi-weed species data. In their reformulated equation, the initial slope parameter ('I') was allowed to vary for each weed species and implicitly served as a competitive index, but the asymptote or maximum yield loss parameter ('A') was fixed for a particular crop. The estimated competition coefficients could be converted into the kind of competitive index values associated with Coble and Mortensen's (1992) competitive load formula by dividing through by the largest 'I' value, i.e., that of the most competitive weed (Swinton et al., 1994b). A somewhat similar approach was taken by Berti and Zanin (1994), who quantified the competitive effect of a mixed weed infestation using the rectangular hyperbolic equation. They transformed the observed density of each weed species into "Density Equivalents" of a hypothetical reference weed that has the 'I' and 'A' parameters both equal to 100.

The term "threshold" is not particularly useful for multi-species weed infestations, because it refers to the abstract "density equivalent" of the most competitive weed species which may not even be present in a particular field. Rather, competitive indices permit the calculation of potential crop yield losses, which can then be compared to the cost of weed control.

## THE IMPORTANCE OF TIME OF WEED EMERGENCE

It has long been recognized that the time of weed emergence relative to the crop is extremely important in determining the magnitude of crop yield losses (Dew 1972; Hagood et al., 1981; O'Donovan et al., 1985). Threshold levels of redroot pigweed (*Amaranthus retroflexus* L.), barnyardgrass [*Echinochloa crus-galli* (L.) Beauv.], common ragweed (*Ambrosia artemisiifolia* L.) and velvetleaf (*Abutilon theophrasti* Medik.) in corn, soybeans, and dry beans (*Phaseolus vulgaris* L.) may be 2 to 10 times higher for weeds emerging 3 to 4 weeks after the crop compared to those emerging with the crop (Knezevic, Weise, and Swanton, 1994; Cardina, Regnier, and Sparrow, 1995; Chikoye, Weise, and Swanton, 1995; Dieleman et al., 1995, 1996; Bosnic and Swanton, 1997a,b). Studies of the critical period of weed control have shown that delaying weed emergence by 3 to 5 weeks was sufficient to prevent significant yield losses in soybeans and corn, respectively (Hall, Swanton, and Anderson, 1992; Van Acker, Swanton, and Weise, 1993). Some authors have even suggested that time of weed emergence relative to the crop may be more important in determining the impact of weeds on crop yield than weed density (Knezevic, Weise, and Swanton, 1994,1995; Chikoye and Swanton, 1995; Dieleman et al., 1995, 1996). Nevertheless, there is clearly an interaction between weed density and duration of weed interference. The length of the critical period or period threshold (Dawson, 1986) varies with weed density, just as density thresholds vary with time of emergence (Weaver, Kropff, and Groenevald, 1992; Dunan et al., 1995).

Empirical models that incorporate time of weed emergence represent a significant step toward improving the accuracy of predictions of crop yield loss. Cousens et al. (1987) introduced a regression equation in which crop yield was estimated as a function of both weed density and time of weed emergence. Several authors formulated a regression model in which crop yield loss was a function of the relative leaf area of the weed with respect to the crop thereby, accounting for the effects of both density and time of emergence with one variable (Kropff, 1988; Kropff and Spitters, 1991; Kropff et al., 1995; Lutman, Risiott, and Ostermann, 1996). Berti et al. (1996) extended the multi-species model of Berti and Zanin (1994) to include time of emergence. They defined the "time density equivalent" as the density of weeds that germinated with the crop, remained until harvest, and caused the same yield loss as that caused by weeds with any given density, time of emergence and time of removal. Although these models improve the accuracy of yield loss estimation over models based on density alone, they suffer from the practical difficulty of quickly estimating time of emergence or leaf area in the field.

Scouting of fields for postemergence weed control decisions usually occurs after crop emergence but before the end of the critical period. Even within this interval, threshold levels or competitive indices based on simultaneous emergence of the crop and weeds will overestimate the need for control of weeds which emerge later. They may also overestimate the need for secondary control in fields which have received soil-applied herbicides. A number of studies have demonstrated that weeds which "escape" soil-applied herbicides to which they are normally sensitive, are less competitive with the crop and cause lower yield losses than untreated weeds (Adcock and Banks, 1991; Weaver, 1991; Black and Dyson, 1993). The competitiveness of weed escapes may be reduced because their time of emergence relative to the crop is delayed, or because of sub-lethal effects of the herbicide. The impact of sub-lethal doses of herbicides on weed growth, competitiveness and seed production in the field has received little attention.

## SEED PRODUCTION FROM SUB-THRESHOLD WEED POPULATIONS

Economic threshold levels are generally based on the costs and benefits of weed control within a single year or crop cycle. Researchers and growers alike, however, are concerned about the implications of weed seed production from sub-threshold populations for future weed management and long-term profit. Such concerns have led to the calculation of "economic optimum thresholds." An economic optimum threshold (EOT) has been defined as the decision threshold that will provide the greatest net benefit over the long term, taking weed seed production into account (Cousens, 1987). These EOT levels are determined by combining weed population dynamic sub-models with crop yield loss functions and economic calculations of discounted net benefit of alternative management strategies over time into what are sometimes referred to as "bioeconomic models" (King et al., 1986; Swinton and King, 1994).

As a rule, EOT's are as much as an order of magnitude lower than thresholds determined on the basis of profits within a single year (Cousens et al., 1986; Bauer and Mortensen, 1992). Cousens et al. (1986) suggested that EOT's can be approximated by dividing the single year economic threshold by the intrinsic rate of increase of a weed species. Studies of the impact of weed seed return by sub-threshold populations of weeds with high rates of increase have concluded that thresholds for these species should be near zero (Norris, 1992; Sattin, Zanin, and Berti, 1992; Cardina and Norquay, 1997). Few studies have examined the impact of seed production from species with lower rates of increase, or considered the ability of IWM systems to manage seed return through changes in tillage practices or crop rotations (Swanton and Murphy, 1996). Bioeconomic models often include parameters characterizing seed production by remaining weeds, seed survival over winter and seedling emergence in the following crop. These parameter values can be modified by crop management practices. VanGessel et al. (1996) demonstrated that in fields with low to moderate weed pressure, IWM systems could prevent an increase in the seed bank without intensive herbicide use.

Estimates of weed population growth and crop yield losses from empirical models are valid only for the conditions under which the models were derived. Many studies have shown that parameter values vary from year to year and site to site, reflecting the influence of weather, time of emergence, and other factors on the relationship between weed density and crop yield (Weaver, 1986; Knezevic, Weise, and Swanton, 1994, 1995; Chikoye and Swanton, 1995; Dieleman et al., 1995, 1996). To be useful for management applications, the parameter values of descriptive models must be based on data drawn from multiple years and sites.

## STABILITY OF WEED THRESHOLD PARAMETERS

A number of studies have examined the stability of the parameter estimates in time and across geographical regions for different functional forms of the crop-weed interference relationship (Bauer et al., 1991; Swinton et al., 1994a; Lindquist et al., 1996). Bauer et al. (1991) used a linear function to relate soybean yield loss to low densities of three weed species and compared parameter estimates over four years and three sites. They found that parameter estimates for velvetleaf varied with environment, with greater yield losses occurring in environments of high yield potential as compared to those with low yield potential. Lindquist et al. (1996) investigated the stability of parameter estimates for velvetleaf in corn across years and locations using the rectangular hyperbolic yield loss function and found that parameter estimates were more stable across years within a location than across locations (South Dakota, Nebraska, Michigan and Colorado). Based on their study, they concluded that parameter values should be based on more than two years of data at a particular location, and that these values cannot be assumed to apply to other locations within a region. Swinton et al. (1994a) compared parameter estimates for their multi-species modification of the rectangular hyperbolic yield function over 13 Minnesota and Wisconsin datasets. They found the parameter values to be stable across years but not locations. Location effects would include differences in crop cultivar and possibly weed biotypes in addition to the effects of weather and time of emergence over years.

The lack of parameter stability across locations and years is of great concern to weed scientists with respect to the accuracy of yield loss predictions. It is not yet clear how important variation in parameter estimates of the yield loss function is for management applications. Studies that have evaluated the performance of bioeconomic models based on these functions have generally shown reasonable performance in terms of profit and level of weed control, in addition to a reduction in herbicide use (Monks et al.,1995; Forcella et al.,1996; Buhler et al., 1997). It may be that these bioeconomic models are sufficiently robust, and/or that yield losses are usually overestimated, so that weed control "failures" are unlikely.

Error is also involved in estimates of weed-free crop yield, crop price, and weed density by the producer or consultant. Monks et al. (1995) suggested that the greatest source of error in recommendations of the threshold-based decision model HERB (Wilkerson, Modena, and Coble, 1991) was overestimation of weed-free crop yield. This overestimation resulted in herbicides being applied more often than necessary. Wiles, Gold, and Wilkerson (1993) concluded that modifying HERB to account for uncertainty about mean weed density due to patchy weed distributions, would improve decision making but would be costly to implement.

The effect of uncertainty about crop price, yield and weed density on economic threshold levels is controversial. Deen et al. (1993) suggested that uncertainty in these variables raises the economic threshold level, under the assumption of risk neutrality, due to the convex nature of the hyperbolic yield loss function. However, Swinton (1991) pointed out that the opposite was true (i.e., uncertainty lowers economic threshold levels) if a sigmoidal yield response function was used, particularly when growers are intolerant of risk. Bauer et al. (1991) calculated confidence intervals for their linear yield loss function, and also demonstrated that economic threshold levels decreased with increasing risk aversion.

Empirical threshold models developed thus far have several limitations. These models cannot be used to explain why the level of observed competition occurred. An empirical approach, although useful for data summarization and for interpolation between data points is however, specific to the data set from which it was derived and thus lacks predictability under different environmental conditions. While it is possible to derive parameter values for all possible conditions, the number of field trials required makes this approach impractical.

## *MECHANISTIC WEED THRESHOLD MODELS*

Mechanistic models of crop-weed competition could potentially address the shortcomings associated with a purely empirical approach. These models, also referred to as explanatory or ecophysiological models, explicitly represent causality between the crop weed competition system variables and consequently, provide a means to explain why observed results occurred. A mechanistic approach to weed crop competition is desirable since relative crop and weed responses to environmental factors, cultural practices and the dynamic nature of competition are considered implicitly (Kropff, 1988; Wilkerson et al., 1990; Kropff, Weaver and Smits, 1992; Weaver, Kropff, and Groeneveld, 1992; Chikoye, Hunt, and Swanton, 1996). Mechanistic models quantify the various physiological and physical processes underlying crop and weed growth and development and the effect of environmental and cultural factors on these processes. Mechanistic models are also better suited for extrapolation outside a data range or to a differing set of conditions. Parameter values associated with many growth and competition processes are stable across location, year, and management practices; consequently, reparameterization is not required for each set of conditions.

Mechanistic competition models essentially consist of coupled weed and crop growth models (Kropff and van Laar, 1993). Competition is simulated dynamically based on resource capture and resource use efficiency by each species which in turn, is based on the underlying physiological processes governing phenology and growth. Resource use efficiency is important because a species that captures a disproportionate amount of a resource but uses it inefficiently will lose its competitive advantage (Berkowitz, 1988). Similarly, a competitive advantage for one resource could infer a competitive advantage for the other resources (Carlson and Hill, 1986; Liebman and Robichaux, 1990).

Mechanistic computer simulation can facilitate an understanding of the complex interrelationships in crop-weed competition (Weaver, 1996). By improving insights into the process of competition and the impact of environmental factors on these processes, cultural practices can be modified in favour of the crop. Dunan et al. (1995), for example, used a mechanistic approach to determine optimal crop density levels that minimized weed competitive effects. Similarly, mechanistic models also increase our understanding of the morphological and physiological traits that contribute to competitive ability (Lindquist and Kropff, 1996). Weaver, Kropff, and Cousens (1994) demonstrated that small differences in the timing of stem extension in winter wheat could have a large effect on outcome of competition with wild oats.

Simulation models also represent a potentially powerful tool for predicting yield loss attributed to weed interference. Predictions can be made over a set of average or extreme environmental conditions for a wide variety of cultural practices and soil types. "What if" scenarios can provide users with alternative answers or hypotheses for further testing. For example, Kropff et al. (1992), used this approach to simulate sugarbeet yield losses over a wide range of lambsquarters (*Chenopodium album* L.) densities and times of emergence.

A number of mechanistic models of weed-crop competition have been developed (Spitters and Aerts, 1983; Graf et al., 1990; Wilkerson et al., 1990; Graf and Hill, 1992; Ball and Shaffer, 1993; Kropff and van Laar, 1993; Grant, 1994; Barbour and Bridges, 1995), but so far most applications of these models have been confined to competition between a single weed and crop within a growing season. Mechanistic models of competition between a crop and multiple weed species that also simulate weed seed production and population growth over time have great potential to contribute to the development of weed management strategies in IWM systems.

## GUIDELINES FOR THE APPLICATION OF WEED THRESHOLDS

The slow acceptance by producers of weed thresholds as a decision making tool may be partly due to the fact that they have been considered in isolation, rather than within an integrated systems approach. The limitations as well as the advantages of thresholds need to be acknowledged. The following guidelines for implementation of a threshold approach to weed management are proposed:

1. Thresholds are an appropriate and useful tool primarily in fields with low weed pressure. Fields with high weed pressure require some sort of weed control prior to or shortly after crop emergence. After these initial control measures, thresholds may be useful for "weed escapes."
2. The importance of thresholds as a tool for weed management will vary with the cropping system. Crops with a high economic value and a lengthy critical period for weed control may have thresholds near zero for some of the more competitive weeds.
3. Thresholds are not an appropriate tool for perennials that need to be controlled before they begin to cause yield losses, or for weeds that reduce crop quality, e.g., eastern black nightshade (*Solanum ptycanthum*

Dun.), are difficult to control in another crop in the rotation, or have become resistant to herbicides.

4. Threshold levels based on simultaneous emergence of the crop and weeds will overestimate the need for control of weeds which emerge later. Scouting of fields should occur during the critical period for weed control. Weeds must be identified, counted, and, if possible, the time of emergence relative to the crop estimated.

5. Threshold levels and crop yield loss parameters determined at one location cannot be applied uncritically without field testing in other geographic regions. Estimates of weed-free crop yield and crop price must be realistic.

6. In fields with multiple weed species, an evaluation of the net benefit of alternative weed control strategies will be more useful than an attempt to identify the threshold for an abstract quantity such as "density equivalent."

7. Thresholds should be supplemented by IWM practices that reduce the impact of seed return, e.g., tillage system, crop rotation, variety/row width selection. For example, for some species the impact of weed seed return is minimized by conservation tillage practices that leave weed seeds in the upper 5 cm of soil, where their survival rate is lower and emergence more uniform than when buried deeply.

Future opportunities for the application of weed thresholds may be enhanced by the introduction of herbicide resistant crops that will provide control opportunities at any crop growth stage, by the registration and labelling of herbicide dose by species matrices, and by the growth of precision farming technologies. Biologically effective (Dieleman et al., 1996) or factor-adjusted (Green, 1991) herbicide doses could lead to lower weed control costs and lower threshold levels. The technologies associated with precision farming may overcome sampling problems associated with weed patchiness, and allow for tracking of weed population changes over time, thereby reducing the risk associated with seed return from sub-threshold weed populations.

## REFERENCES

Aarts, H. F. M. and C. L. M. De Visser. (1985). A management information system for weed control in winter wheat. *Proc. British Crop Protection Conf. Weeds.* pp. 679-686.

Adcock, T. E. and P. A. Banks. (1991). Effects of preemergence herbicides on the competitiveness of selected weeds. *Weed Sci.* 39:54-56.

Ball, D. A. and M. J. Shaffer. (1993). Simulating resource competition in multispecies agricultural plant communities. *Weed Res.* 33:299-310.

Barbour, J. C. and D. C. Bridges. (1995). A model of competition for light between peanut *(Arachis hypogaea)* and broadleaf weeds. *Weed Sci.* 43:247-257.

Bauer, T. A. and D. A. Mortensen. (1992). A comparison of economic and economic optimum thresholds for two annual weeds in soybeans. *Weed Technol.* 6:228-235.

Bauer, T. A., D. A. Mortensen, G. A. Wicks, T. A. Hayden, and A. R. Martin. (1991). Environmental variability associated with economic thresholds for soybeans. *Weed Sci.* 39:564-569.

Beckett, T. H., E. W. Stoller, and L. W. Wax. (1988). Interference of four annual weeds in corn (*Zea mays*). *Weed Sci.* 36:764-769.

Berkowitz, A. R. (1988). Competition for resources in weed-crop mixtures. In *Weed Management in Agroecosystems: Ecological Approaches*, eds. M. A. Altieri and M. Liebman, CRC Press, Boca Raton, FL. pp. 89-119.

Berti, A. and G. Zanin. (1994). Density equivalent: A method for forecasting yield loss caused by mixed weed populations. *Weed Res.* 34:327-332.

Berti, A., C. Dunan, M. Sattin, G. Zanin, and P. Westra. (1996). A new approach to determine when to control weeds. *Weed Sci.* 44:496-503.

Black, I. D. and C. B. Dyson. (1993). An economic threshold model for spraying herbicides in cereals. *Weed Res.* 33:279-290.

Blackshaw, R. E. and G. B. Schaalje. (1993). Density and species ratio effects on interference between redstem filaree (*Erodium cicutarium*) and round-leaved mallow (*Malva pusilla*). *Weed Sci.* 41:594-599.

Blackshaw, R. E., G. W. Anderson, and J. Dekker. (1987). Interference of *Sinapis arvensis* L. and *Chenopodium album* L. in spring rapeseed (*Brassica napus* L.). *Weed Res.* 27:207-213.

Bosnić A. Č. and C. J. Swanton. (1997a). Influence of barnyardgrass (*Echinochloa crus-galli* ) time of emergence and density on corn (*Zea mays*). *Weed Sci.* 45:276-282.

Bosnić A. Č. and C. J. Swanton. (1997b). Economic decision rules for postemergence herbicide control of barnyardgrass (*Echinochloa crus-galli*) in corn (*Zea mays*). *Weed Sci.* (In press).

Buchanan, G. A. and E. R. Burns. (1971). Weed competition in cotton: I. Cocklebur and redroot pigweed. *Weed Sci.* 19:580-582.

Buchanan, G. A., R. H. Crowley, J. E. Street, and J. A. McGuire. (1980). Competition of sicklepod (*Cassia obtusifolia*) and redroot pigweed (*Amaranthus retroflexus*) with cotton (*Gossypium hirsutum*). *Weed Sci.* 28:258-262.

Buhler, D. D., R. P. King, S. M. Swinton, J. L. Gunsolus, and F. Forcella. (1997). Field evaluation of a bioeconomic model for weed management in soybean (*Glycine max*). *Weed Sci.* 45:158-165.

Cardina, J., E. Regnier, and D. Sparrow. (1995). Velvetleaf (*Abutilon theophrasti*) competition and economic thresholds in conventional- and no-tillage corn (*Zea mays*). *Weed Sci.* 43:81-87.

Cardina, J. and H. M. Norquay. (1997). Seed production and seed bank dynamics in subthreshold velvetleaf (*Abutilon theophrasti*) populations. *Weed Sci.* 45:85-90.

Carlson, H. L. and J. E. Hill. (1986). Wild oat (*Avena fatua*) competition with spring wheat: Effects of nitrogen fertilizer. *Weed Sci.* 34:29-33.

Chikoye, D. and C. J. Swanton. (1995). Evaluation of three empirical models depicting *Ambrosia artemisiifolia* competition in white bean. *Weed Res.* 35:421-428.

Chikoye, D., S. F. Weise, and C. J. Swanton. (1995). Influence of common ragweed

(*Ambrosia artemisiifolia*) time of emergence and density on white bean (*Phaseolus vulgaris*). *Weed Sci.* 43:375-380.

Chikoye, D., L. A. Hunt, and C. J. Swanton (1996). Simulation of competition for photosynthetically active radiation between common ragweed (*Ambrosia artemisiifolia*) and dry bean (*Phaseolus vulgaris*). *Weed Sci.* 44:545-554.

Clements, D. R., S. F. Weise, and C. J. Swanton. (1994). Integrated weed management and weed species diversity. *Phytoprotection* 75:1-18.

Coble, H. D. (1985). Development and implementation of economic thresholds for soybean. In *CIPM: Integrated Pest Management on Major Agricultural Systems*, eds. R. E. Frisbie and P. L. Adkisson. Texas A & M University, pp. 295-307.

Coble, H. D. and D. A. Mortensen. (1992). The threshold concept and its application to weed science. *Weed Technol.* 6:191-195.

Combellack, J. H. and G. Friesen. (1992). Summary of outcomes and recommendations from the First International Weed Control Congress. *Weed Techn.* 6:1043-1058.

Connolly, J. (1988). Experimental methods in plant competition research in crop-weed systems. *Weed Res.* 28:431-436.

Cousens, R. (1985). An empirical model relating crop yield to weed and crop density and a statistical comparison with other models. *J. Agric. Sci.* 105:513-521.

Cousens, R. (1987). Theory and reality of weed control thresholds. *Plant Prot. Quart.* 2:13-20.

Cousens, R., B. J. Wilson, and G. Cussans. (1985). To spray or not to spray: The theory behind the practice. *Proc. Br. Crop Prot. Conf. Weeds,* vol. 3:671-678.

Cousens, R., C. J. Doyle, B. J. Wilson, and G. W. Cussans. (1986). Modeling the economics of controlling *Avena fatua* in winter wheat. *Pestic. Sci.* 17:1-12.

Cousens, R., P. Brain, J. T. O'Donovan, and P. A. O'Sullivan. (1987). The use of biologically realistic equations to describe the effects of weed density and relative time of emergence on crop yield. *Weed Sci.* 35:720-725.

Czapar, G. F., M. P. Curry, and M. E. Gray. (1995). Survey of integrated pest management practices in central Illinois. *J. Prod. Agric.* 8: 483-486.

Dale, M. R. T., D. J. Blundon, D. A. MacIsaac, and A. G. Thomas. (1991). Multiple species effects and spatial autocorrelation in detecting species associations. *J. Vegetation Sci.* 2:635-642.

Dawson, J. H. (1986). The concept of period thresholds. Proc. Eur. Weed Res. Soc. Symp., *Economic Weed Control.* pp. 327-331.

Deen, W., A. Weersink, C. G. Turvey, and S. E. Weaver. (1993). Weed control decision rules under uncertainty. *Rev. Agric. Econ.* 15:39-50.

Dew, D. A. (1972). Index of competition for estimating crop loss due to weeds. *Can. J. Plant Sci.* 52:921-927.

Dieleman, A., A. S. Hamill, S. F. Weise, and C. J. Swanton. (1995). Empirical models of pigweed (*Amaranthus* spp.) interference in soybean (*Glycine max*). *Weed Sci.* 43:612-618.

Dieleman, A., A. S. Hamill, G. C. Fox, and C. J. Swanton. (1996). Decision rules for postemergence control of pigweed (*Amaranthus* spp.) in soybean (*Glycine max*). *Weed Sci.* 44:126-132.

Dunan, C. M., P. Westra, E. E. Schweizer, D. W. Lybecker, and F. D. Moore III.

(1995). The concept and application of early economic period threshold: The case of DCPA in onions. *Weed Sci.* 43:634-639.

Flint, E. P and D. T. Patterson. (1983). Interference and temperature effects on growth in soybean (*Glycine max*) and associated $C_3$ and $C_4$ weeds. *Weed Sci.* 31:193-199.

Forcella, F., R. P. King, S. M. Swinton, D. D. Buhler, and J. L. Gunsolus. (1996). Multi-year validation of decision aide for integrated weed management in row crops. *Weed Sci.* 44:650-661.

Gerowitt, B. and R. Heitefuss. (1990). Weed economic thresholds in cereals in the Federal Republic of Germany. *Crop Prot.* 9:323-331.

Graf, B. and Hill, J. E. (1992). Modeling the competition for light and nitrogen between rice and *Echinochloa crus-galli*. *Agric. Systems* 40:345-359.

Graf, B., O. Rakotobe, P. Zahner, V. Delucchi, and A. P. Gutierrez. (1990). A simulation model for the dynamics of rice growth and development: Part I - The carbon balance. *Agric. Systems*. 32:341-365.

Grant, R. (1994). Simulation of competition between barley and wild oats under different managements and climates. *Ecological Modeling*. 71:269-287.

Green, J. M. (1991). Maximizing herbicide efficiency with mixtures and expert systems. *Weed Tech.* 5: 894-897.

Hagood, E. S., Jr., T. T. Bauman, J. L. Williams, Jr., and M. M. Schreiber. (1981). Growth analysis of soybeans (*Glycine max*) in competition with jimsonweed (*Datura stramonium*). *Weed Sci.* 29:500-504.

Håkansson, S. (1988). Competition in stands of short-lived plants: Density effects measured in three-component stands. Crop Production Science. 3, Dept. of Crop Production Science, Swedish University of Agricultural Science, Uppsala, Sweden. pp. 181.

Hall, M. R., C. J. Swanton, and G. W. Anderson. (1992). The critical period of weed control in grain corn (*Zea mays*). *Weed Sci.* 40:441-447.

Hume, L. (1989). Yield losses in wheat due to weed communities by green foxtail (*Setaria viridis*): a multi-species approach. *Can. J. Plant Sci.* 69:521-529.

Ikerd, J. E. (1993). The need for a systems approach to sustainable agriculture. *Agriculture, Ecosystems and Environments*. 46:147-160.

Jensen, P. K. (1991). Weed size hierarchies in Denmark. *Weed Res.* 31:1-7.

King, R. P., D. W. Lybecker, E. E. Schweizer, and R. L. Zimdahl. (1986). Bioeconomic modeling to simulate weed control strategies for continuous corn (*Zea mays*). *Weed Sci.* 34:972-976.

Knezevic, S. Z., S. F. Weise, and C. J. Swanton. (1994). Interference of redroot pigweed (*Amaranthus retroflexus*) in corn (*Zea mays*). *Weed Sci.* 42:568-573.

Knezevic, S. Z., S. F. Weise, and C. J. Swanton. (1995). Comparison of empirical models depicting density of *Amaranthus retroflexus* L., and relative leaf area as predictors of yield loss in maize (*Zea mays* L.). *Weed Res.* 35:207-215.

Kroh, G. C. and S. N. Stephenson. (1980). Effects of diversity and pattern on the relative yield of four Michigan first year fallow field species. *Oecologia*. 45:366-371.

Kropff, M. J. (1988). Modeling the effects of weeds on crop production. *Weed Res.* 28:465-471.

Kropff, M. J. and L. A. P. Lotz. (1992). Systems approaches to quantify crop-weed interactions and their application in weed management. *Agric. Sys.* 40:265-282.

Kropff, M. J. and C. J. T. Spitters. (1991). A simple model of crop loss by weed competition from early observations on relative leaf area of the weeds. *Weed Res.* 31:97-105.

Kropff, M .J. and H. H. van Laar (1993). Empirical models for crop-weed competition. In *Modeling Crop-Weed Interactions*, eds. M.J. Kropff and H.H. van Laar. CAB International, Wallingford, U.K. pp. 9-24.

Kropff, M. J., S. E. Weaver, and N. A. Smits. (1992). Use of ecophysiological models for crop- weed interference: Relations amongst weed density, relative time of weed emergence, relative leaf area, and yield loss. *Weed Sci.* 40:296-301.

Kropff, M. J., C. J. T. Spitters, B. J. Schnieders, W. Joenje, and W. DeGroot. (1992). An eco- physiological model for interspecific competition applied to the influence of *Chenopodium album* L. on sugarbeet: II. Model evaluation. *Weed Res.* 32:451-463.

Kropff, M. J., L. A. P. Lotz, S. E. Weaver, H. J. Bos, J. Wallinga, and T. Migo. (1995). A two parameter model for prediction of crop yield loss by weed competition from early observation of relative area of the weeds. *Annals of Applied Biol.* 126:329-346.

Liebman, M. and R. H. Robichaux. (1990). Competition by barley and pea against mustard: Effect on resource acquisition, photosynthesis and yield. *Agric. Ecosystems Environ.* 31:155- 172.

Lindquist, J. L. and M. J. Kropff. (1996). Applications of an ecophysiological model for irrigated rice (*Oryza sativa*)-*Echinochloa* competition. *Weed Sci.* 44:52-56.

Lindquist, J. L., D. A. Mortenson, S. A. Clay, R. Schmenk, J. J. Kells, K. Howatt, and P. Westra. (1996). Stability of corn (*Zea mays*)-velvetleaf (*Abutilon theophrasti*) interference relationships. *Weed Sci.* 44:309-313.

Lutman, P. J. W., R. Risiott, and H. P. Ostermann. (1996). Investigations into alternative methods to predict the competitive effects of weeds on crop yields. *Weed Sci.* 44:290-297.

Lybecker, D. W., E. E. Schweizer, and P. Westra. (1991). Computer aided decisions for weed management in corn. Abst. in W. *J. Agric. Econ.* 16:456.

Marra, M. C. and G. A. Carlson. (1983). An economic threshold model for weeds in soybeans (*Glycine max*). *Weed Sci.* 31:604-609.

Milroy, S. P., D. W. Turner, and N. A. Goodchild. (1990). Competition in binary mixtures of three grass weed species and their impact on wheat yield. Proc. of the EWRS Symposium. *Integrated Weed Management in Cereals*:211-216.

Monks, C. D., D. C. Bridges, J. W. Woodruff, T. R. Murphy, and D. J. Berry. (1995). Expert system evaluation and implementation for soybean (*Glycine max*) weed management. *Weed Technol.* 9:535-540.

Moolani, M. K., E. L. Knake, and F. W. Slife. (1964). Competition of smooth pigweed with corn and soybean. *Weeds* 12:126-128.

Norris, R. F. (1992). Case history for weed competition/population ecology: Barnyardgrass (*Echinochloa crus-galli*) in sugarbeets (*Beta vulgaris*). *Weed Technol.* 6:220-227.

O'Donovan, J. T. (1996). Weed economic thresholds: Useful agronomic tool or pipe dream? *Phytoprotection* 77:13- 28.

O'Donovan, J. T., E. A. de St. Remy, P. A. O'Sullivan, D. A. Dew, and A. K. Sharma. (1985). Influence of the relative time of emergence of wild oat (*Avena fatua*) on yield loss of barley(*Hordeum vulgare*) and wheat (*Triticum aestivum*). *Weed Sci.* 33:498-503.

Oliver, L. R. (1988). Principles of weed threshold research. *Weed Technol.* 2:398-403.

Onofri, A. and F. Tei. (1994). Competitive ability and threshold levels of three broadleaf weed species in sunflower. *Weed Res.* 34:471-479.

Roush, M. L., S. R. Radosevich, R. G. Wagner, B. D. Maxwell, and T. D. Peterson. (1989). A comparison of methods for measuring effects of density and ratio in plant competition experiments. *Weed Sci.* 37:268-275.

Sattin, M., G. Zanin, and A. Berti. (1992). Case history for weed competition/population ecology: Velvetleaf (*Abutilon theophrasti*) in corn (*Zea mays*). *Weed Technol.* 6:213-219.

Schweizer, E. E. and T. C. Lauridson. (1985). Powell amaranth *(Amaranthus powelli)* interference in sugarbeet (*Beta vulgaris*). *Weed Sci.* 33:518-520.

Shurtleff, J. L. and H. D. Coble. (1985). Interference of certain broadleaf weed species in soybeans (*Glycine max*). *Weed Sci.* 33:654-657.

Sibuga, K. P., and J. D. Bandeen. (1980). Effects of green foxtail (*Setaria viridis*) and common lambsquarter (*Chenopodium album*) interference in field corn (*Zea mays*). *Can. J. Plant Sci.* 60:1419-1425.

Sims, B. D. and L. R. Oliver. 1990. Mutual influence of seedling johnsongrass *(Sorghum halepense)*, sicklepod (*Cassia obtusifolia*), and soybean (*Glycine max*). *Weed Sci.* 38:139-147.

Spitters, C. J. T. (1983). An alternative approach to the analysis of mixed cropping experiments 1 Estimation of competition effects. *Netherlands J. Agric. Sci.* 31:1-11.

Spitters, C. J. T. and R. Aerts. (1983). Simulation of competition for light and water in crop- weed association. *Aspects of Applied Biol.* 4:467-483.

Stoller, E. W., S. K. Harrison, L. M. Wax, E. E. Regnier, and E. D. Nafziger. (1987). Weed interference in soybeans (*Glycine max*). *Rev. Weed Sci.* 3:155-181.

Streibig, J. C. (1989). The herbicide dose-response curve and the economics of weed control. *Brit. Crop Prot. Conf. - Weeds.* pp. 927-935.

Swanton, C. J. and S. D. Murphy. (1996). Weed science beyond the weeds: The role of integrated weed management (IWM) in agroecosystem health. *Weed Sci.* 44:437-445.

Swanton, C. J. and S. F. Weise. (1991). Integrated weed management: The rationale and approach. *Weed Technol.* 5:657-663.

Swinton, S. M. (1991). Response to risk in weed control decisions under expected profit maximization: Comment. *Journal of Agric. Economics* 42:404-406.

Swinton, S. M. and R. P. King. (1994). A bioeconomic model for weed management in corn and soybean. *Agric. Syst.* 44:313-335.

Swinton, S. M., D. D. Buhler, F. Forcella, J. L. Gunsolus, and R. P. King. (1994a).

Estimation of crop yield loss due to interference by multiple weed species. *Weed Sci.* 42:103-109.

Swinton, S. M., J. Stern, K. Renner, and J. Kells. (1994b). Estimating weed-crop interference parameters for weed management models. Res. Rpt. 538. Michigan Agricultural Experiment Station, Michigan State University. 20 p.

Thill, D. C., J. M. Lish, R. H. Callihan, and E. J. Bechinski. (1991). Integrated weed management–a component of integrated pest management: A critical review. *Weed Technol.* 5:648-656.

Toler, J. E., B. Guice, and E. C. Murdock. (1996). Interference between johnsongrass (*Sorghum halepense*), smooth pigweed (*Amaranthus hybridus*), and soybean (*Glycine max*). *Weed Sci.* 44:331-338.

Vail, G. D. and L. R. Oliver. (1993). Barnyardgrass (*Echinochloa crus-galli*) interference in soybeans (*Glycine max*). *Weed Technol.* 7:220-225.

Van Acker, R. C. (1996). Multiple-weed species interference in broadleaved crops: Evaluation of yield loss prediction and competition models. Ph.D. Thesis, University of Reading, Reading, UK. p. 180.

Van Acker, R. C., C. J. Swanton, and S. F. Weise. (1993). The critical period of weed control in soybean (*Glycine max*). *Weed Sci.* 41:194-220.

Van Acker, R. C., P. J. W. Lutman, and R. J. Froud-Williams. (1997). Predicting yield loss due to interference from two weed species using early observations of relative weed leaf area. *Weed Res.* 37: (*in press*).

VanGessel, M. J. and K. A. Renner. (1990). Redroot pigweed (*Amaranthus retroflexus*) and barnyardgrass (*Echinochloa crus-galli*) interference in potatoes (*Solanum tuberosum*). *Weed Sci.* 38:338-343.

VanGessel, M. J., E. E. Schweizer, D. W. Lybecker, and P. Westra. (1996). Integrated weed management systems for corn (*Zea mays*) production in Colorada–A case study. *Weed Sci.*44:423-428.

Weaver, S. E. (1986). Factors affecting threshold levels and seed production of jimsonweed (*Datura stramonium* L.) in soybeans (*Glycine max* (L.) Merr.). *Weed Res.* 26:215-223.

Weaver, S. E. (1991). Size-dependent economic thresholds for three broadleaf weed species in soybeans. *Weed Technol.* 5: 674-679.

Weaver, S. E. (1996). Simulation of crop-weed competition: Models and their applications. *Phytoprotection* 77: 3-11.

Weaver, S .E., M. J. Kropff, and R. M. W. Groeneveld. (1992). Use of ecophysiological models for crop-weed interference: The critical period of weed interference. *Weed Sci.* 40:302-307.

Weaver, S. E., M. J. Kropff, and R. Cousens. (1994). A simulation model of competiton between winter wheat and *Avena fatua* for light. *Ann. Appl. Biol.* 124:315-331.

Wiles, L. J., H. J. Gold, and G. G. Wilkerson. (1993). Modelling the uncertainty of weed density estimates to improve postemergence herbicide control decisions. *Weed Res.* 33:241-252.

Wilkerson, G. G., J. W. Jones, H. D. Coble, and J. L. Gunsolus. (1990). SOYWEED: A simulation model of soybean and common cocklebur growth and competition. *Agron. J.* 82:1003-1010.

Wilkerson, G. G., S. A. Modena, and H. Coble. (1991). HERB: Decision model for postemergence weed control in soybean. *Agron. J.* 83:413-417.

Wilson, B. J. (1986). Yield responses of winter cereals to the control of broadleaved weeds. *Proc. of European Weed Res. Symposium, Economic Weed Control.* pp. 75-82.

Wilson, B. J. and K. J. Wright. (1990). Predicting the growth and competitive effects of annual weeds in wheat. *Weed Res.* 30:201-211.

Zanin, G., A. Bertini, and L. Toniolo. (1993). Estimation of economic thresholds for weed control in winter wheat. *Weed Res.* 33:459-467.

Zimdahl, R. L. (1980). *Weed-Crop Competition: A Review.* International Plant Protection Center. Corvallis, OR: Oregon State University. 196 p.

RECEIVED: 06/10/97
ACCEPTED: 10/30/97

# Ecological Implications
# of Using Thresholds for Weed Management

## Robert F. Norris

**SUMMARY.** Various types of thresholds have been developed for weed management in an attempt to provide a more rational approach to decision making. The economic threshold concept was originally developed for management of arthropod pests, and is based on an understanding of arthropod population biology. Adoption of a management strategy for weeds that was developed for maintaining arthropod populations below a damaging level, referred to as the economic injury level or EIL, is not ecologically sound. Many of the factors that regulate populations of the two types of pest are different. For arthropod management the economic threshold (ET) is defined as the pest population at which treatment should be initiated to stop the population from increasing to the EIL. Weed science has adopted the ET to be the same as the EIL; this leads to maintenance of a relatively high seed bank as weeds at or below the ET density are allowed to produce seed. Research where weed seed production was accurately determined in corn, sugarbeets, alfalfa and other crops is now suggesting that several important weed species should be managed so that they do not produce seed. I am proposing that a new threshold called a no seed threshold, or NST, should be established for such weeds. Application of the ET concept to an invading weed species is disastrous as it leads to establishment of the seedbank before any control action is taken; for an invading species that is expanding its range the use of NST seems more appropriate. Progressive farmers in California have adopted NST for management of weeds. These farmers claim that the strategy is economically superior to that

Robert F. Norris, Associate Professor, Weed Science Program, University of California, Davis, CA 95616 (E-mail: rfnorris@ucdavis.edu).

[Haworth co-indexing entry note]: "Ecological Implications of Using Thresholds for Weed Management." Norris, Robert F. Co-published simultaneously in *Journal of Crop Production* (Food Products Press, an imprint of The Haworth Press, Inc.) Vol. 2, No. 1 (#3), 1999, pp. 31-58; and: *Expanding the Context of Weed Management* (ed: Douglas D. Buhler) Food Products Press, an imprint of The Haworth Press, Inc., 1999, pp. 31-58. Single or multiple copies of this article are available for a fee from The Haworth Document Delivery Service [1-800-342-9678, 9:00 a.m. - 5:00 p.m. (EST). E-mail address: getinfo@haworthpressinc. com].

© 1999 by The Haworth Press, Inc. All rights reserved.

*31*

using single season ET. In one case adoption of NST for weed management has resulted in decreased reliance on herbicides as weed control can be attained using non-chemical techniques. There is urgent need for weed science to develop improved data on weed population dynamics that is coupled to economics of weed control and crop production; until such data become available it is not feasible to accurately assess the use of thresholds for weed management. *[Article copies available for a fee from The Haworth Document Delivery Service: 1-800-342-9678. E-mail address: getinfo@haworthpressinc.com]*

**KEYWORDS.** Economic thresholds, invasion, no seed thresholds, pest management, population dynamics

## WEEDS AND THE THRESHOLD CONCEPT

The concept of utilizing economic thresholds to provide a more rational way of making pest control decisions originated in the 1950s (Stern et al., 1959). The first efforts were all directed towards more rational management of arthropod pests, and the concept of economic thresholds is now well established for integrated pest management (IPM) programs for many arthropod pests (see Stern, 1973; Pedigo, Hutchins and Higley, 1986; Pedigo, 1996). In its simplest form the concept attempts to relate the population development of an arthropod pest (Figure 1) to the anticipated single season crop loss from such pest attack (Figure 2). Economic thresholds have also been developed as a component of the decision making for nematodes (Ferris, 1978; Osteen, Moffitt and Johnson, 1988), and pathogens (Zadoks, 1985). This paper addresses utilization and applicability of thresholds for weed management, proposes a new threshold, reviews published data supporting the concept of not letting weeds set seed, briefly explores the problem of using thresholds to manage invading weed species, and concludes with two examples of farming operations that use stopping weeds from producing seed as a management philosophy.

In response to mounting pressure to reduce herbicide use, and as a component of IPM programs in crop production, there has been considerable effort made to develop economic thresholds for weed management (Coble and Mortensen, 1992). There are now several examples of economic thresholds being developed for use in cereal production (Cousens et al., 1986; Heitefuss, Gerowitt and Wahmoff, 1987; Gerowitt and Heitefuss, 1990; Black and Dyson, 1993; Zanin, Berti and Toniolo, 1993; Kwon et al., 1995). There has also been considerable effort to develop economic thresholds for decision making for weed management in soybeans [*Glycine max* (L.) Merr.] (Coble, 1985;

FIGURE 1. Hypothetical example of pest arthropod populations dynamics in relation to time, showing relative position of economic threshold and economic injury levels. GEP is the general population position at which the population stabilizes between outbreaks. Dashed line = population without control, solid line = population with control.

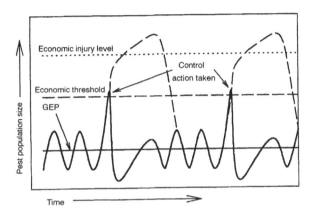

FIGURE 2. Generalized relationship between pest density and economic loss. Dotted line with open circles represents the population following a control action applied when the economic threshold had been exceeded.

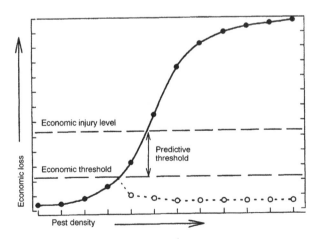

Renner and Black, 1991; Weersink, Deen and Weaver, 1991; Wilkerson, Modena and Coble, 1991; Bauer and Mortensen, 1992) and corn (*Zea mays* L.) (Lybecker, Schweizer and King, 1991; Sattin, Zanin and Berti, 1992; Cardina, Regnier and Sparrow, 1995; Berti et al., 1996; Bosnic and Swanton, 1997). Examples of other crops in which researchers have attempted to establish single-season economic thresholds include alfalfa (*Medicago sativa* L.) (Légère and Deschenes, 1989), onions (*Allium cepa* L.) (Dunan et al., 1995), sugarbeets (*Beta vulgaris* L.) (Norris, 1992a) and tomatoes (*Lycopersicon esculentum* L.) (Akey et al., 1995a). The general principles of applying thresholds for weed management have been extensively discussed (e.g., Cousens, Wilson and Cussans, 1985; Cussans, Cousens and Wilson, 1986; Auld, Menz and Tisdell, 1987; Cousens, 1987; Thornton et al., 1990; Doyle, 1991; Jordan, 1992; O'Donovan, 1996; Wallinga and van Oijen, 1997) and the reader is referred to these papers.

There are several fundamental problems with the way that the economic threshold concept has been applied to weed management. Weed science seems to have equated the economic threshold with the economic injury level; this creates a fundamental problem in relation to timing of initiation of control measures. The second problem is that the economic threshold concept as applied to weed management exacerbates development of herbicide resistant weeds. If we continue to use economic thresholds, as currently defined for weed management, I feel that we can assure ourselves of an ever increasing problem with herbicide resistant weeds. The use of economic thresholds for managing invasive weeds is another fundamental strategic error, as the approach allows the plant to become established before any control action is initiated. Utilizing economic thresholds requires high levels of weed management in each crop, and I argue that the whole concept is thus dependent on availability of herbicides to control the inevitable weed population that *will be* present in the field each year. Without herbicides it is doubtful if the economic threshold concept could be utilized. The modeling approach used by Wallinga and van Oijen (1997) suggests that the economic underpinning of the economic threshold concept is, in fact, flawed in relation to long-term weed management.

There are compelling biological and ecological reasons why the adoption of single season economic thresholds for weed management should be questioned. This paper expands and updates ideas presented at the First International Weed Control Congress in Melbourne (Norris, 1992b). I argue that the economic threshold paradigm as utilized for arthropod management does not lead to sound long-term economic weed management. This paper will present arguments based on the biology and ecology of weeds to support this position; others have presented mathematical arguments supporting the same position (Wallinga and van Oijen, 1997). At the outset I realize that some of the positions I am taking in this paper will be controversial; if I stimulate discussion and thinking about the utilization of thresholds in weed management, then I feel that my efforts will have been useful.

## DEFINITIONS

In reviewing literature on economic thresholds it is apparent that different authors define thresholds in different ways. Cousens (1987) reviewed various types of thresholds that have been used, and interpreted them in relation to weed management. I have used these definitions and incorporated ideas presented by Weersink et al. (1991), Coble and Mortensen (1992) and Donovan (1996). I have also included information from entomological definitions (Stern et al., 1959; Stern, 1973; Pedigo, Hutchins and Higley, 1986; Pedigo, 1996).

*Threshold:* This unqualified term is often used, but is of very little value because it is too vague. It simply means the lowest level of stimulus to which there is a reaction. Its use should be avoided in relation to weed management.

*Competition threshold:* This implies that as weed density decreases there is a level below which there is no further loss. Biologically this does not make sense, and Cousens (1987) argued, and I concur, that the term should not be used.

*Statistical threshold:* This is referred to as a damage threshold in most of the entomological literature. It is the point at which statistically valid yield (or other parameter) losses can be determined. It has been given various other names by weed scientists (see Cousens, 1987).

*Economic injury level (EIL):* This threshold is usually not discussed in relation to weed management, and was not included by Cousens (1987) or by Coble and Mortensen (1992). It is defined as "the lowest population density which can cause economic damage." It is sometimes also referred to as the "damage threshold" (Figure 2). As far as I can determine this is the threshold that most weed scientists call an economic threshold (see below). Weersink, Deen, and Weaver (1991) note this discrepancy in their discussion of economic thresholds for weed management. Lack of application of the EIL concept is a fundamental problem in the way that thresholds are being used for weed management.

*Economic threshold (ET):* This is typically defined by weed science as the point at which losses equal cost of control (Cousens, Wilson and Cussans, 1985; Auld and Tisdell., 1987; Cousens, 1987; O'Donovan, 1996). There is, however, a fundamental difference between how this threshold is used by weed scientists and by entomologists. The definition presented by Cousens (1987) states that the ET is "that weed density at which the cost of control measures equals the increased return in yield which could result" and is widely accepted by weed science. The entomological definition of ET states that it is "that pest density at which control measures should be initiated to prevent an increasing pest population from reaching the EIL" (Figure 2). When ET is equated with EIL the population is treated much later in its development and economic loss will have occurred. This is exactly the situation in which weed science finds itself when adopting the ET definition given above. This will be discussed at greater length below.

The definition given above implies a single cropping cycle. Throughout this paper the use of the term economic threshold implies the single-season nature of the definitions used above.

*Predictive threshold:* This is the difference between the ET and the EIL (Figure 2). For insects and pathogens the ET is always lower than the EIL. The predictive threshold thus implies that management action must be taken *before* attaining the EIL. For weeds this question has not been addressed due to the equating of EIL and ET, but attempts to develop EOT (see below) suggest that the ET is perhaps 2- to 10-fold lower than the EIL. Another way to state this is to say that for a weed population, the ET occurs one to several *years* prior to achieving the EIL.

*Economic optimum threshold (EOT):* This threshold was proposed by Cousens (1987) and is now seeing limited use in weed science. The EOT attempts to include the economic impacts of the multiyear population dynamics characteristic of weeds (Jordan, 1992). It is not used in the other pest management disciplines. Modeling suggests that the EOT is lower than the (single season) ET (Table 1). There are, however, no actual examples of weed population dynamics data for different threshold management levels to support these model predictions.

If weed science were to adopt the use of EIL and ET as defined by entomologists there would be no need for EOT. The use of ET to prevent a population from achieving an EIL would automatically mean that the long-term nature of weed population dynamics would be considered. This would require that weed science reevaluate how it is currently using the term ET.

*Safety threshold:* This threshold attempts to allow a safety margin due to uncertainty about economics and the actual losses that will occur. The result is lower values than for ET, but reduction is not well defined.

*Visual threshold:* This is an intuitive threshold, and is basically what is visually acceptable to the land manager. It is probably the threshold that many farmers and pest control advisors use (see Czapar, Curry and Wax, 1997). It is difficult to quantify.

*Action threshold:* This threshold is defined as the population at which a grower decides to institute a control tactic (Coble and Mortensen, 1992). This is a combination of economic threshold, safety threshold and visual threshold.

*No seed threshold (NST):* This threshold is defined on the basis that weeds present or remaining in a field should not be permitted to set seed (Norris,

TABLE 1. Comparison of predicted economic threshold (ET) and economic optimum thresholds (EOT) for weeds in several crops.

| Weed | Crop system | ET | EOT | Reference |
|---|---|---|---|---|
| Wild oat | Cereals | 8 to 10/m$^2$ | 2 to 3/m$^2$ | Cousens et al., 1986 |
| Blackgrass | Cereals | around 30/m$^2$ | 7.5/m$^2$ | Doyle, Cousens, and Moss, 1986 |
| Velvetleaf | Soybeans | 2.1/m$^2$ | 0.3 to 0.4/m$^2$ | Bauer and Mortensen, 1992 |
| Sunflower | Soybeans | 1.2/m$^2$ | 0.4/m$^2$ | Bauer and Mortensen, 1992 |
| Velvetleaf | Corn | Probably zero | – | Cardina, Regnier and Sparrow, 1995 |
| | | | | Cardina and Norquay, 1997 |
| | | Zero | – | Zanin and Sattin, 1988; Sattin, Zanin, and Berti, 1992 |
| Hempnettle | Alfalfa and oats | Zero | – | Légère and Deschenes, 1989 |
| Barnyardgrass | Tomatoes | Zero | – | Akey et al., 1995a |
| Barnyardgrass | Sugarbeets | Zero | – | Norris, 1992a |

1995). Weeds that will not be producing seed by harvest (e.g., emerged too late) and which do not reduce crop yield (or cause other problems, such as harvest difficulty) would be permitted to grow. It is feasible that this threshold may not be applicable to broadcast low-value crops but rather be applied to relatively high value row crops. This concept will be developed further in this paper.

*Zero threshold:* This implies that weeds are not allowed to grow. It is widely discussed as a management option, but was not included by Cousens (1987). This concept invokes a negative connotation due to the absence of plants other than the crop.

## ARTHROPOD AND WEED POPULATION ECOLOGY

The threshold concept is based on understanding pest population biology. Stern et al. (1959) and virtually all publications since then place the different thresholds (EIL, ET, etc.) for arthropod populations in relation to population development (Figures 1 and 2). Although many of the current threshold concepts were initially developed for arthropod management, they have been adopted for weed management with little change. There are, however, fundamental differences in the ecology and population biology of weeds and arthropods that lead to different interpretations of how these principles can be applied to using threshold management concepts for the two classes of pests (Table 2). The following discussion expands the comparison between the two types of pest organism.

*Trophic level:* Weeds are producers in an ecosystem trophic dynamics sense. They are green plants that carry out photosynthesis and can manufacture sugars from carbon dioxide and water using energy from sunlight. Arthropods are consumers, and therefore must ingest complex organic molecules as food. Pest arthropods that feed on crop plants, for which most thresholds have been developed, are thus primary consumers, or herbivores, feeding on the producers. In many instances weeds can actually serve as a food source for plant pest arthropods.

Resource supply and the factors regulating the population of the consumer are not the same as those for the producer. The resources for weeds are water, light, carbon dioxide and mineral nutrients. Weed population dynamics can be directly altered by changes in these resources. Light availability under a canopy, for instance, can substantially alter weed population dynamics (seed output). The resources for arthropods are the plants or other consumer organisms. Any effect of primary ecosystem resources on arthropods is indirect; changing light intensity has almost no direct effect on arthropod population dynamics. If the concepts of bottom-up and top-down driven trophic dynamics are used then the plants (weeds?) drive the system in the former, and in the

TABLE 2. Aspects of the population ecology of weeds and arthropods compared.

| Population parameter | Weeds | Arthropods | Implications for thresholds |
|---|---|---|---|
| Trophic level | Producers | Consumers | Use different resources |
| Longevity | Persistent seedbank | No long-term carryover for most spp. | Long-term implications |
| Population decrease | Populations decline slowly; do not "crash" | Populations usually "crash" at some time of the year | Population controls and loss rates differ |
| Generation time | 1 or less generations/yr | 1 to multiple generations/yr | Time scales differ |
| Population synchronicity | Population not synchronous | Population often synchronous | Prediction reliability |
| Population initiation | Delayed emergence | Repeated generations | Prediction difficulty |
| Organism mobility | Most are stationary (but moved by human activity) | Mobile | Prediction differences; invasion processes different |
| Off-season survival | Dormancy; survive within the managed ecosystem | Hibernation/aestivation; often not within managed ecosystem | Prediction difficulty |
| Fecundity per individual per generation | Relatively high fecundity: typically 1:1000 to 1:1000000+ per generation cycle | Relatively low fecundity: 1:100 to 1:1000+ typical per generation cycle | Rapidity of population regeneration |
| Damage potential | 'Damage/individual varies | Damage/individual ± fixed | Prediction of damage/loss |

latter the herbivores drive the system. Whichever system is functioning it is not reasonable to expect that the economic thresholds for the consumer will operate in a manner similar to those for the producer.

*Longevity:* The longevity of the population affects population size, genetic turn-over, and size of initial population at the start of the season. Weeds typically produce seeds which, in turn, will produce the next generation. Weed seeds can remain viable in the soil for many years. A short-lived weed might have seed that remains viable for 2 to 5 years, while many long-lived weed seeds can persist for in excess of 20 years, and reports of over 100 years have been made for several species. This leads to considerable overlap of generations, with a phenomenon in the seed bank that has been referred to as "genetic memory." For most pest arthropods there is no equivalent to the seedbank, and typical longevity of any one generation is from a few days to about one year. There are some noted exceptions, such as cicadas and shrimps. There is little population overlap between generations except within a season, and thus there is no "genetic memory" in the sense that there is for a weed seedbank containing seeds produced in several different years. The ability of weeds to establish a seedbank dictates that the management paradigm for weeds should be different from that used for arthropods.

· *Population decrease:* Weed populations decline slowly over time due to long-lived seeds with varying levels of dormancy (see below). Unless catastrophic events (e.g., flooding, fire) kill plants, the seedbank decline typically follows an exponential decay. Arthropod populations often "crash"; they experience a sudden large decrease in numbers due to natural events like an epizootic, hot or cold weather, and seasonal changes. Only those members of the population that are in the suitable stage to survive the crash will survive to the next favorable season. This means that an arthropod population can decrease from damaging (above EIL) to inconsequential (below ET), within a day or two to a few weeks. This type of population decrease *does not occur* for weeds, for which population decrease occurs over a few to many years (Wilson and Lawson, 1992; Thompson, Band and Hodgson, 1993; Burnside et al., 1996; Radosevich, Holt and Ghersa, 1997). The threshold concept works well for a population of multivoltine consumer organisms that crashes each year, but is not readily applicable to univoltine consumer pests (Pedigo, Hutchins and Higley, 1986; Pedigo, 1996). This difference between utility of economic thresholds for univoltine and multivoltine arthropods raises a question about applicability of economic thresholds for producer organisms for which the population continues from year to year.

*Generation time:* The typical generation time for most weeds is one year; some perennial species may, however, require several years before any offspring are produced. The generation time for univoltine arthropods is once per year. In contrast, multivoltine arthropods have generation times that range from a month or two to as short as a few weeks. Most arthropod thresholds are for those species that possess multivoltine population dynamics; this means that the population can increase rapidly *within* the season. Examples include aphids, mites, Lygus bugs, *Helicoverpa zea* Bodie (bollworms, corn earworms, tomato fruitworm), numerous other Lepidopteran pests, leafminers (*Liriomyza* spp.), whiteflies (*Bemisia* spp.), etc. Controlling one of these types of arthropod pest at the ET delays or stops the population from attaining the EIL. The use of thresholds is considered much less feasible for univoltine arthropods (Pedigo, Hutchins and Higley, 1986; Pedigo, 1996). On the basis of generation time weed population dynamics are not suitable for use of economic thresholds as employed for management of multivoltine arthropods.

*Population synchronicity:* Populations of weeds are not synchronous. Due to population overlap for perennials, and varying levels of seed dormancy in annuals, a weed population is typically made of individuals of differing chronological ages. The difference in age of individuals in the population range from a few weeks to several years. This can create difficulty in obtaining effective control even if decision making is not affected. Populations of many arthropod populations are synchronous. All individuals hatch within a few days, the adults mate and lays eggs within a few days, and the adults often die within a few days of each other. Population synchronicity improves the ability to make economic threshold decisions for many arthropod pests as it improves the reliability of predicting events and damage. Lack of population synchronicity in weeds makes economic threshold decisions much less reliable, and decreases the chances that a control strategy will be successful.

*Organism mobility:* Populations of most important weeds are not mobile (unless aided by animals or humans). Weed populations are typically fairly stable in a particular field (specific agroecosystem), and thus their size can be predicted over time. Many pest arthropods are mobile, at least in one phase of their life-cycle (usually adults). Also, many pest arthropods are not present in the managed ecosystem during the off-season. This means that they must migrate back to the field in, or must build-up to, damaging numbers. In either case using an economic threshold can suggest treatment before the EIL is attained. For most weeds the population is already present in the field at the start of the season even if it is not readily visible. On the basis of organism mobility using an economic threshold concept for weed management that was originally developed for management of mobile arthropods does not make sense. The case of weeds with wind-borne seeds [such as groundsel (*Senecio vulgaris* L.), prickly lettuce (*Lactuca serriola* L.), willowherbs (*Epilobium* spp.)] is special as they can invade previously non-infested fields without human or animal intervention, and is further discussed under the topic of thresholds in relation to invasion of weeds.

*Off-season survival:* Most weed seeds possess from one to several mechanisms of dormancy that regulate the percentage of the population that will germinate at any particular germination event (Radosevich, Holt and Ghersa, 1997). Arthropod populations can also exhibit dormancy, and either hibernate or aestivate to minimize the impacts of the non-favorable season (Pedigo, 1996). Arthropod dormancy typically lasts for only one off-season, whereas weed dormancy can last for many years. These two different survival strategies result in different ways in which populations are regulated.

*Fecundity per generation:* Many weeds produce more than 1000 seeds/plant when permitted to grow in crops that are weak competitors, and recent evidence is showing that several common weeds can produce over 100,000 seeds/plant when growing in less competitive crops. Most arthropods produce from 10s to 100s of offspring, but few exceed 1000 per individual. In conjunction with the multivoltine nature of many arthropod populations the fecundity per individual leads to different rates of population increase between weeds and arthropods. Many common weed populations can explode *in a single generation* (and do not crash). Most arthropods require three or four generations to build up to similar population levels. If a fecundity of 100 offspring per individual is assumed, and there is no population regulation between generations (extremely unlikely) it will take between three and four generations to achieve the same population that a single parent producing 100,000 offspring per generation achieves. This difference in population increase between weed and arthropod populations leads to a different interpretation of economic threshold time-frame. In the case of many arthropods the time-frame is less than a growing season; for most weeds the time-frame is many years.

*Damage potential:* The amount of yield loss per individual weed plant is variable over a wide range depending on conditions such as time of year at emergence (day length), soil fertility, proximity and type of neighbors, and soil moisture. A mature redroot pigweed (*Amaranthus retroflexus* L.) plant that is 10 cm tall does not cause the same competitive loss as one that is 150 cm tall, yet this range of size can commonly occur in a single field. In contrast, an arthropod of equivalent developmental stage of a single species can be expected to cause essentially the same amount of damage as another individual of the same species and developmental stage. A numerical estimate of the pest arthropod thus provides a fairly reliable estimate of the amount of damage that will be sustained. A numerical count of weeds present may, however, have little relationship with the amount of damage that will be sustained. This necessitates questioning the meaning of economic thresholds for weeds that are based on numerical counts of individuals present.

These differences in biology/ecology lead to very different population dynamics between a multivoltine arthropod and an annual weed. A typical population development of an arthropod population is shown in Figure 1. The population at the beginning of the season is relatively low and said to be at the general equilibrium position or population (GEP). When conditions are correct the population increases; if it exceeds the economic threshold control action is taken to keep the population from increasing to a damaging level.

The population dynamics of any weed that reproduces by seed is different (Figure 3). Due to the presence of the seedbank, the GEP is initially high. A portion of the population germinates under favorable conditions, and becomes seedlings. At this point there is little decline in the overall population, but the size of the seedbank has decreased proportionally to the number of seeds that germinated. Following a control action the total population is

FIGURE 3. Theoretical population dynamics of an annual weed with seedbank showing the proportions of the population represented by the seedbank and the growing plants. Abbreviations: GEP = general equilibrium position; EIL = economic injury level; ET = economic threshold; NST = no seed threshold.

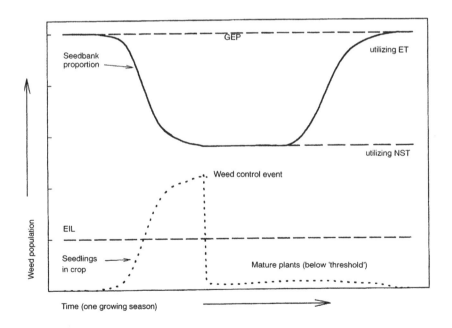

reduced in proportion to the number of germinated seeds that were killed, and the above ground population is reduced to whatever is the accepted "threshold." For an economic threshold, as defined earlier for weed science, seed production by the weeds judged below the threshold has the potential to return the seedbank to its original size when the system is operating with equilibrium populations. Utilization of NST would result in a seedbank that is lower than the original value by the amount of germination, plus any predation and death that occurred. Continued annual use of single season economic thresholds for weeds results in a relatively high GEP, with a more or less stable seedbank. The use of NST is predicted to result in a stepwise decline in the seedbank each time there is a germination event. The use of EOT would result in a seedbank that is lower than that maintained if ET is used, but will still be adequate to assure a rapid increase in weed population if control strategies were relaxed. Use of NST raises a question regarding the difference in cost between implementing EOT vs. NST. I argue that the slight increase in cost in going from EOT to NST is more than offset by the reduction in weed population in future years. Resolution of this question awaits results of long-term management experiments.

The foregoing discussions show that there is a fundamental difference in management strategy used against arthropod pests and against weeds. The aim of arthropod management is to stop the population from *increasing* to the EIL. The aim of weed management, using economic thresholds as currently defined, is to temporarily *decrease* the population down to a level that is acceptable for crop production. I argue, for this reason and all those discussed above, that using the same management strategy, namely economic thresholds, for weeds and pest arthropods does not make ecological sense.

## ADDITIONAL CONCEPTS

There are several other issues that are relevant to discussing the utility of thresholds for weed management. O'Donovan (1996) pointed out several of these and they are included here in abbreviated format.

Most arthropod economic thresholds are for a single species. Some authors have suggested that the interactions between different arthropod pests needs to be considered (e.g., Newsom, 1980). In the case of weeds, multiple species interactions are the norm and several authors have noted this problem (e.g., Cousens, 1987; O'Donovan, 1996). Calculating economic thresholds for a single species may not have much utility when most fields have mixed species present. Adopting NST reduces the problem of multi-species interactions at the decision making level.

In order for thresholds to be used the predictions they make must be reliable over years and locations. The weed population and yield loss components of threshold decisions are typically based on some form of regression model. Several researchers have pointed out that the various parameters in such models are proving to be unstable (Cardina, Regnier and Sparrow, 1995; Johnson et al., 1995; McGiffen et al., 1997). Utilization of an NST management philosophy reduces the significance of lack of stability of predictions.

The use of thresholds implies that the pest population can be assessed in an easy and timely manner before deciding to employ a control tactic. O'Donovan (1996) pointed out the problem of weed population assessment. Sim (1987) considered difficulty of population assessment a major reason for lack of adoption of thresholds for weed management in cereals in England, and Kwon et al. (1995) specifically excluded multi-year consideration from the PALWEED-WHEAT model due to difficulty of seedbank assessment. The actual techniques for weed population assessment are difficult (e.g., seedbank analysis), time consuming, and often require considerable expertise (recognition of seeds and/or young seedlings of several species). Associated with this are the actual costs of equipment and personnel to make the assessment. As noted previously, there is a serious problem in relation to what actually constitutes a unit of measurement. The number of seeds or seedlings present may really not be adequate for predictive purposes due to the plasticity of plant growth and the clonal nature of perennials. Biomass of weed vegetation may be a better predictor of competition and seed production but the data are difficult to obtain. Relative leaf area of crop and weeds may also be a useful way to assess weed impacts on crops but may have limited utility in predicting long-term population dynamics. Adoption of NST would probably result in reduced costs for population assessment as evaluation is on a presence/absence basis, whereas ET or EOT require development of information on density (or other parameter) per unit area. The costs of population assessment should be considered when judging the utility of using ET versus NST.

Weeds are normally present in patchy distributions (e.g., van Groenendael, 1988; Hughes, 1990; Thornton et al., 1990; O'Donovan, 1996; Cardina, Johnson and Sparrow, 1997). All authors discussed the implication of patchiness in relation to competition, crop losses, and weed population assessment. As adoption of NST requires only presence/absence information the need for evaluation of weed patchiness is reduced.

The adoption of any form of thresholds for weed management has serious implications in terms of development of resistance to herbicides in weed populations. When weeds remain in the field they produce seed, which then pass the genetic make-up of the population to the next generation. If the weeds at or below the threshold have been selected by herbicide application, then allowing them to produce seed will perpetuate the resistance characteristic. Full adoption of NST would eliminate the problem of weeds developing resistance to herbicides as no seeds are produced and thus no genes are passed on. The utility of this approach to managing a herbicide resistant weed problem has been demonstrated by Powles, Tucker and Morgan (1992) who were able to eliminate paraquat-resistant *Hordeum glaucum* Steud. in three years by eliminating seed production.

Adoption of an NST strategy implies an integrated approach to weed management, and will probably require use of hand labor to remove weeds that escape other management tactics. There are essentially no data currently available for the cost of hand weeding when weed densities are at or below the economic threshold. Estimated costs of hoeing low weed densities, and grower experiences are presented in following sections.

Land ownership may be an important reason for use of single season economic thresholds. If land is not owned but is rather farmed on short-term leases there is little incentive to consider weed seed production and the costs of weed control on a long-term basis. It is thus feasible that land-ownership is in fact a serious impediment to adoption of the sound long-term weed management programs. Landlord perception of long-term weed management problems were, however, listed as a major reason for lack of adoption of single season economic thresholds for weed management by farmers in Illinois (Czapar, Curry and Wax, 1997). The socio-economics of land ownership probably impinges on adoption of economic thresholds for weed management.

Weeds remaining at harvest were listed by farmers in Illinois as the most important reason for lack of adoption of single season economic threshold is used for weed management (Czapar, Curry and Wax, 1997). Adoption of a NST philosophy removes this concern.

## *POPULATION DYNAMICS RESEARCH EXAMPLES*

Most research that relates crop loss to weed presence has provided little or no information on seed production, seedbank dynamics, and other aspects of the population biology of the weeds in the system (see Zimdahl, 1980). In the absence of such information I argue that it is not feasible to draw valid conclusions about what might constitute an economic threshold for the weed, because it is not possible to place the value into a pest population context in the way that entomologists have done when developing economic thresholds for arthropod management.

Elliott (1972) suggested that if a farm has a very low population of wild oats (*Avena fatua* L.) then the best management strategy would be to hand rogue to stop the formation of seeds. In 1984 I concluded, from using a simple weed population dynamics model, that the weed threshold should be zero (Norris, 1984). During the last 10 years, there have been several examples of competition studies where data for weed population dynamics were also developed and which support this concept (Table 1).

Using a corn/velvetleaf [*Abutilon theophrasti* (L.) Medic.] system in the Po valley of Italy, Zanin and Sattin (1988) and Sattin, Zanin and Berti (1992) concluded that the only logical management strategy for the weed, based on a relatively simple model of population dynamics, was to not permit the weed to produce seed. This conclusion has recently been supported for velvetleaf management in corn in the central USA (Cardina, Regnier and Sparrow, 1995; Cardina and Norquay, 1997). Légère and Deschenes (1989), working with alfalfa and oats (*Avena sativa* L.) in Quebec, Canada, concluded on the basis of weed population dynamics that hempnettle (*Galeopsis tetrahit* L.) should not be allowed to produce seed. Norris (1992a) likewise concluded that, based on population dynamics, barnyardgrass [*Echinochloa crus-galli* (L.) Beauv.] should not be permitted to set seed in sugarbeet fields in California, USA. A recent study (Akey et al., 1995b), also in California, concluded that barnyardgrass should not be allowed to set seed in processing tomato fields.

The conclusion by the authors of the papers noted above was that seed production by weeds at or below economic threshold densities was so high that it would perpetuate the need for weed control in the next crop. For velvetleaf, it was concluded that a single year of seed production by a sub-economic threshold density of the weed would result in seedling populations above the economic threshold for several years (Cardina and Norquay, 1997). From my own work (Norris, 1992a), I concluded that barnyardgrass in sugarbeets at the single season economic threshold of about 1 plant per 10 m of crop row would return about 18,000 seeds/m$^2$ to the seedbank, of which we now estimate that about 80% will survive to the following spring. This far exceeds the economic threshold for the weed in any crop in California. One barnyardgrass plant in a hectare of sugarbeets or tomatoes in California produces sufficient seeds to reinfest, with dispersal by human activity, the entire hectare at about 10 seeds/m$^2$. At 80% survival this will mandate weed control in the following crops.

The argument is made by all the above-noted authors that stopping seed production by a relatively low number of weeds in the current crop is more cost effective than controlling a large population in the next crop. I have attempted to predict costs of handweeding using current hoeing costs in California. If it is assumed that the cost of labor for hoeing is $7.50/hr, that the crew walks at about 5 km/hr when not hoeing, and that it takes 15 seconds to remove a weed, then the cost of hoeing weed densities below about 1 every 20 m of crop row does not exceed $25.00/ha. At weed densities below 1 per 100 m of crop row the cost does not exceed $10.00/ha. Even using the threshold density of 1 plant every 10 m the cost of hand weeding would only be about $50.00/ha. These calculated values are in close agreement with $7.50/ha provided by Mark Grewall of the J. G. Boswell Company (see below) for hoeing escape weeds in cotton (*Gossypium hirsutum* L.). Seed rain from weeds at the densities discussed here will result in the need for weed control in the subsequent crop which would cost more than the cost of hand weeding.

The reader is referred to the papers noted above for details of the experiments, but I feel that it is striking that when accurate data on seed production and knowledge of seedbank behavior were combined with data on economic impacts that the researchers all concluded that utilizing any form of economic thresholds as a management strategy did not make sense. I argue that current suggestions for using economic thresholds for weed management are based on crop yield loss data in the absence of weed population biology information, and thus do not properly evaluate the economic impacts of the weeds.

Wallinga and van Oijen (1997) also concluded that economic thresholds are not a sound weed management strategy. They state that "the economic underpinning of the threshold concept is deceptive and does not provide a base for rational use of weed control in the long term." They arrived at this conclusion using a strictly economic modeling approach. Their conclusion was thus the same as that derived from the field research data discussed above. This strengthens the argument that the use of economic thresholds for weed management is not appropriate.

A further set of field data has been developed that also strengthens the reason for not adopting economic thresholds. Sugarbeets were grown at the research farm at the University of California at Davis in the presence of varying densities of barnyardgrass, which was grown as cohorts initiated at varying dates after establishment of the crop. The agronomic techniques used have been described elsewhere (Norris, 1992a). Barnyardgrass was germinated with the crop, or at 2, 4, 6, 9, and 12 weeks after crop germination.

Sugarbeet yield loss resulting from the 0-delay cohort of barnyardgrass that emerged with the crop was similar to that reported previously (Norris, 1992a), and indicated that the single-season economic threshold would be about 1 weed/10 m of crop row (Figure 4A). A 2-week delay in barnyardgrass emergence in relation to the sugarbeets substantially reduced the magnitude of the crop loss, which resulted in a single-season economic threshold of about 1 plant/m. The difference between the threshold for the 0-delay versus the 2-week delay cohorts was a 10-fold increase in barnyardgrass density. A 4-week delay in barnyardgrass emergence relative to the sugarbeets increased the economic threshold density to about 3 or 4 plants/m. There were no yield losses when barnyardgrass emerged after 9 or more weeks delay regardless of weed density. In the absence of seed production information these data would suggest that after about 4 to 6 weeks there is no

FIGURE 4. Impact of increasing barnyardgrass density in relation to time of emergence on 'A' sugarbeet root yield and 'B' barnyardgrass seed production. The value listed as 'delay' indicate the time between initial sugarbeet irrigation and the first irrigation for delayed cohorts of barnyardgrass.

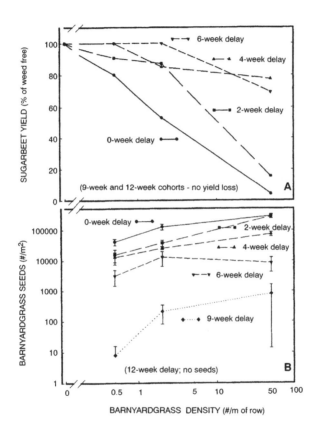

single-season requirement for further weed management action unless the density of the late emerging cohort was still very high.

Barnyardgrass seed production was estimated at harvest by removing inflorescences from randomly selected tillers and measuring the inflorescence lengths. This permitted prediction of seed rain based on inflorescence length and numbers (Norris, 1992c). Seed production for the 0-delay barnyardgrass cohort ranged from 40,000/m$^2$ for the 0.5/m density to nearly 300,000/m$^2$ at the high density (Figure 4B); these figures were consistent with those reported previously (Norris, 1992a). With each delay in cohort initiation the number of seeds produced decreased. At the six week delay the 0.5 and 2.0/m density were estimated to have produced 3100 and 13000 seeds/m$^2$, respectively. At these densities for this delay in initiation there was no economic yield loss, yet the seed production was sufficient to cause major problems in the next crop. Even at the 9-week initiation delay there was sufficient seed production to resupply much of the seedbank yet there was no economic impact on the current crop yield. These data for seed production by late emerging cohorts of barnyardgrass again suggest that the single season economic thresholds would be a poor way to manage the weed due to the number of seeds produced by sub-threshold densities of the weed.

## WEED INVASIONS AND THE THRESHOLD CONCEPT

Dewey, Jenkins and Tonioli (1995) proposed that invading noxious weed species should be treated like a "wildfire." The implication is that control should be carried out while the "fire" is small, rather than attempting control when it has become a large conflagration. This is the accepted paradigm for fighting wildfires. Control costs are much lower when a wildfire is controlled early. One could say that we have a very low threshold for wildfires. In the context of weed management this could be rephrased to state that the threshold for invading species is low; controlling a few plants is much less costly than controlling many well established plants.

Non-critical acceptance of the single season economic threshold concept is *the* most important problem in relation to weed invasion. Most land managers do not act until the problem caused by an invading weed reaches the EIL (Figure 5), at which time a seedbank has already been established. Even using EOT, or ET as defined entomologically, would mean treatment when the population is low. The NST approach would suggest that newly established individual plants should be removed prior to seed production. Such action would stop the plant from developing a seedbank.

Velvetleaf makes a good example to illustrate the point made above. Hand rogueing of the weed is fairly easy in most crops when the populations are low. Cardina and Norquay (1997) determined that a single year of seed production from a sub-threshold velvetleaf population results in seedling populations well above the threshold level in following years. Allowing the plant to produce seed results in a very long lived weed problem (Figure 6); even with no reseeding there were adequate seeds remaining in the soil to create a new infestation even after 17 years of corn and soybean production, and even more if alfalfa had been grown (Lueschen et al., 1993). The use of ET, as defined by weed science, is a disastrous approach to managing invading species that leads to increased costs and reduced production. The use of NST is the logical threshold to use for management of invading species that is

FIGURE 5. Hypothetical increase in population for an invading weed species, shown in relation to economic injury level and economic threshold used in the accepted entomological sense. Economic injury level is population at which economic loss starts to occur; economic threshold (?) is the population at which control measures should be initiated.

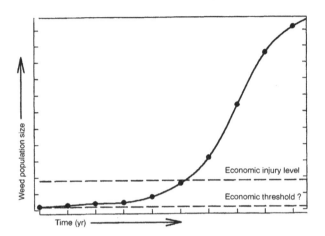

spread by human activity. Adoption of ET may, however, be the only logical approach to managing an invading species with wind born seeds.

## USE OF NO SEED THRESHOLDS FOR WEED MANAGEMENT: FARM EXAMPLES IN CALIFORNIA

Although anecdotal in nature, weed management programs used by two California farms serve as practical examples of the advantages of adopting policies that do not permit weeds to produce seed. The farm managers refuse to use thresholds for weed management, although they subscribe to and use thresholds for management of other types of pests. They argue that the use of thresholds for weed management would not be economically acceptable due to the long-term problems created by letting weeds set seed, and which thus perpetuate the seedbank.

One example is the largest arable corporate farm in California (J. G. Boswell Co., 60,000 ha of arable crop production) (see article by Horstmeier, 1995). The whole farm has been operated on a policy of not letting weeds set seed for over 40 years. They still use hand weeding in 30,000 ha of cotton two or three times a season to ensure that weeds like nightshade (*Solanum* spp.) and annual morning glory (*Ipomoea* spp.) do not produce seed. When I

FIGURE 6. Decline of velvetleaf seedbank over time in relation to different cultural practice. Graph is redrawn from data presented by Lueschen et al. (1993).

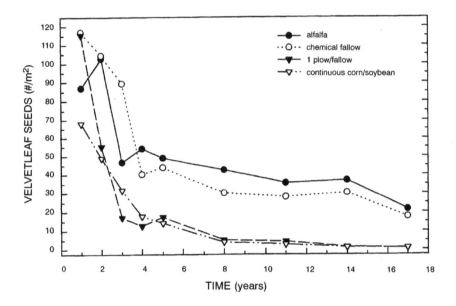

asked Mr. Grewal, the manager described in the above-mentioned article (Horstmeier, 1995), how he could afford the hand weeding his response was "how can I afford not to?" What Mr. Grewal meant was that he judged the $5 or $6 per hectare cost of hand weeding to be cheaper than controlling a large weed population in the next season. This is a very different paradigm than that of the visual economic threshold that most farmers appear to use. Mr. Grewal states that weed management costs for the Boswell farming operation are lower than those for their competitors. The Boswell Company philosophy emphasizes a holistic approach to weed management that most farms never actually practice. The following are some examples of such practices: farm machinery *is* cleaned when it leaves fields in which weeds were present; weeds are not allowed to set seed when present in crop stubble following harvest; weeds are not permitted to grow on field margins, roadsides and ditchbanks; all equipment storage areas are maintained free of weeds; drivers of harvesters are instructed to go around isolated infestations of weeds so that they do not spread the seed.

Another example is a 180 ha intensive vegetable farm in the Salinas valley.

Louis Manzoni felt that weed management was becoming so difficult in vegetable production that he decided to kill the weed seedbank by treating the whole ranch with methyl bromide (done over several years). The single treatment coupled with the change in weed management philosophy to NST has allowed Mr. Manzoni to grow vegetable crops without using herbicides since that time. Seven years after using methyl bromide, Mr. Manzoni had essentially no weeds present on the ranch. He is so adamant that weeds should not set seed that he even uses a hand crew, at a cost that he estimated to be over $150/ha, to pull weeds out of a berseem clover (*Trifolium alexandrinum* L.) cover crop! Mr. Manzoni informed me that he can obtain contracts with packer/shippers for his vegetable production much more easily now that his weed management has changed to NST. Mr. Manzoni states that he recovered the cost of the methyl bromide in four crops (2 years) and that since that time his weed control costs have been much lower than before he adopted NST. Mr. Manzoni considers that his neighbors who permit weeds to set seed in their crops or along the margins of fields are "crazy."

These farms represent two different arable agro-ecosystems, yet both claim that their weed management costs are lower than their neighbors (competitors?) who use one or other form of thresholds (or let at least some weeds go to seed). The weed management programs for the two farm examples have changed from a concept of control to achieve a successful crop, to a strategy aimed at stopping weed reinfestation. Both farms are owned, not leased, and both managers point out that there is incentive for using long-term weed management strategies. This suggests that the weed management paradigm used by most farmers is substantially contributing to their continuing weed problems and relatively high levels of weed management required each year, and strengthens the argument proposed by the researchers who have arrived at the same conclusions through combination of crop loss and weed population biology research (Table 1).

Until the research community develops reliable data on the long-term impacts of different weed management paradigms on weed population dynamics and farm economics, we will have to be satisfied with accepting that the strategies used by some of our progressive farmers may be correct.

## *CONCLUSION*

In light of ecological principles of population biology, in light of mounting weed population biology research evidence in relation to economic losses, in light of revised thinking about weed invasion processes, and in light of progressive farmer experiences there is strong reason to reject use of single season economic thresholds for weed management. The only debate that should be going on is between using multi-year economic thresholds (EOT) as proposed by Cousens (1987) and exemplified by work by Doyle, Cousens and Moss (1986) and by Bauer and Mortensen (1992), or adopting the policy of not letting weeds set seed as proposed for velvetleaf in corn (Zanin and Sattin, 1988; Sattin, Zanin and Berti, 1992; Cardina, Regnier and Sparrow, 1995; Cardina and Norquay, 1997), hempnettle in alfalfa and oats (Légère and Deschenes, 1989), barnyardgrass in sugarbeets (Norris, 1992a), and barnyardgrass in tomatoes (Akey et al., 1995a). The ultimate utility of using thresholds for weed management cannot be resolved without data from long-term research that provides reliable economic information in relation to weed management practice coupled with information on the population dynamics of the weeds. I therefore argue that there is urgent need for weed science to develop reliable data that couples weed population biology with crop losses caused by weeds.

I also argue that there is need to develop a weed management paradigm, and not to blindly accept that developed for management of arthropod pests. If we continue to pursue a weed management paradigm that uses single-season economic thresholds we are probably doomed to continue using large quantities of herbicides and to face ever increasing problems of herbicide resistant weeds. If a no seed threshold weed management paradigm is adopted, weed control costs should ultimately decrease, and reliance on herbicides to "put out the fire" each year will also decrease. I also strongly believe that weed science also needs to reevaluate the trend away from use of hand labor for weed management, as it is this tactic which makes NST work. The development of "smart" sprayers that can detect the presence of a weed have the potential to greatly increase the ease of implementing an NST philosophy in the future. If we adopt no seed thresholds based on integrated weed management practices, the whole problem of resistant weeds should decrease; if no seed are produced, then no resistance genes are passed to the next generation.

## REFERENCES

Akey, W.C., R.F. Norris, M. Rejmanek, and C.L. Elmore. (1995a). Does it matter where the weeds are? The impact of plant aggregation on interspecific competition. *Weed Sci. Soc. Am. Abst.* 35: 48.

Akey, W.C., R.F. Norris, M. Rejmanek, and C.L. Elmore. (1995b). Influence of density and spatial distribution of barnyardgrass [*Echinochloa crus-galli* (L.) Beauv.] on its growth and seed production in competition with direct seeded tomatoes. *Weed Sci. Soc. Am. Abst.* 35: 48.

Auld, B.A., K.M. Menz, and C.A. Tisdell (1987). *Weed Control Economics.* London, NW1 7DX, UK: Academic Press (London) Ltd.

Auld, B.A. and C.A. Tisdell. (1987). Economic thresholds and response to uncertainty in weed control. *Agricultural Systems* 25: 219-227.

Bauer, T.A. and D.A. Mortensen. (1992). A comparison of economic and economic optimum thresholds for two annual weeds in soybeans. *Weed Technology* 6: 228-235.

Berti, A., C. Dunan, M. Sattin, G. Zanin, and P. Westra. (1996). A new approach to determine when to control weeds. *Weed Science* 44: 496-503.

Black, I.D. and C.B. Dyson. (1993). An economic threshold model for spraying herbicides in cereals. *Weed Research* 33: 279-290.

Bosnic, A.C. and C.J. Swanton. (1997). Economic decision rules for postemergence herbicide control of barnyardgrass (*Echinochloa crus-galli*) in corn (*Zea mays*). *Weed Science* 45: 557-563.

Burnside, O.C., R.G. Wilson, S. Weisberg, and K.G. Hubbard. (1996). Seed longevity of 41 weed species buried 17 years in eastern and western Nebraska. *Weed Sci* 44: 74-86.

Cardina, J., G.A. Johnson, and D.H. Sparrow. (1997). The nature and consequence of weed spatial distribution. *Weed Science* 45: 364-373.

Cardina, J. and H.M. Norquay. (1997). Seed production and seedbank dynamics in subthreshold velvetleaf (*Abutilon theophrasti*) populations. *Weed Science* 45: 85-90.

Cardina, J., E. Regnier, and D. Sparrow. (1995). Velvetleaf (*Abutilon theophrasti*) competition and economic thresholds in conventional- and no-tillage corn (*Zea mays*). *Weed Science* 43: 81-87.

Coble, H.D. (1985). Development and implementation of economic thresholds for soybean. *Integrated Pest Management on Major Agricultural Systems,* eds. R.E. Frisbie and P.L. Adkisson. College Station, TX: Texas A&M University. MP-1616, pp. 285-307.

Coble, H.D. and D.A. Mortensen. (1992). The threshold concept and its application to weed science. *Weed Technology* 6: 191-195.

Cousens, R. (1987). Theory and reality of weed control thresholds. *Plant Protection Quarterly* 2: 13-20.

Cousens, R., C., J. Doyle, B.J. Wilson, and G.W. Cussans. (1986). Modelling the economics of controlling *Avena fatua* in winter wheat. *Pesticide Science* 17: 1-12.

Cousens, R., B.J. Wilson, and G.W. Cussans (1985). *To Spray or Not to Spray: The Theory Behind the Practice.* British Crop Protection Conference-Weeds, Brighton, U.K., pp. 671-678.

Cussans, G.W., R.D. Cousens, and B. Wilson, J. (1986). Thresholds for weed control–the concepts and their interpretation. *Proc. EWRS Symp. 1986. Economic Weed Control* pp:253-260.

Czapar, G.F., M.P. Curry, and L.M. Wax. (1997). Grower acceptance of economic thresholds for weed management in Illinois. *Weed Technology* 11: 828-831.

Dewey, S.A., M.J. Jenkins, and R.C. Tonioli. (1995). Wildfire suppression–a paradigm for noxious weed management. *Weed Technology* 9: 621-627.

Doyle, C.J. (1991). Mathematical models in weed management. *Crop Protection* 10:432-444.

Doyle, C.J., R. Cousens, and S.R. Moss. (1986). A model of the economics of controlling *Alopecurus myosuroides* Huds. in winter wheat. *Crop Protection* 5: 143-150.

Dunan, C.M., P. Westra, E.E. Schweizer, D.W. Lybecker, and F.D. Moore. (1995). The concept and application of early economic period threshold–the case of DCPA in onions (*Allium cepa*). *Weed Science* 43: 634-639.

Elliott, J.G. (1972). *Wild Oats, Where Next?* Proceedings British Weed Control Conference, Brighton, U.K., pp. 965-976.

Ferris, H. (1978). Nematode economic thresholds: derivations, requirements, and theoretical considerations. *Journal of Nematology* 10: 341-350.

Gerowitt, B. and R. Heitefuss. (1990). Weed economic thresholds in cereals in the Federal Republic of Germany. *Crop Protection* 9: 323-331.

Heitefuss, R., B. Gerowitt, and W. Wahmoff (1987). *Development and Implementation of Weed Economic Thresholds in the F. R. Germany.* British Crop Protection Conference–Weeds., pp. 1025-1033.

Horstmeier, G. (1995). Not one weed. *Top Producer*: 26-27.

Hughes, G. (1990). The problem of weed patchiness.\ *Weed Research* 30:223-224.

Johnson, G.A., D.A. Mortensen, L.J. Young, and A.R. Martin. (1995). The stability of weed seedling population models and parameters in Eastern Nebraska corn (*Zea mays*) and soybean (*Glycine max*) fields. *Weed Science* 43: 604-611.

Jordan, N. (1992). Weed demography and population dynamics: implications for threshold management. *Weed Technology* 6: 184-190.

Kwon, T.J., D.L. Young, F.L. Young, and C.M. Boerboom. (1995). PALWEED-WHEAT: a bioeconomic decision model for postemergence weed management in winter wheat (*Triticum aestivum*). *Weed Science* 43: 595-603.

Légère, A. and J.-M. Deschenes. (1989). Effects of time of emergence, population density and interspecific competition on hemp-nettle (*Galeopsis tetrahit*) seed production. *Canadian Journal of Plant Science* 69: 185-194.

Lueschen, W.E., R.N. Anderson, T.R. Hoverstad, and B.K. Kann. (1993). Seventeen years of cropping systems and tillage affect velvetleaf (*Abutilon theophrasti*) seed longevity. *Weed Science* 41: 82-86.

Lybecker, D.W., E.E. Schweizer, and R.P. King. (1991). Weed management decisions in corn based on bioeconomic modeling. *Weed Science* 39:124-129.

McGiffen, M.E., F. Forcella, M.J. Lindstrom, and D.C. Reicosky. (1997). Covariance of cropping systems and foxtail density as predictors of weed interference. *Weed Science* 45: 388-396.

Newsom, L.D. (1980). The next rung up the integrated pest management ladder. *Entomological Society of America Bulletin* 26: 369-374.

Norris, R.F. (1984). Weed thresholds in relation to long-term population dynamics. *Proceedings of the Western Society of Weed Science* 37: 38-44.

Norris, R.F. (1992a). Case history for weed competition/population ecology: barn-yardgrass (*Echinochloa crus-galli*) in sugarbeets (*Beta vulgaris*). *Weed Technology* 6: 220-227.

Norris, R.F. (1992b). *Have Ecological and Biological Studies Improved Weed Control Strategies?* Proceedings of the First International Weed Control Congress, Melbourne, Australia, pp. 7-33.

Norris, R.F. (1992c). *Predicting Seed Rain in Barnyardgrass (Echinochloa crus-galli).* IX Colloquium on Weed Biology and Ecology, Dijon, France, European Weed Research Society, pp. 377-386.

Norris, R.F. (1995). Thresholds: detrimental for weed management? *European Journal of Plant Pathology, Abstracts*: 969.

O'Donovan, J.T. (1996). Weed economic thresholds: useful agronomic tool or pipe dream? *Phytoprotection* 77: 13-28.

Osteen, C.D., J. Moffitt, and A.W. Johnson. (1988). Risk efficient action thresholds for nematode management. *Journal of Production Agriculture* 1: 332-338.

Pedigo, L.P. (1996). *Entomology & Pest Management*. Upper Saddle River, NJ: Prentice Hall.

Pedigo, L.P., S.H. Hutchins, and L.G. Higley. (1986). Economic injury levels in theory and practice. *Annual Review of Entomology* 31: 341-368.

Powles, S.B., E.S. Tucker, and T.R. Morgan. (1992). Eradication of paraquat-resistant *Hordeum glaucum* Steud. by prevention of seed production for 3 years. *Weed Research* 32: 207-211.

Radosevich, S., J. Holt, and C. Ghersa (1997). *Weed Ecology: Implications for Management*. New York, NY: John Wiley and Sons, Inc.

Renner, K.A. and J.R. Black. (1991). SOYHERB–A computer program for soybean herbicide decision making. *Agronomy Journal* 83: 921-925.

Sattin, M., G. Zanin, and A. Berti. (1992). Case history for weed competition/population ecology: velvetleaf (*Abutilon theophrasti*) in corn (*Zea mays*). *Weed Technology* 6: 213-219.

Sim, L.C. (1987). *The Value and Practicality of Using Weed Thresholds in the Field*. British Crop Protection Conference–Weeds., Brighton, U.K., pp. 1067-1071.

Stern, V.M. (1973). Economic thresholds. *Annual Review of Entomology* 18: 259-280.

Stern, V.M., R.F. Smith, R. van den Bosch, and K.S. Hagen. (1959). The integrated control concept. *Hilgardia* 29: 81-99.

Thompson, K., S.R. Band, and J.G. Hodgson. (1993). Seed size and shape predict persistence in the soil. *Functional Ecology* 7: 236-241.

Thornton, P.K., R.H. Fawcett, J.B. Dent, and T.J. Perkins. (1990). Spatial weed distribution and economic thresholds for weed control. *Crop Protection* 9: 337-342.

van Groenendael, J.M. (1988). Patchy distribution of weeds and some implications for modelling population dynamics. *Weed Research* 28: 437-441.

Wallinga, J. and M. van Oijen. (1997). Level of threshold weed density does not affect the long-term frequency of weed control. *Crop Protection* 16: 273-278.

Weersink, A., W. Deen, and S. Weaver. (1991). Defining and measuring economic threshold levels. *Canadian Journal of Agricultural Economics* 39: 619-625.

Wilkerson, G.G., S.A. Modena, and H.D. Coble. (1991). HERB: decision model for postemergence weed control in soybean. *Agronomy Journal* 83: 413-417.

Wilson, B.J. and H.M. Lawson. (1992). Seedbank persistence and seedling emergence of seven weed species in autumn-sown crops following a single years seeding. *Annals of Applied Biology* 120: 105-116.

Zadoks, J.C. (1985). On the conceptual basis of crop loss assessment: the threshold theory. *Review of Phytopathology* 23: 455-473.

Zanin, G., A. Berti, and L. Toniolo. (1993). Estimation of economic thresholds for weed control in winter wheat. *Weed Research* 33: 459-467.

Zanin, G. and M. Sattin. (1988). Threshold level and seed production of velvetleaf (*Abutilon theophrasti* Medicus) in maize. *Weed Research* 28: 347-352.

Zimdahl, R.L. (1980). *Weed-Crop Competition: A Review.* Corvallis, OR.: International Plant Protection Center, Orgeon State University.

RECEIVED: 08/15/97
ACCEPTED: 03/13/98

# Increasing Crop Competitiveness to Weeds Through Crop Breeding

Todd A. Pester
Orvin C. Burnside
James H. Orf

**SUMMARY.** Increasing the ability of crops to compete against weeds, through either enhancing crop tolerance or crop interference to weeds, provides an attractive addition to current weed control practices and could be an integral component of weed management systems. Research has shown that considerable variability exists among crop cultivars with respect to their ability to compete with weeds. Despite this evidence, directed research on competitive crops has been minimal. Reasons for this lack of emphasis in plant breeding programs include the effectiveness of current weed management with tillage and herbicides, and the lack of easily identifiable crop characteristics that are indicative of weed competitiveness. Expanded knowledge of specific crop-weed interactions would facilitate crop competitiveness to weeds through either crop management practices or plant breeding. Plant breeders need basic and applied information to identify favorable crop-weed competitive traits in order to enhance or incorporate those traits into crop cultivars. Accelerated research on weed competitive crops

---

Todd A. Pester, Graduate Research Assistant, Department of Bioagricultural Science and Pest Management, Colorado State University, Fort Collins, CO 80523. Orvin C. Burnside and James H. Orf, Professors, Department of Agronomy and Plant Genetics, University of Minnesota, St. Paul, MN 55108.

Address correspondence to: Todd A. Pester, Department of Bioagricultural Science and Pest Management, Colorado State University, Fort Collins, CO 80523 (E-mail: tpester@lamar.colostate.edu).

[Haworth co-indexing entry note]: "Increasing Crop Competitiveness to Weeds Through Crop Breeding." Pester, Todd A., Orvin C. Burnside, and James H. Orf. Co-published simultaneously in *Journal of Crop Production* (Food Products Press, an imprint of The Haworth Press, Inc.) Vol. 2, No. 1 (#3), 1999, pp. 59-76; and: *Expanding the Context of Weed Management* (ed: Douglas D. Buhler) Food Products Press, an imprint of The Haworth Press, Inc., 1999, pp. 59-76. Single or multiple copies of this article are available for a fee from The Haworth Document Delivery Service [1-800-342-9678, 9:00 a.m. - 5:00 p.m. (EST). E-mail address: getinfo@haworthpressinc.com].

© 1999 by The Haworth Press, Inc. All rights reserved.

should lead to more economical, effective, and feasible integrated weed management programs for all crops. *[Article copies available for a fee from The Haworth Document Delivery Service: 1-800-342-9678. E-mail address: getinfo@haworthpressinc.com]*

**KEYWORDS.** Crop morphology, integrated weed management, plant breeding

## INTRODUCTION

Before the widespread use of selective herbicides, farmers relied on an integrated weed management system consisting of crop interference, crop rotations, selective tillage, hand weeding (Parish, 1990), and preventive weed control methods (Burnside, 1970). Over the past 50 years, however, these integrated practices have been increasingly displaced by herbicide and tillage treatments that have reduced management and labor inputs required for crop production. While herbicides and tillage have proven to be highly effective methods of weed management, current crop production concerns over high input costs, negative environmental impacts, and food safety have spurred renewed interest in alternative methods of weed management (Burnside, 1992; McWhorter, 1984; Swanton and Weise, 1991; Wyse, 1992). Enhancing the ability of crop species to compete with weeds is an attractive control option that could be an integral component of future weed management systems (McWhorter and Barrentine, 1975; Rose et al., 1984; Urwin, Wilson, and Mortensen, 1996). Weed scientists can follow the lead of plant pathologists and plant breeders, who have successfully used host resistance as a major method of controlling or mitigating plant diseases, by developing genotypes that are more competitive to weeds.

Increasing crop competitiveness to weeds could be accomplished through both improved crop management practices and directed plant breeding efforts. Crop production practices should be directed at increasing crop competitiveness to weeds by providing the most suitable management technology possible to favor crop growth. For example, narrow row spacing, increased plant density, and time of planting are capable of reducing weed pressures (Buhler and Gunsolus, 1996; Malik, Swanton, and Michaels, 1993). Plant breeding should be directed at developing crop cultivars that are genetically superior competitors to weeds, either through crop tolerance or crop interference to weeds. Researchers have documented differential weed competitive ability among commonly grown cultivars of many crops indicating potential weed management progress through plant breeding (Berkowitz, 1988; Callaway, 1992; Callaway and Forcella, 1992; Lanning et al., 1997; Rose et al.,

1984). Characteristics commonly identified to make crops more competitive to weeds include rapid germination and root development, early above ground growth and vigor, rapid leaf area and canopy establishment, large leaf area development and duration, and greater plant height (Callaway, 1992; Challaiah et al., 1986).

Increasing weed competitiveness of crops should be considered as only a part of the total weed management strategy. No weed management method has proven to be a panacea, so one must integrate various weed control technologies into a comprehensive weed management system in order to achieve economic weed control. Additionally, we need to determine the associated costs of increasing crop competitiveness to weeds. These may include increased production expenses (e.g., seed, fertility, and equipment) and costs in plant fitness such as resource allocation and possible lower yield potential.

The objectives of this paper are to (a) review the history of crop competitiveness to weeds, (b) evaluate reasons for slow development of weed competitiveness in crop cultivars and needed stimulation for future research, (c) enumerate appropriate breeding methods and biotechniques to enhance crop competitiveness to weeds, and (d) discuss implications of adding crop competitiveness into future systems of weed management.

## HISTORY OF CROP COMPETITIVENESS TO WEEDS

Research on crop competitiveness to weeds dates back to the 1930's or earlier (Crafts and Robbins, 1962; Pavlychenko and Harrington, 1934). Over the past 60 years many crop-weed competition studies have been conducted throughout the world (Berkowitz, 1988; Crafts and Robbins, 1962; Kropff and van Laar, 1993; Rice, 1984). This research revealed considerable variation in weed competitiveness among crops and crop cultivars. Some crop cultivars are more capable of maintaining yield despite high weed populations (crop tolerance to weeds), while other crop cultivars are able to suppress weed growth and weed seed production (crop interference to weeds) (Ahmed and Hoque, 1981; Bridges and Chandler, 1988; Callaway, 1992; Callaway and Forcella, 1992; McWhorter and Hartwig, 1972; Rose et al., 1984).

Differences in crop interference to weeds among cultivars have been established by many researchers examining plant variables such as growth attributes, light interception, or seed production of crops and weeds (Callaway, 1992). Vigorous growth attributes including early emergence, high leaf area index (LAI), dense canopy, and plant height were identified as superior weed competitive traits among several crops. In rice (*Oryza sativa* L.), LAI and increased biomass at an early stage of growth were the traits most closely related to weed competitive ability (Garrity, Movillon, and Moody, 1992; Kawano, Gonzalez, and Lucena, 1974). In dry bean (*Phaseolus vulgaris* L.), LAI and leaf size accounted for 73% of the total variation in weed biomass (Wortmann, 1993). Malik, Swanton, and Michaels (1993) also found that high LAI, enhanced by narrow row spacing and increased crop density, increased dry bean weed competitive ability. Similar results have been reported for soybean [*Glycine max* (L.) Merr.] where early emergence, leaf area expansion rate, and rapid canopy closure were associated with reduced growth of several annual grass and broadleaf weed species (Bussen et al., 1997; Pester, 1996; Rose et al., 1984).

Researchers have also identified profuse tillering, branching, and vine growth habits as favorable weed competitive characteristics in rice, dry bean, and soybean (Akobundu and Ahissou, 1985; Rose et al., 1984; Soto and Gamboa,1984; Urwin, Wilson, and Mortensen, 1996). Many of these vigorous growth characteristics enhance weed competitiveness by reducing light quality and quantity beneath the crop canopy thereby impeding weed seed germination and weed seedling growth. This has been demonstrated in wheat (*Triticum aestivum* L.) and barley (*Hordeum vulgare* L.) (Challaiah et al., 1983; Lanning et al., 1997). The ease of measuring plant height in crop-weed competition experiments permits its frequent recording. In studies with rice, wheat, cotton, and soybean, increased plant height has been the character most closely related to weed biomass reduction (Ahmed and Hoque, 1981; Balyan et al., 1991; Blackshaw, 1994; Bridges and Chandler, 1988; Garrity, Movillon, and Moody, 1992; Wicks et al., 1986).

In many situations crop interference also acts to reduce weed seed production (Blackshaw, 1994; Friesen, Nickel, and Morrison, 1992). Increasing crop interference could be part of an integrated approach to weed management aimed at minimizing future yield losses to crops by reducing weed seed banks. If competitive crop cultivars were able to reduce the amount of weed seed being returned to the soil each year, this lowering of the weed seed bank would reduce weed density during future weed seed germination events. However, the ability of current crop cultivars to consistently reduce weed seed production is highly variable (Blackshaw, 1994; Koscelny et al., 1990; Pester, 1996). Given the typical existing high weed seed bank populations (Crafts and Robbins, 1962) and seed longevities (Burnside et al., 1996), it would be more realistic to concentrate research on reducing weed biomass during each crop production year. This may change someday when feasible and economic methods are developed to destroy the soil weed seed bank.

The results of these crop-weed competition studies illuminate the advantages of aggressive, early plant growth in suppressing weed biomass, and in some cases, weed seed production. Interestingly, many crop cultivars currently being grown have been selected by plant breeders for early, vigorous growth in optimal environments. This indicates that plant breeders may already be inadvertently selecting for weed competitiveness in crop cultivars.

## REASONS FOR THE SLOW DEVELOPMENT OF WEED COMPETITIVE CROPS

Past research reveals that differences in weed competitiveness exist among cultivars of crops that could markedly reduce weed biomass production in future crop production systems (Callaway, 1992). With this vision of future weed management, one might ask why has the development of crop cultivars with increased competitiveness toward weeds been so slow to reach fruition?

A primary reason for the slow development of weed competitive crops has been the effectiveness of current herbicide and tillage methods in controlling weeds. Farmers have grown accustomed to 90% plus efficacy at controlling weeds from one or two herbicide applications, making it difficult to justify a search for other methods. Similarly, selective tillage within a growing crop is often very effective at relatively low cost. In the past, crop production specialists have focused on high yields in optimum environments. Weed competitive crop cultivars have not been able to reduce weeds to acceptable levels by themselves and therefore have not received much consideration.

Selecting plants for competitiveness against weeds has also been impeded by a lack of suitable selection criteria for evaluating plant progeny. Despite the fact that many traits have been identified that correlate to weed competitive ability, the logistics and feasibility of selecting for these traits in a breeding program have not been resolved. As scientists gain a better understanding of the characteristics and mechanisms of weed competitive ability among crop cultivars, and develop the appropriate techniques for exploiting them, we can then breed for such characteristics in crop plants (Baldwin and Santleman, 1980). Mechanistic research such as path analysis could help identify such specific crop traits that interfere strongly with early weed growth (Jordan, 1993; Pantone, Baker, and Jordan, 1992). However, if a plant breeder can screen and select for a single trait (e.g., seed germination, vigorous early growth, rapid leaf area expansion, or days to maturity) that increases weed competitiveness and thus can make consistent improvements in subsequent filial generations, the trait's direct or indirect relationship to weed competitiveness may be irrelevant.

The complexity of crop-weed interactions has also hindered the development of weed competitive crops. Weed competitive crop traits may vary among weed densities and production systems. Crop production is often impeded by a wide range of weed species and weed densities which are most often spatially and temporally variable (Mortensen, Johnson, and Young, 1993; Thornton et al., 1990; Wiles et al., 1992). These variable weedy conditions will likely influence crop competitive performance across a landscape. Additionally, specific crop traits believed to be associated with weed competitiveness may vary in their response among crop species (Callaway and Forcella, 1992). For example, later maturity was important in increasing weed tolerance of rice cultivars (Smith, 1974), but early maturity increased weed tolerance in corn hybrids (Staniforth, 1961).

Finally, the associated costs in plant fitness of increasing weed competitive ability of crop cultivars are not well understood. Can increased crop competitiveness economically replace other weed management practices? To answer this question we need more information on the effects of enhanced crop competitiveness on weed population dynamics (e.g., weed seed bank). We need to know if weed competitive cultivars are less resistant to growth stresses from water, nutrients, light, temperature, plant diseases, and insects (Blum, 1988). Also, we need to better understand any associated yield costs from increasing weed competitiveness of crops. However, considering the high costs of weed management, increasing crop competitiveness to weeds would be an economical and effective addition to integrated weed management systems.

## STIMULATION FOR DEVELOPMENT OF WEED COMPETITIVE CROPS

Many factors are currently driving the demand for alternative methods of weed management. First, crop production is becoming more competitive on a worldwide basis every year. Modest increases in crop yield and market value have not kept up with rising input costs for crop production in most countries. Second, the continuous loss of farm labor in developed countries coupled with increasing farm size has created the need for weed management methods that require less of the farmer's time. Third, the impacts of tillage and herbicides on soil erosion and water quality have aroused environmental concerns. Developing weed competitive crops would help alleviate these problems and reduce the burden currently placed on tillage and chemical weed control methods. Countries in earlier developmental stages could also benefit from incorporating this technology as a component of integrated weed management.

The roles of tillage and herbicides in weed management systems are changing. Weed control is usually achieved in conventionally-tilled crops with an integration of cultural, mechanical, and chemical practices. Increases in reduced-till and no-till is often associated with increased expenditures for chemical weed control (Buhler, 1995). Controlling weeds has become more of a challenge in no-till crop production because the farmer has given up the tillage option (Worsham, 1991). Consumer concerns regarding the effects of pesticides on the environment and food and feed quality and safety have resulted in re-evaluation of current herbicides (Goldburg, 1992; Wauchope, 1994). All herbicides in the United States registered prior to November 1984 must be re-registered or not used beyond 2001. Company personnel often decide to discontinue producing herbicides used on minor acreage crops because it is not cost effective to keep them registered. Also, weed species are developing resistance to herbicides at an accelerated rate (Holt, 1992).

Weed management needs to utilize a system of integrated technologies to reduce the effects of weeds on the crop while addressing the concerns of current production systems. Enhancing the ability of crop species to compete with weeds offers a new weed management option that would complement integrated weed management systems with little additional cost or effort to the farmer. However, in years of poor crop establishment or other conditions not conducive to crop competitiveness, rescue treatments of tillage, selective postemergence herbicides, or even hand weeding may be required. Development of new integrated weed management systems, including weed competitive crops, will require a concerted effort of many agronomic disciplines.

## BASIC RESEARCH NEEDS OF WEED COMPETITIVE CROPS

There are many aspects of crop-weed competition that are not well understood. Basic research must be implemented to address these gaps in knowledge before significant progress can be made toward developing crops with increased weed competitiveness. Mechanistic research should be conducted to identify specific relationships between crop and weed interactions (Jordan, 1993; Pantone, Baker, and Jordan, 1992). As in many areas of crop-weed ecology, our understanding of these issues is limited by insufficient weed biology research (Aldrich, 1984; Radosevich and Holt, 1984). Modeling efforts could also help determine the effects of the competitive trait or mechanism on weed population dynamics and identify critical events in crop and weed life cycles to maximize the effects of crop competition (Dotray and Young, 1993).

Weed competitive traits identified through basic weed biology and ecology research could stimulate effort and progress in plant breeding programs. To facilitate rapid, cost-effective screening of hundreds of crop biotypes, these traits should be morphological and easily identified by plant breeders. Past research that identified vigorous early growth as a competitive advantage typically included measurements of morphological characteristics. Estimates of many of these morphological characteristics could be obtained visually and therefore lend themselves to rapid evaluation by plant breeders. Measurements of other characteristics such as light interception and below ground interactions may be equally valuable to increasing weed competitive ability but are more difficult to obtain. It will also be necessary to determine the heritability of such traits. Callaway and Forcella (1992) describe a soybean breeding experiment that showed inherited crop weed competitiveness through the $F_4$ generation, indicating potential for successful incorporation of weed competitive traits through breeding.

Crop tolerance and crop interference represent two approaches to increase weed competitiveness through plant breeding. Currently there is disagreement over which approach is most appropriate to pursue (Callaway and Forcella, 1992; Jordan, 1993). Crop tolerance is associated with reducing yield loss in the presence of weeds, thereby ensuring some crop yield even in unfavorable growing conditions (e.g., drought, shade, and limited nutrients). While maintaining crop yield is important, this approach seems counter productive as it would allow weeds to continue growing unabated, potentially returning large amounts of seed to the weed seed bank and interfering with crop harvest. In contrast, crop interference is associated with weed suppression. Weber and Staniforth (1957) found that soybean yield losses from weeds are usually proportional to the amount of water, nutrients, and light used by weeds at the expense of soybean. Selecting for crop interference to weeds would help crops aggressively compete with weeds early in the growing season to minimize their water and nutrient consumption and to negatively impact weed biomass and weed seed production. However, crop interference may or may not be obtained at the expense of other fitness factors such as crop yield. Recent research has shown that current crop cultivars selected for rapid establishment and high yield under weed-free conditions may also be superior competitors against weeds due to their inherent aggressive growth (Bussan, 1997; Pester, 1996). Regardless of which approach for increasing crop competitiveness is selected, there is a requirement for establishing definite plant breeding objectives prior to implementing the research, as each approach employs different selection methods and ecological principles.

Determination of when and how long weeds compete with a crop is important in developing a breeding program for increasing crop competitiveness to weeds (Aldrich, 1984; Burnside, 1979; Crafts and Robbins, 1962; Hall, Swanton, and Anderson, 1992). It is also important to know how long weeds have to be controlled in a crop before the crop itself can effectively compete with late-emerging weeds. Early weed removal has been shown to aid crop stand establishment, and there is an inverse relationship between crop stand and production of above-ground weed biomass (Hall, Swanton, and Anderson, 1992; Knake and Slife, 1965). The influence of early weed removal has also been identified in maintaining crop quality and yield. Burnside (1979) found that soybean seed weight and seed number per plant both increased when weeds were controlled. Soybean weeded at two and four weeks after planting did not show significantly reduced soybean yields from later emerging weeds. Thus, weed control during three to four weeks after planting was the most critical period in obtaining high soybean yields. Studies designed to identify critical weed-free periods should be conducted across a range of environments because the critical period will change somewhat with different growing conditions (e.g., weed competition to crops occurs earlier when moisture is limiting). Knowledge of these critical periods can help determine the potential effectiveness of a competitive cultivar and help producers develop appropriate weed management systems to enhance weed control.

A third approach to breeding for increased competitiveness is developing crop cultivars with resistance to herbicides (Duke, 1996). Research to develop herbicide resistant crops using conventional breeding methods or biotechnology is currently being conducted, mainly by private industry (Wilcut et al., 1996). These newly developed crop cultivars will lead us into future use of herbicides with greater selectivity. However, exploitation of this technology has resulted in many seed companies being purchased by herbicide manufacturers with the intent of developing crop cultivars that promote the use of their herbicides. Increasing crop competitiveness through interference to weeds would reduce the need for herbicides and therefore would not be a priority research objective for herbicide companies (Burnside, 1992). Consequently, breeding for increased crop competitiveness will likely be initiated by or even relegated to public plant breeders and seed companies without herbicide company affiliations.

## APPLIED RESEARCH NEEDS OF WEED COMPETITIVE CROPS

Genetic variability is critical to making progress in a breeding program (Moll and Stuber, 1974). Much of the past weed competitiveness research has been conducted with a limited amount of crop genetic variability. Callaway (1992) found that in most studies fewer than 10 cultivars were evaluated. Once true characteristics have been identified that correlate to weed competitiveness, a survey of the germplasm collections should be conducted to establish the extent of the variability and identify lines with the highest competitive ability within each crop species. For example, Dilday et al. (1991) has completed such a worldwide search of 10,000 rice accessions for allelopathic activity, and 347 weed competitive accessions were identified from 30 different countries. Similar searches are needed for weed competitive traits within each crop species to expedite plant breeding progress.

New crop cultivars displaying enhanced weed competitive ability should be evaluated in field scale integrated weed management systems. These field studies should focus on weed population dynamics and associated costs of crop fitness. Experiments of this nature would apply more to crop management variables that would synergistically support crop competitiveness to weeds. In particular, better understanding of weed control costs related to agricultural production, such as increased soil erosion from cultivation and water-quality effects from herbicides, would be critical to weighing costs and benefits of crop competitiveness to weeds.

Several cultural practices should be evaluated that could work in concert with weed competitive cultivars (e.g., cultivation, reduced tillage, crop rotation, narrowed row spacing, optimum planting date and rate, fertility, and chemical and biological control of insects and diseases). For example, plant residue left on the soil surface provides many benefits to crop production such as erosion control, water conservation, reduced temperature fluctuations, increased soil organic matter levels, improved soil structure, and suppression of weeds (Crutchfield, Wicks, and Burnside, 1985). Plant residue management is an important and effective method of weed control in no-till cropping systems, horticultural crops, and gardens. Crop cultivars selected for reduced tillage methods may differ from those used in conventional production systems because they may need to grow under cooler, wetter soil conditions (Burnside and Moomaw, 1984). Malik, Swanton, and Michaels (1993) found the ability of dry bean cultivars to reduce weed biomass was enhanced in medium (46 cm) and narrow rows (23 cm) as compared to traditional wide rows (69 cm). Cultivar, row spacing, and seeding density combinations that maximized LAI when grown under weedy conditions also significantly reduced weed biomass. Weed biomass may also be reduced by planting a mixture of cultivars in specific weedy areas of a field. Mumaw and Weber (1957) grew composites of unlike soybean cultivars and found this planting method exceeded the mean yield of their pure lines more than composites of similar soybean cultivars. When Buhler and Gunsolus (1996) delayed soybean planting from mid-May to early-June, weed densities and yield losses from weeds were reduced. Combinations of these types of cultural practices could fabricate more effective integrated weed management systems.

In breeding for crops with more competitiveness to weeds it will be important to consider future trends in cropping systems. Several traits were identified by Francis (1985) as being particularly important in developing cultivars for multiple cropping systems such as crop maturity, photoperiod sensitivity, temperature sensitivity, morphology, root system, seedling growth rate, and density response. Similar interactions apply to crop-weed competition. Field experiments should be designed to evaluate crop-weed interactions across various environments, locations, and years. Applied research that advances our knowledge and understanding of crop competitiveness to weeds is critical to successfully breeding crop cultivars that will provide a useful alternative to weed management through crop interference to weeds.

## STANDARD BREEDING METHODS FOR WEED COMPETITIVE CROPS

As noted earlier, there are many different crop traits or characteristics that can contribute to a genotype being more competitive to weeds. Plant breeding methods used for developing cultivars more competitive to weeds or more tolerant of weeds depends on the traits or characteristics that would be under selection. Breeders need crop traits or characteristics that are relatively easy to identify or measure and can be determined on thousands of genotypes rapidly and inexpensively. The key to success in breeding for crop competitiveness to weeds is the correct identification of those genotypes that are more competitive.

The breeding method or methods most appropriate for development of competitive crop cultivars depends on whether the trait is qualitative or quantitative in nature. Traits such as plant height and maturity could be considered qualitative in nature and are relatively easy to select for since major genes for these traits exist in most crop species. Most of the traits described as contributing to competitiveness such as early vigor, LAI, canopy establishment, leaf expansion rate, and yield are quantitative and thus require extensive measurements in order to identify the best genotypes. Since quantitative traits can be influenced significantly by the environment, measurements need to be made over years and locations to determine the extent of genotype-environment interaction (Hicks, Stucker, and Orf, 1992).

Breeding methods most frequently used for qualitative traits are the pedigree method or single seed descent (Poehlman and Sleper, 1995). Selection could begin as early as the $F_2$ generation or be done later as dictated by the plant breeding method. For quantitative traits, crop evaluation and selection would need to be carried out in order to achieve a greater level of homozygosity, and would probably involve some type of recurrent selection (Poehlman and Sleper, 1995). Data for selection of quantitative traits would need to be collected over several environments.

Progress through breeding is most rapid when crop genotypes with high levels of competitiveness are identified. This is best accomplished by screening the germplasm collections of crop species. Screening for traits like plant height, maturity, leaf area, or vigor may be carried out under weed-free conditions and then verified in field trials with weeds present. Traits such as yield of the crop under weedy conditions or reduction of weed biomass by crop genotype require that the trait be measured with weeds present. Basic and applied research needs to identify appropriate methods to facilitate cultivar selection when grown with weeds. For proper evaluation of weed competitive crop cultivars, plant breeders need information on weed population dynamics, crop-weed interactions, density effects, row spacing, and cultural methods. Understanding these variables will make selection of competitive genotypes and development of competitive cultivars by plant breeders more effective.

## IMPACT OF BIOTECHNOLOGY ON WEED COMPETITIVE CROPS

Biotechnology is beginning to have a significant impact on weed management. The development of herbicide resistant cultivars and hybrids provides farmers with additional options for weed management, but it also requires that they pay more attention to details. Some herbicide resistant cultivar and herbicide combinations provide control of a broad spectrum of weed species while others only are effective on broadleaf weeds, grass weeds, or certain species of weeds. Thus, the use of biotechnology to develop herbicide resistant cultivars does not necessarily preclude the need for competitive genotypes, but it may result in the need for competitiveness for only certain species or types of weeds. In one sense this may make the development of competitive genotypes easier, since there may be fewer weeds to consider, but in another sense it may require more work, since genotypes may need to be developed for specific cropping systems.

Although the use of biotechnology to develop genotypes more competitive with weeds is theoretically possible, the practical application of biotechnology for increasing crop competitiveness is far off. In order to use the tools of biotechnology, such as gene cloning and gene transfer, scientists first need to know what traits or characteristics contribute to crop competitiveness. Then they need to be able to accurately measure the phenotype of plants in a population or populations that have variability for crop competitiveness. Molecular markers associated with crop competitiveness then need to be identified and used to either clone the gene(s) for competitiveness (so they can be transferred to other genotypes or species) or to follow the incorporation of competitiveness into new cultivars. Since it appears that many traits associated with crop competitiveness are quantitative, the use of biotechnological techniques will be challenging. It is also likely that with these complex traits there will be interactions that will further complicate the process of incorporating competitiveness into productive cultivars.

The tools of biotechnology offer the potential to contribute to enhancing crop competitiveness to weeds. However, before there are major contributions to crop competitiveness to weeds through biotechnology, other than herbicide resistant cultivars, much work needs to be done to further understand crop competitiveness and the traits or characteristics underlying competitive crop genotypes.

## *FARMER UTILIZATION OF WEED COMPETITIVE CROPS*

No single weed control method is a panacea for weed management. However, crop competitiveness to weeds can provide a partial, but economical and effective addition to weed management systems. Some crop competitive characteristics may affect a specific weed or group of weeds such as grass or broadleaf species. Weed species shifts are a concern if crops are more competitive against a single weed or group of weeds; however, this is true of any weed management technology (Swanton, Clements, and Derksen, 1993). Farmers will be better served if weed competitive crops are thought of as just another component of their integrated weed management system.

Crop cultivars that are more competitive to weeds would be the simplest and most easily utilized weed management technology that one could provide farmers. Farmers are used to evaluating and selecting new crop cultivars that may provide increased yield or better plant disease, nematode, or insect control. Once farmers realize that a new crop cultivar can control 30 to 50% of their weeds, acceptance of this technology would be rapid. Furthermore, farmers who experience the value of weed competitive cultivars will have the confidence to attempt more high risk production practices such as narrow row crop production or reduced-till and no-till cropping systems.

As farmers begin integrating weed management components they can incorporate new technologies to enhance their weed control program in each crop (Buhler, 1995; Burnside et al., 1993; Mt. Pleasant, Burt, and Frisch, 1994). Precision agriculture techniques will allow farmers to describe the spatial variability of many biotic and abiotic factors such as weeds that are often spatially aggregated across a field (Mortensen, Johnson, and Young, 1993; Thornton et al., 1990; Wiles et al., 1992). Spatial statistical analysis techniques are continually improving to facilitate scientific quantification of spatial variability across a landscape. Technologies such as differential application of fertilizer and herbicides will benefit from information pertaining to this spatial variability. Once identified, areas of high weed infestation can be geographically referenced (Johnson, Mortensen, and Gotway, 1996) using tools such as global positioning systems (GPS) and geographical information systems (GIS) to generate prescription maps and facilitate site specific weed management options. These site specific management options may eventually include planting specific weed competitive crop cultivars in areas of high weed density based on weed seed bank analysis. Crop yield monitors further help farmers quantify the negative effects of weeds and also help evaluate effectiveness of crop cultivars with enhanced weed competitiveness. The future of weed management and crop production will become even more exciting and challenging, but the end result of utilizing crop competitiveness should be more economical, dependable, and productive weed management and higher crop yields and profits for the farmer.

## FUTURE IMPACT OF CROP COMPETITIVENESS ON WEED MANAGEMENT

Researchers have shown that differences in weed competitiveness of crop cultivars are the rule rather than the exception (Callaway, 1992). Economic costs to farmers from additional seed expenses of crop cultivars with enhanced weed competitive ability should be minimal; similar to cultivars with insect or disease resistance. Reduced herbicide and tillage costs will be realized because of less herbicide used and fewer applications or trips across the field. Reduced applications of herbicides resulting from crops being more competitive to weeds will further aid management of herbicide resistant weeds or variable growing conditions (Holt, 1992). These options will also reduce the environmental impact from crop production practices. Finally, specific planting of problem areas using competitive cultivars may provide an alternative to herbicide and tillage treatments in cropping situations.

As farmers and researchers examine previous weed management practices and exploit current technologies to develop new tools, it is important to realize that there will never be a permanent weed control program. Weed management in the future must always incorporate many methods working together as integrated components of a weed control system. This will reduce current reliance on individual approaches and provide opportunities for synergy among system components. Nowadays the weed scientist and farmer is limited in the number of feasible weed management methods available. Thus, public weed scientists must undertake accelerated research on alternative weed management methods such as preventive, cultural, mechanical, biological, and integrative approaches or these methods will not be exploited. This research should lead to more economical, effective, and feasible integrated weed management programs for all crops and non-cropland situations as we move into the twenty-first century.

## REFERENCES

Ahmed, N.U. and M.Z. Hoque. (1981). Plant height as a varietal characteristic in reduce weed competition in rice. *International Rice Research Newsletter* 6(3):20.

Akobundu, O. and A. Ahissou. (1985). Effect of interrow spacing and weeding frequency on the performance of selected rice varieties on hydromorphic soils of West Africa. *Crop Protection* 4:71-76.

Aldrich, R.J. (1984). *Weed-Crop Ecology: Principles in Weed Management*. Breton, North Scituate, MA.

Baldwin, F.L. and P.W. Santleman. (1980). Weed science and integrated pest management. *Bioscience* 30:675-678.

Balyan, R.S., R.K. Malik, R.S. Panwar, and S. Singh. (1991). Competitive ability of winter wheat cultivars with wild oat (*Avena ludoviciana*). *Weed Sci.* 39:154-158.

Berkowitz, A.R. (1988). Competition for resources in weed-crop mixtures. In *Weed Management in Agroecosystems: Ecological Approaches,* eds. M.A. Altieri and M. Liebman. Boca Raton, FL: CRC Press, pp. 89-120.

Blackshaw, R.E. (1994). Differential competitive ability of winter wheat cultivars against downy brome. *Agron. J.* 86:649-654.

Blum, A. (1988). *Plant Breeding for Stress Environments*. Boca Raton, FL: CRC Press.

Bridges, D.C. and J.M. Chandler. (1988). Influence of cultivar height on competitiveness of cotton (*Gossypium hirsutum*) with johnsongrass (*Sorghum halepense*). *Weed Sci.* 36:616-620.

Buhler, D.D. (1995). Influence of tillage systems on weed population dynamics and management in corn and soybean in the central USA. *Crop Sci.* 35:1247-1258.

Buhler, D.D. and J.L. Gunsolus. (1996). Effect of date of preplant tillage and planting on weed populations and mechanical weed control in soybean (*Glycine max*). *Weed Sci.* 44:373-379.

Burnside, O.C. (1970). Progress and potential for nonherbicidal control of weed through preventive weed control. In *FAO International Conference on Weed Control,* ed. J.T. Holstun, Jr. Urbana, IL: Weed Sci. Soc. America, pp. 464-483.

Burnside, O.C. (1979). Soybean (*Glycine max*) growth as affected by weed removal, cultivar, and row spacing. *Weed Sci.* 27:562-565.

Burnside, O.C. (1992). Weed science–the step child. *Weed Technol.* 7:515-518.

Burnside, O.C., N.H. Krause, M.J. Wiens, M.M. Johnson, and E.A. Ristau. (1993). Alternative weed management for the production of kidney beans (*Phaseolus vulgaris*). *Weed Technol.* 7:940-945.

Burnside, O.C. and R.S. Moomaw. (1984). Influence of weed control treatments on soybean cultivars in an oat-soybean rotation. *Agron. J.* 76:887-890.

Burnside, O.C., R.G. Wilson, S. Weisberg, and K.G. Hubbard. (1996). Seed longevity of 41 weed species buried 17 years in eastern and western Nebraska. *Weed Sci.* 44:74-86.

Bussan, A.J., O.C. Burnside, J.H. Orf, E.A. Ristau, and K.J. Puettmann. (1997). Field evaluation of soybean (*Glycine max*) genotypes for weed competitiveness. *Weed Sci.* 45:31-37.

Callaway, M.B. (1992). A compendium of crop varietal tolerance to weeds. *Am. J. Alt. Agric.* 7(4):169-180.

Callaway, M.B. and F. Forcella. (1992). Crop tolerance to weeds. In *Crop Improvement for Sustainable Agriculture Systems,* eds. M.B. Callaway and C.A. Francis. Lincoln, NE: University of Nebraska Press, pp. 100-131.

Challaiah, O.C. Burnside, G.A. Wicks, and V.A. Johnson. (1986). Competition between winter wheat (*Triticum aestivum*) cultivars and downy brome (*Bromus tectorum*). *Weed Sci.* 34:689-693.

Challaiah, R.E. Ramsel, G.A. Wicks, O.C. Burnside, and V.A. Johnson. (1983). Evaluation of the weed competitive ability of winter wheat cultivars. *North Central Weed Control Conf. Proc.* 38:85-91.

Crafts, A.S. and W.W. Robbins. (1962). *Weed Control.* New York, NY: McGraw-Hill Book Co., Inc.

Crutchfield, D.A., G.A. Wicks, and O.C. Burnside. (1985). Effect of winter wheat (*Triticum aestivum*) straw mulch level on weed control. *Weed Sci.* 34:110-114.

Dilday, R.H., P. Nastasi, J. Lin, and R.J. Smith, Jr. (1991). Allelopathic activity in rice (*Oryza sativa* L.) against ducksalad [*Heteranthera limosa* (Sw.) Willd.]. In *Sustainable Agriculture for the Great Plains,* eds. J.D. Hanson, M.J. Shaffer, D.A. Ball, and C.V. Cole, Symposium Proceedings. USDA, ARS, ARS-89, pp. 255.

Dotray, P.A. and F.L. Young. (1993). Characterization of root and shoot development of jointed goatgrass (*Aegilops cylindrica*). *Weed Sci.* 41:353-361.

Duke, S.O., ed. (1996). *Herbicide-Resistant Crops: Agricultural, Environmental, Economic, Regulatory, and Technical Aspects.* Boca Raton, FL, CRC Press.

Francis, C.A. (1985). Cultivar development for multiple cropping systems. *CRC Crit. Rev. Plant Sci.* 3:133-168.

Friesen, L.F., K.P. Nickel, and I.N. Morrison. (1992). Round-leaved mallow (*Malva pusilla*) growth and interference in spring wheat (*Triticum aestivum*) and flax (*Linum usitatissimum*). *Weed Sci.* 40:448-454.

Garrity, D.P., M. Movillon, and K. Moody. (1992). Differential weed suppression ability in upland rice cultivars. *Agron. J.* 84:586-591.

Goldburg, R.J. (1992). Environmental concerns with the development of herbicide-tolerant plants. *Weed Technol.* 6:647-652.

Hall, M.R., C.J. Swanton, and G.W. Anderson. (1992). The critical period of weed control in grain corn (*Zea mays*). *Weed Sci.* 40:441-447.

Hicks, D.R., R.E. Stucker, and J.H. Orf. (1992). Choosing soybean varieties from yield trials. *J. Prod. Agric.* 5(3):303-307.

Holt, J.S. (1992). History of identification of herbicide-resistant weeds. *Weed Technol.* 6:615-620.

Johnson, G.A., D.A. Mortensen, and C.A. Gotway. (1996). Spatial and temporal analysis of weed seedling populations using geostatistics. *Weed Sci.* 44:704-710.

Jordan, N. (1993). Prospects for weed control through crop interference. *Ecol. Appl.* 3:84-91.

Kawano, K., H. Gonzalez, and M. Lucena. (1974). Intraspecific competition, competition with weeds, and spacing response in rice. *Crop Sci.* 14:841-845.

Knake, E.L. and F.W. Slife. (1965). Giant foxtail seeded at various times in corn and soybeans. *Weeds* 13:331-334.

Koscelny, J.A., T.F. Peeper, J.B. Solie, and S.G. Solomon, Jr. (1990). Effect of wheat (*Triticum aestivum*) row spacing, seeding rate, and cultivar on yield loss from cheat (*Bromus secalinus*). *Weed Technol.* 4:487-492.

Kropff, M.J. and H.H. van Laar, eds. (1993). *Modelling Crop-Weed Interactions.* Wallingford, Oxon OX10 8DE, United Kingdom: CAB International.

Lanning, S.P., L.E. Talbert, J.M. Martin, T.K. Blake, and P.L. Bruckner. (1997). Genotype of wheat and barley affects light penetration and wild oat growth. *Agron. J.* 89:100-103.

Malik, V.S., C.J. Swanton, and T.E. Michaels. (1993). Interaction of white bean (*Phaseolus vulgaris* L.) cultivars, row spacing, and seeding density with annual weeds. *Weed Sci.* 41:62-68.

McWhorter, C.G. (1984). Future needs in weed science. *Weed Sci.* 32:850-855.

McWhorter, C.G. and W.L. Barrentine. (1975). Cocklebur control in soybeans as affected by cultivars, seeding rates, and methods of weed control. *Weed Sci.* 23:386-390.

McWhorter, C.G. and E.E. Hartwig. (1972). Competition of johnsongrass and cocklebur with six soybean varieties. *Weed Sci.* 20:56-59.

Moll, R.H. and C.W. Stuber. (1974). Quantitative genetics–empirical results relevant to plant breeding. *Adv. Agron.* 26:277-313.

Mortensen, D.A., G.A. Johnson, and L.J. Young. (1993). Weed distribution in agricultural fields. In *Site Specific Crop Management,* eds. P. Robert and R.H. Rust. Am. Soc. Agron., pp. 113-124.

Mt. Pleasant, J., R.F. Burt, and J.C. Frisch. (1994). Integrating mechanical and chemical weed management in corn (*Zea mays*). *Weed Technol.* 8:217-223.

Mumaw, C.R. and C.R. Weber. (1957). Competition and natural selection in soybean varietal composites. *Agron. J.* 49:154-160.

Pantone, D.J., J.B. Baker, and P.W. Jordan. (1992). Path analysis of red rice (*Oryza sativa* L.) competition with cultivated rice. *Weed Sci.* 40:313-319.

Parish, S. (1990). A review of non-chemical weed control techniques. *Biol. Agric. Hort.* 7:117-137.

Pavlychenko, T.K. and J.B. Harrington. (1934). Competitive efficiency of weeds and cereal crops. *Can. J. Res.* 10:77-94.

Pester, T.A. (1996). *Quantifying Soybean Weed Competitiveness in Greenhouse and Field Experiments.* MS Thesis, Univ. of Minnesota, St. Paul, MN.

Poehlman, J.M. and D.A. Sleper. (1995). *Breeding Field Crops.* Ames, IA: Iowa State University Press.

Radosevich, S.R. and J.S. Holt. (1984). *Weed Ecology: Implications for Vegetation Management.* New York, NY: John Wiley & Sons.

Rice, E.L. (1984). *Allelopathy.* 2nd ed. Academic Press, New York, NY.

Rose, S.J., O.C. Burnside, J.E. Specht, and B.A. Swisher. (1984). Competition and allelopathy between soybeans and weeds. *Agron. J.* 76:523-528.

Smith, R.J. (1974). Competition of barnyardgrass with rice cultivars. *Weed Sci.* 22:423-426.

Soto, A., and C. Gamboa. (1984). Competencia entre malas hierbas y el frijol en funcion del cultivar, la poblacion y la distancia entre hileras. *Agron. Costarricense* 8:45-52.

Staniforth, D.W. (1961). Responses of corn hybrids to yellow foxtail competition. *Weeds* 9:132-136.

Swanton, C.J. and S.F. Weise. (1991). Integrated weed management: the rationale and approach. *Weed Technol.* 5:657-663.

Swanton, C.J., D.R. Clements, and D.A. Derksen. (1993). Weed succession under conservation tillage: a hierarchical framework for research and management. *Weed Technol.* 7:286-297.

Thornton, P.K., R.H. Fawcett, J.B. Dent, and T.J. Perkins. (1990). Spatial weed distribution and economic thresholds for weed control. *Crop Protection* 9:337-344.

Urwin, C.P., R.G. Wilson, and D.A. Mortensen. (1996). Late season weed suppression from dry bean (*Phaseolus vulgaris*) cultivars. *Weed Technol.* 10:699-704.

Wauchope, R.D. (1994). Herbicides in runoff and surface waters: environmental and regulatory impacts introducton. *Weed Technol.* 8:850-851.

Weber, C.R. and D.W. Staniforth. (1957). Competitive relationships in variable weed and soybean stands. *Agron. J.* 49:440-444.

Wicks, G.A., R.E. Ramsel, P.T. Nordquist, J.W. Schmidt, and Challaiah. (1986). Impact of wheat cultivars on establishment and suppression of summer annual weeds. *Agron. J.* 78:59-62.

Wilcut, J.W., H.D. Coble, A.C. York, and D.W. Monks. (1996). The niche for herbicide-resistant crops in U.S. agriculture. In *Herbicide-Resistant Crops,* ed. S.O. Duke. Boca Raton, FL: Lewis Publishers, CRC Press, pp. 213-230.

Wiles, L.J., G.W. Oliver, A.C. York, H.J. Gold, and G.G. Wilkerson. (1992). Spatial distribution of broadleaf weeds in North Carolina soybean (*Glycine max*) fields. *Weed Sci.* 40:554-557.

Worsham, A.D. (1991). Allelopathic cover crops to reduce herbicide input. *Proc. South. Weed Sci. Soc.* 44:58-69.

Wortmann, C.S. (1993). Contribution of bean morphological characteristics to weed suppression. *Agron. J.* 85:840-843.

Wyse, D. L. (1992). Future of weed science research. *Weed Technol.* 6:162-165.

RECEIVED: 05/14/97
ACCEPTED: 10/27/97

# Genetic Approach to the Development of Cover Crops for Weed Management

## Michael E. Foley

**SUMMARY.** Economic and environmental issues are driving efforts to improve cover crops for weed management. Cover crop residues on the soil surface interfere with weeds by releasing allelochemicals and by physical suppression. Optimizing allelopathic potential, biomass production, and other desirable cover crop characteristics using classical and molecular genetic approaches holds great promise for improving the efficacy and selectivity of cover crops. Likewise, investigating allelopathy at the genetic and molecular level should aid in understanding the biochemical basis for allelopathy in plants. *[Article copies available for a fee from The Haworth Document Delivery Service: 1-800-342-9678. E-mail address: getinfo@haworthpressinc.com]*

**KEYWORDS.** Allelopathy, crop breeding, weed suppression

### *INTRODUCTION*

The discovery of synthetically produced organic chemicals with utility as herbicides revolutionized weed management in the second-half of the 20th

Michael E. Foley, Associate Professor, Department of Botany and Plant Pathology, Purdue University, West Lafayette, IN 47907-1155 (E-mail: foley@btny. purdue.edu).

The support of the U.S. Department of Agriculture (96-34103-3097) and the Purdue University Office of Agricultural Research is gratefully acknowledged. This is journal series paper No. 15367.

[Haworth co-indexing entry note]: "Genetic Approach to the Development of Cover Crops for Weed Management." Foley, Michael E. Co-published simultaneously in *Journal of Crop Production* (Food Products Press, an imprint of The Haworth Press, Inc.) Vol. 2, No. 1 (#3), 1999, pp. 77-93; and: *Expanding the Context of Weed Management* (ed: Douglas D. Buhler) Food Products Press, an imprint of The Haworth Press, Inc., 1999, pp. 77-93. Single or multiple copies of this article are available for a fee from The Haworth Document Delivery Service [1-800-342-9678, 9:00 a.m. - 5:00 p.m. (EST). E-mail address: getinfo@ haworthpressinc.com].

© 1999 by The Haworth Press, Inc. All rights reserved.

century. The improved efficacy of herbicides and the application of bio-technology to develop herbicide resistant crops have sustained this revolution (Beyer et al., 1988; Delannay et al., 1995). Synthetic herbicides continue to be one of the most important tools for weed management in the United States. Nevertheless, several factors have stimulated interest in alternative approaches to weed management. Economic and regulatory considerations sometimes preclude development or registration of herbicides for crops with limited acreage, value, or for human consumption. Environmental considerations, such as herbicides in surface and groundwater, have raised concerns about public health. And the incidence of herbicide resistant weeds raises questions about the sustainability of existing weed management practices.

One alternative approach to the uses of synthetic herbicides is management of plant residues on the soil surface to "smother" weeds (Swanton and Weise, 1991). Residues can be a product from the preceding crop, as in no-tillage production systems, and from cover crops. Interest in cover crop residues for weed management has come about partly through the use of conservation tillage systems that are compatible with the use of fall or spring-planted cover crops. Conservation tillage practices have been mandated to reduce soil erosion. Plant residues on the soil surface reduce soil erosion, reduce nutrient leaching and runoff, enhance soil structure and organic matter content, and suppress weeds (Barnes and Putnam, 1983; Putnam and De-Frank, 1983; Mohler, 1991; Liebl et al., 1992; Ateh and Doll, 1996; Swanton and Murphy, 1996).

The use of crops for weed management is an old concept. There is renewed interest in this concept as it relates to enhancing cultivar competitiveness and developing cover crops (Swanton and Weise, 1991; Jordan, 1993; Wyse, 1994; Lemerle et al., 1996; Bastiaans et al., 1997; Bussan et al., 1997). Before organic herbicides came into prominent use, "smother crops" were used in crop rotations to suppress the growth of weeds and break cycles of weed infestation associated with the production of specific crops (Robbins, Crafts, and Raynor, 1942). Smother crops shade out weeds through rapid growth and thick stands. Some common smother crops are alfalfa (*Medicago sativa*), buckwheat (*Fagopyrum esculentum*), foxtail millet (*Setaria italica*), rye (*Secale cereale*), and sorghum (*Sorghum* spp.). Because products from smother crops provide net income, they are still useful as main, rotational or catch crops in existing production systems (Schreiber, 1992; Young et al., 1994). The concept of cover crops originates from smother crops. Smother and cover crops are sometimes one in the same (e.g., sorghum, rye) and both are grown between the main or economically important crops. The main function of a cover crop is biomass production with management for suppression of weeds (Barnes and Putnam, 1983; Swanton and Murphy, 1996). This paper will focus on allelopathy as a means to enhance the capability of cover crops to suppress weeds. Allelopathy is a concept that needs further verification. It is often difficult repeat in the field results of laboratory bioassays demonstrating allelopathy. Determining a genetic basis for allelopathy and genetically manipulating allelopathic potential in some plant is critical for proving the concept.

## INTERFERENCE

Interference refers to the various negative interactions of one plant with another (Cousens and Mortimer, 1995). As it relates to weed management, interference is often approached from the perspective of interspecific effects of weeds on crops. With cover crops however, the potential for interference must be considered for both weeds and the crop. The components of interference are competition for common resources, such as light, space, nutrients and water, and deleterious allelopathic effects (Muller, 1969). Allelopathy refers to both beneficial and deleterious biochemical interactions between all types of plants (Molisch, 1937). In a classical botanical context, plants also denote bacteria, fungi, and algae. The emphasis in weed management is on deleterious effects of allelochemicals (Rice, 1974; Putnam, 1986). More complex indirect sources of interference, for example, one plant affecting the nutrient status of another, are probably not applicable in frequently disturbed agroecosystems (Fuerst and Putnam, 1983). In less frequently disturbed agroecosystems, e.g., no-tillage systems, and when cover crops are involved, indirect sources of interference such as nutrient availability, need to be considered (Waddington, 1978; Waddington and Bowren, 1978; Purvis Jessop, and Lovett, 1985). Some indirect types of interference are unpredictable. For example, in one year a wheat (*Triticum aestivum*) straw mulch for weed suppression increased frost injury to emerged soybeans (*Glycine max*) (Vidal and Bauman, 1996). The frost injury interacted with giant foxtail (*Setaria faberi*) interference to decrease soybean yields relative to areas without residue. Ideally, cover crops should not interfere with the main crop such that yield, quality, or net return is reduced. In practice, however, cover crop residue sometimes adversely affects the main crop (Raimbault, Vyn, and Tollenaar, 1990; Johnson, Defelice, and Helsel, 1993; De Haan et al., 1994; Ateh and Doll, 1996; Brecke and Shilling, 1996). Approaching the ideal will require research in individual crop production systems: (1) to determine the actions and interactions of cover crop and environmental factors on crop yield (Schonbeck et al., 1993; Dwyer et al., 1995; Mangan et al., 1995), and (2) to ascertain the type and level of management factors required to optimize suppression of weeds without adversely impacting the crop (Masiunas, Weston, and Weller, 1995; Mwaja, Masiunas, and Weston, 1995; Dabney et al., 1996; Smeda and Weller, 1996).

Cover crop biomass suppresses weeds through competition and allelopathy. Biomass killed shortly before planting the crop is termed mulch or residue. Residue may include the root system. Living cover crop biomass is termed living mulch. Green-manure crops, which are cover crops grown to be incorporated into the soil, can also suppress weeds (Boydston and Hang, 1995; Al-Khatib, Libbey, and Boydston, 1997). The physical presence of cover crop mulch on the soil surface can greatly reduce weed density and biomass (Putnam and DeFrank, 1983; Masiunas, Weston, and Weller, 1995; Ateh and Doll, 1996). Likewise, allelochemicals arising directly or indirectly from mulch or residue suppress weeds (Bell and Muller, 1973; Barnes and Putnam, 1983; Putnam and DeFrank, 1983; Barnes and Putnam, 1986; Brown and Morra, 1995). Increasing levels of biomass with uniform soil coverage enhances weed suppression (Teasdale, Beste, and Potts, 1991; Buhler, 1995; Teasdale, 1996; Vidal and Bauman, 1996). However, the relative contribution of physical and allelochemical components of biomass to weed suppression is poorly defined.

## COMPONENTS OF INTERFERENCE

Separating the physical and allelopathic components of interference is problematic (Fuerst and Putnam, 1983; Weidenhamer, 1996). Nonetheless, as cover crops for different crop and management systems are developed (Wyse, 1994), understanding these components and their interactions will be imperative to maximize and minimize impacts on weeds and crops, respectively. At one time, the success of barley (*Hordeum vulgare*) as a smother crop was attributed to physical competition. Overland (1966) devised experiments that eliminated the competition component of weed suppression by barley and determined that allelochemicals also contribute to the suppression of weeds. However, she was not able to quantify the relative contribution of each component to the suppression of weeds. A stair step apparatus, which passes and recycles nutrient solution through the rhizosphere of two plants, has been used to separate competitive and allelopathic mechanisms (Rice, 1974). Despite some limitation with this method, Bell and Koeppe (1972) determined that 35% of the inhibition of corn (*Zea mays*) growth by mature giant foxtail was through an allelopathic mechanism.

Barnes and Putnam (1983) determined that poplar wood shavings could be used as a nontoxic mulch control in experiments designed to separate the physical and allelochemical effects of rye cover crops. Rye residues reduced weed biomass in no-tilled peas by 73% compared to their control. About one-half the weed suppression from rye was attributed to allelochemicals. The remainder was due to physical conditions imposed by the rye residue (Putnam and DeFrank, 1983). The allelochemicals were derived primarily from rye shoots, and secondarily from roots (Barnes and Putnam, 1986). But in a study designed to control indirect sources of interference, e.g., moisture and fertilizer availability, by capillary mat subirrigation, rye shoot tissue did not suppress weeds but root tissue did (Hoffman et al., 1996). In a field study, retention or removal of rye shoot residue in no-tilled corn had little influence on several variables related to corn growth and development (Raimbault, Vyn, and Tollenaar, 1990). It was inferred that allelochemicals from rye roots inhibited growth and development of corn. Varietal differences in allelochemicals in both roots and shoots of rye may explain the conflicting reports related to the role of allelochemicals in the suppression of weeds (Copaja, Berria, and Niemeyer, 1991; Perez and Ormeno-Nunez, 1991). The rye variety 'Wheeler' used by Barnes and Putnam (1983) was selected for its high biomass production and allelochemical potential. As much as anything, the use of different varieties and blends of rye may account for the conflicting reports on the efficacy of rye and rye tissues for weed suppression (Putnam and DeFrank, 1983; Moore, Gillespie, and Swanton, 1994; Hoffman et al., 1996).

## GENETIC-BASED DEVELOPMENT OF COVER CROPS

Cover crops that produce allelochemicals are being used for weed management with variable levels of success (Swanton and Murphy, 1996; Weston, 1996). There are many reasons for this. Some relate to economic, environmental, and management factors. Others relate to a lack of knowledge about the biology of characteristics regarded as fundamental to suppression of weeds. Allelopathy, which is regarded as fundamental to weed suppression by cover crops, has been investigated using biochemical approaches. Yet many questions remain unanswered. For example, with rye, definitive proof is lacking that hydroxamic or phenolic acids are fundamental to allelopathy (Shilling, Liebl, and Worsham, 1985; Barnes and Putnam, 1987; Niemeyer, 1988a). Alternative approaches are clearly necessary to ascertain the fundamentals of allelopathy in plants.

The importance of genetics to understanding the biological basis for allelopathy and improving cover crops has been stressed repeatedly (Rice, 1974; Rice, 1984; Putnam, 1986; Rice, 1995). Putnam (1986) envisioned plant geneticists exploiting allelopathic variability in germplasm to enhance the efficacy of crops and cover crops. Although Putnam's vision has not come to pass (Rice, 1995; Weston, 1996), fundamental advances in genetics and genetic technology in the last decade have provided new opportunities to develop knowledge-based pest management strategies (Martin et al., 1993; Padgette et al., 1995). Enhancing the efficacy of cover crops for weed management will require basic information on genetic factors associated with allelopathy and other important characteristics (Table 1). Then crop and

TABLE 1. Cover crop characteristics amenable to genetic manipulation.

| Characteristic | Genetic attributes |
|---|---|
| Allelopathic potential | Increase or decrease |
| Biomass production | High, moderate or low |
| Allelochemical specificity | Weed specificity, crop selectivity |
| Germination | Rapid germination and emergence under a variety of conditions, seedling vigor |
| Life cycle | Short or long, rapid growth, programmed necrosis |
| Morphology | . Optimize height, leaf angle, leaf size, rooting depth, rooting pattern, etc. |
| Agronomic | Nondormant seed, resistant to pests, good seed production when desired, winter hardiness |

cover crop germplasm can be exploited to optimize weed suppression for different crops and production systems. Beyond that, genetic approaches could verify the nature of putative allelochemicals, reveal their biosynthetic pathways, and facilitate development of molecular markers to aid in breeding cover crops for weed suppression.

*Brassica* spp., rice, rye, and sorghum are amenable to genetic investigations because these species are relatively well characterized at the genetic and molecular levels (Chittenden et al., 1994; Phillips, Wehling, and Wricke, 1994; Meyerowitz and Somerville, 1994; Loarce, Hueros, and Ferrer, 1996). Initially the germplasm should be characterized for allelopathic characteristics and biomass production. The focus has been on characterizing allelopathic potential in the germplasm of crops and cover crops (Putnam and Duke, 1974; Leather, 1983; Fay and Duke, 1977; Massantini, Caporali, and Zellini, 1977; Niemeyer, 1988b; Hinen and Worsham, 1991; Dilday, Lin, and Yan, 1994; Ben-Hammouda, Kremer, and Minor, 1995; Olofsdotter, Navarez, and Moody, 1995; Nimbal et al., 1996; Wahle, 1996), because this trait holds great promise for enhancing the suppression of weeds (Putnam, 1986; Rice, 1995). Considerable differences among cultivars and accessions in some collections have been reported (Putnam and Duke, 1974; Fay and Duke, 1977; Dilday, Lin, and Yan, 1994). For other crops and cover crops, e.g., rye, the amount of germplasm screened may be insufficient to detect variability in allelopathic potential (Hinen and Worsham, 1991; Wahle, 1996; Burgos and Talbert, 1997). It is particularly important to include landraces and wild relatives in these screens because domestication of crops may have sacrificed traits like allelopathy for grain yield (Niemeyer, 1988b; Copaja, Berria, and Niemeyer, 1991; Hoult and Lovett, 1993; Olofsdotter, Navarez, and Moody, 1995).

Determining biomass production by cover crops has not been a high priority. Biomass should be considered however, because it can physically suppress weeds. Thus, collectively manipulating both biomass production and allelopathic potential might be highly desirable (De Haan et al., 1994). My colleagues and I are screening rye germplasm for allelopathic potential of shoots using standard bioassay procedures (Barnes and Putnam, 1983). The sheer magnitude of growing and screening approximately 1900 rye accessions in the USDA's world collection is a laborious task that will not be complete until shoot biomass production and allelochemical production by roots have been determined.

Screening cover crop germplasm for allelopathic potential will accomplish several things. First, varieties adapted to local or regional conditions that have equal or greater allelopathic potential than current selections could be identified. These varieties could be evaluated under local conditions, and if superior to existing blends and varieties, quickly adopted. Second, investigating allelopathic potential in germplasm grown under uniform environmental conditions will determine if allelopathy is a heritable trait. If allelopathic potential is heritable, conventional plant breeding schemes could be used to introgress this trait into locally adapted varieties. Finally, identifying the extreme phenotypes within the germplasm will set the stage for genetic and molecular genetic analyses.

Genetic analyses in part relies on identifying accessions with high and low, or no allelopathic potential. Plants from accessions with high and low allelopathic potential would be cross pollinated to produce $F_1$ seeds. $F_1$ seeds would be germinated and plants grown to produce $F_2$ seeds. Following germination of $F_2$ seeds, a large population of $F_2$ seedlings would be grown, seedling tissue harvested, and bioassayed to evaluate segregation for allelopathy within the $F_2$ population. Assuming allelopathic potential is a heritable trait and differences are distinguishable by bioassay, then $F_2$ seedlings representing the extreme phenotypes would be taken to seed. These seeds would be germinated and seedling tissue harvested and bioassayed to determine allelopathic potential of the $F_3$ families. This step, called progeny testing, is necessary to verify that individuals selected for the extremes of allelopathic potential breed true. Similar experiments using the entire $F_2$ population might be done to estimate heritability for allelopathic potential.

Molecular genetic analyses can also utilize individuals with high and low levels of allelopathic potential. Deoxyribonucleic acid (DNA) from individual $F_2$ plants that breed true for high and low allelopathic potential, respectively can be pooled, i.e., bulked, and used for bulked segregant analysis (Michelmore, Paran, and Kesseli, 1991). Bulked segregant analysis in conjunction with random amplified polymorphic DNA (RAPD) analysis (Williams et al., 1990) is used to identify molecular markers linked to simple or quantitative traits of interest when the phenotype, e.g., allelopathic potential, can be readily identified. Identifying molecular markers linked to genes that determine allelopathic potential would be useful for breeding crops and cover crops using marker assisted selection, and in initiating the arduous process of cloning genes associated with this trait. Cloning genes from any species is not a trivial undertaking. However, new techniques and unified genetic maps of higher plants should increase opportunities for identifying, manipulating and cloning genes for complex traits (Lovett, 1994; Collins, 1995; Tanksley, Ganal, and Martin, 1995; Vos et al., 1995; Hill et al., 1996; Jansen, 1996; Paterson et al., 1996).

Another useful genetic tool to use in understanding allelopathy and developing cover crops for weed suppression would be near-isogenic lines (NIL) of an allelopathic plant. Near-isogenic lines are an uncommon but vital tool in investigations of weed management (McCloskey and Holt, 1990). After identifying individuals with the extreme phenotypes in the germplasm and gaining insight into the genetics, e.g., dominant or recessive trait, reciprocal crosses using an inbred allelopathic and non-allelopathic parent could be done. The primary selection criteria would be the presence or absence of allelopathic potential. Secondary traits like level of biomass production could be considered; however, this would complicate the task of selection at each generation. After several generations of backcrossing using parents with contrasting phenotypes as the recurrent parents, NILs in one or more genetic backgrounds would be available. Tissue from the allelopathic and non-allelopathic NILs could be used as treatment and control in subsequent investigations aimed at separating physical and allelochemical components of weed suppression. Near-isogenic lines would also be useful in determining the effect of environmental variables on cover crop-induced suppression of weeds and in biochemical, genetic and molecular genetic investigations of allelopathy.

## PROSPECTS

Cover crops are being used in selected situations for weed management (Mangan et al., 1995; Teasdale, 1996; Weston, 1996). Further acceptance by growers will depend on such things as increasing cover crop efficacy and selectivity (Table 1). Although biomass production should not be ignored, increasing the efficacy of cover crops for weed suppression will largely depend on exploiting their allelopathic potential. Utilizing allelopathic potential may involve breeding for intrinsically higher levels of a particular allelochemical. For example, if genetic approaches verify that a hydroxamic acid in rye is the primary allelochemical responsible for suppression of weeds, rye might be bred for higher hydroxamic acid levels or rates of exudition (Dunn, Long, and Routley, 1981; Simcox and Weber, 1985). Glucosinolates are a family of putative allelochemicals found in many members of Brassicaceae (Brown and Morra, 1995; Brown and Morra, 1996). If structural variants of glucosinolates display differential activity on weeds or selectivity on crops, genetic and molecular approaches could be used immediately to develop cover crops. Germplasm could be selected for production of specific types of glucosinolates (Giamoustaris and Mithen, 1996) and for other characteristics for the suppression of weeds (De Haan et al., 1994).

Nearly all secondary phytochemicals arise from three metabolic pathways (Conn, 1995). As genes and gene products involved in the synthesis of phytochemicals are identified (Chapple, 1994), new cloning techniques may offer the potential to introduce into crops "cassettes of genes" that encode for multiple steps of a biochemical pathway (Michelmore, 1996). Inter- and intraspecific transfer of genes using conventional and contemporary methods should lead to the development of a cornucopia of cover crops. These cover crops would be designed for specific crops and cropping systems and targeted to specific weeds and other pests (Nicol et al., 1992; Giamoustaris and Mithen, 1996; Givovich and Niemeyer, 1996; Halbrendt, 1996).

Enhancing the allelopathic potential of cover crops will undoubtedly affect crop selectivity. Therefore, an integral component of cover crop development will be evaluating their effect on crops, i.e., stand establishment and yield (Dabney et al., 1996; Smeda and Weller, 1996; Vidal and Bauman, 1996). Just as the efficacy of herbicides has been tested against germplasm of important crops (Hardcastle, 1974; Fleming, Banks, and Legg, 1988), so to will the efficacy of cover crops be evaluated (Hicks et al., 1989). Intraspecific variation in the response of crops to allelochemicals might lead to breeding for resistance just as has been done for herbicides (Sebastian et al., 1989; Dotray et al., 1993). Modification of natural products by crops may rely on the transfer of genes that confer tolerance in one organism to another. For example, resistance to glufosinate herbicide has been conferred on many species of higher plants by introducing the *pat* or *bar* gene isolated from *Streptomyces* spp. (De Block et al., 1987; Strauch, Wohlleben, and Puhler, 1988). The product of these genes, a phosphinothricin acetyltransferase, catalyzes the modification and thus detoxification glufosinate (Droge-Laser et al., 1994). From this biotechnology is coming a plethora of glufosinate-resistant crops

(Mickleson and Harvey, 1997; Sankula et al., 1997; Shankle, Shaw, and Medlin, 1997). The abundance of secondary compounds in plants and other organisms and their role in defense suggests the potential to discover and clone genes involved in detoxification of allelochemicals that arise directly or indirectly from cover crops (Balke, Davis, and Lee, 1987; Friebe, Wieland, and Schulz, 1996; Schuler, 1996). Perhaps through biotechnology will come a plethora of allelochemical-resistant crops.

In no-tillage systems, relative emergence times between crop and weeds shifts in favor of the weeds (Buhler, 1995). Instead of tillage, a mixture of herbicides is frequently used to shift the equilibrium of interference back to the crop. As critical periods of weed control and weed economic thresholds are better understood and utilized, cover crop mulches, alone or with herbicides (Yenish, Worsham, and York, 1996), may be used to shift the equilibrium of interference back to the crop. Implementing the scenario just described will require a concerted effort by everyone who has vested interest in the safe and efficient production of food and fiber.

## REFERENCES

Al-Khatib, K., C. Libbey, and R. Boydston. (1997). Weed suppression with *Brassica* green manure crops in green peas. *Weed Science* 45: 439-445.

Ateh, C.M. and J.D. Doll. (1996). Spring-planted winter rye (*Secale cereale*) as a living mulch to control weeds in soybean (*Glycine max*). *Weed Technology* 10: 347-353.

Balke, N.E., M.P. Davis, and C.C. Lee. (1987). Conjugation of allelochemicals by plants. In *Allelochemicals-Role in Agriculture and Forestry*, ed. G.R. Waller. Washington, DC: American Chemical Society, pp. 214-227.

Barnes, J.P. and A.R. Putnam. (1986). Evidence for allelopathy by residues and aqueous extracts of rye (*Secale cereale*). *Weed Science* 34: 384-390.

Barnes, J.P. and A.R. Putnam. (1987). Role of benzoxazinones in allelopathy by rye (*Secale cereale* L.). *Journal of Chemical Ecology* 13: 889-906.

Barnes, J.P. and A.R. Putnam. (1983). Rye residues contribute weed suppression in no-tillage cropping systems. *Journal of Chemical Ecology* 9: 1045-1057.

Bastiaans, L., M.J. Kropff, N. Kempuchetty, A. Rajan, and T.R. Migo. (1997). Can simulation models help design rice cultivars that are more competitive against weeds. *Field Crops Research* 51:101-111.

Bell, D.T. and D.E. Koeppe. (1972). Noncompetitive effects of giant foxtail on growth of corn. *Agronomy Journal* 64: 321-325.

Bell, D.T. and C.H. Muller. (1973). Dominance of California annual grasslands by *Brassica nigra. American Midland Naturalist.* 90: 277-299.

Ben-Hammouda, M., R.J. Kremer, and H.C. Minor. (1995). Phytotoxicity of extracts from sorghum plant components on wheat seedlings. *Crop Science* 35: 1652-1656.

Beyer, E.M., M.J. Duffy, J.V. Hay, and D.D. Schlueter. (1988). Sulfonylurea herbicides. In *Herbicides: Chemistry, Degradation, and Mode of Action.* Vol. 3. eds. P.C. Kearney and D.D. Kaufman. New York, NY: Marcel Dekker, Inc., pp. 117-189.

Boydston, R.A. and A. Hang. (1995). Rapeseed (*Brassica napus*) green manure crop suppresses weeds in potato (*Solanum tuberosum*). *Weed Technology* 9: 669-675.

Brecke, B.J. and D.G. Shilling. (1996). Effects of crop species, tillage, and rye (*Secale cereale*) mulch on sicklepod (*Senna obtusifolia*). *Weed Science* 44: 133-136.

Brown, P.D. and M.J. Morra. (1995). Glucosinolate-containing plant tissues as bio-herbicides. *Journal of Agricultural and Food Chemistry* 43: 3070-3074.

Brown, P.D. and M.J. Morra. (1996). Hydrolysis products of glucosinolates in *Brassica napus* tissues as inhibitors of seed germination. *Plant and Soil* 181: 307-316.

Buhler, D.D. (1995). Influence of tillage systems on weed population dynamics and management in corn and soybean in the central USA. *Crop Science* 35: 1247-1258.

Burgos, N.R. and R.E. Talbert. (1997). Response of seedling vegetables and weeds to allelochemicals from rye (*Secale cereale* L.). *Proceedings of the Weed Science Society of America* 37: 18.

Bussan, A.J., O.C. Burnside, J.H. Orf, E.A. Ristau, and K.J. Puettmann. (1997). Field evaluation of soybean (*Glycine max*) genotypes for weed competitiveness. *Weed Science* 45: 31-37.

Chapple, C. (1994). Genetic characterization of secondary metabolism in *Arabidopsis*. In *Genetic Engineering of Plant Secondary Metabolism*, eds. B.E. Ellis, G.W. Kuroki, and H.A. Stafford. New York, NY: Plenum Press, pp. 251-274.

Chittenden, L.M., K.F. Scherz, Y. Lin, R.A. Wing, and A.H. Paterson. (1994). A detailed RFLP map of *Sorghum bicolor* × *S. propinquum*, suitable for high-density mapping, suggests ancestral duplication of *Sorghum* chromosomes or chromosomal segments. *Theoretical and Applied Genetics* 87: 925-933.

Collins, F.S. (1995). Positional cloning moves from perditional to traditional. *Nature Genetics* 9: 347-350.

Conn, E.E. (1995). The world of phytochemicals. In *Phytochemicals and Health*, eds. D.L. Gustine and H.E. Flores. Rockville, MD: American Society of Plant Physiologists, pp. 1-14.

Copaja, S.V., B.N. Berria, and H.M. Niemeyer. (1991). Hydroxamic acid content of perennial Triticeae. *Phytochemistry* 30: 1531-1534.

Cousens, R. and M. Mortimer. (1995). *Dynamics of Weed Populations.* Cambridge, UK: Cambridge University Press.

Dabney, S.M., J.D. Schreiber, C.S. Rothrock, and J.R. Johnson. (1996). Cover crops affect sorghum seedling growth. *Agronomy Journal* 88: 961-970.

De Block, M., J. Bottermann, M. Vandewiele, J. Dockx, C. Thoen, V. Gossele, N.R. Movva, C. Thompson, M. Van Montagu, and J. Leemans. (1987). Engineering herbicide resistance in plants by expression of a detoxifying enzyme. *The EMBO Journal* 6: 2513-2518.

De Haan, R.L., D.L. Wyse, N.J. Ehlke, B.D. Maxwell, and D.H. Putnam. (1994). Simulation of spring-seeded smother plants for weed control in corn (*Zea mays*). *Weed Science* 42: 35-43.

Delannay, X., T.T. Bauman, D.H. Beighley, M.J. Buettner, H.D. Coble, M.S. Defe-
lice, C.W. Derting, T.J. Diedrick, J.L. Griffin, E.S. Hagood, F.G. Hancock, S.E.
Hart, B.J. Lavallee, M.M. Loux, W.E. Lueschen, K.W. Matson, C.K. Moots, E.
Murdock, A.D. Nickell, M.D.K. Owen, E.H. Paschall, L.M. Prochaska, P.J. Ray-
mond, D.B. Reynolds, W.K. Rhodes, F.W. Roeth, P.L. Sprankle, L.J. Tarochione,
C.N. Tinius, R.H. Walker, L.M. Wax, H.D. Weigelt, and S.R. Padgette. (1995).
Yield evaluation of a glyphosate-tolerant soybean line after treatment with gly-
phosate. *Crop Science* 35: 1461-1467.
Dilday, R.H., J. Lin, and W. Yan. (1994). Identification of allelopathy in the USDA-
ARS rice germplasm collection. *Australian Journal of Experimental Agriculture*
34: 907-910.
Dotra, P.A., L.C. Marshall, W.B. Parker, D.L. Wyse, D.A. Somers, and B.G. Gengen-
bach. (1993). Herbicide tolerance and weed control in sethoxydim-tolerant corn
(*Zea mays*). *Weed Science* 41: 213-217.
Droge-Laser, W., U. Siemeling, A. Puhler, and I. Broer. (1994). The metabolites of
the herbicide L-phosphinothricin (Glufosinate). Identification, stability, and mo-
bility in transgenic, herbicide-resistant and untransformed plants. *Plant Physiology* 105: 159-166.
Dunn, G.M., B.J. Long, and D.G. Routley. (1981). Inheritance of cyclic hydroxa-
mates in *Zea mays* L. *Canadian Journal of Plant Science* 61: 583-593.
Dwyer, L.M., B.L. Ma, H.N. Hayhoe, and J.L.B. Culley. (1995). Tillage effects on
soil temperature, shoot dry matter accumulation and corn grain yield. *Journal of
Sustainable Agriculture* 5: 85-99.
Fay, P.K. and W.B. Duke. (1977). An assessment of allelopathic potential in *Avena*
germplasm. *Weed Science* 25: 224-228.
Fleming, A.A, P.A. Banks, and J.G. Legg. (1988). Differential response of maize
inbreds to bentazon and other herbicides. *Canadian Journal of Plant Science* 68:
501-507.
Friebe, A., I. Wieland, and M. Schulz. (1996). Tolerance of *Avena sativa* to the
allelochemical benzoxazolinone. Degradation of BOA by root-colonizing bacteria.
*Vereinigung fur Angewandte Botanik* 70: 150-154.
Fuerst, E.P. and A.R. Putnam. (1983). Separating the competitive and allelopathic
components of interference: Theoretical principles. *Journal of Chemical Ecology*
9: 937-944.
Giamoustaris, A. and R. Mithen. (1996). Genetics of aliphatic glucosinolates. IV.
Side-chain modification in *Brassica oleracea*. *Theoretical and Applied Genetics*
93: 1006-1010.
Givovich, A. and H.M. Niemeyer. (1996). Role of hydroxamic acids in the resistance
of wheat to the Russian Wheat Aphid, *Diuraphis noxia* (Mordvilko) (Hom.,
Aphididae). *Journal of Applied Entomology* 120: 537-539.
Halbrendt, J.M. (1996). Allelopathy in the management of plant-parasitic nematodes.
*Journal of Nematology* 28: 8-14.
Hardcastle, W.S. (1974). Differences in the tolerance of metribuzin by varieties of
soybeans. *Weed Research* 14: 181-184.
Hicks, S.K., C.W. Wendt, J.R. Gannaway, and R.B. Baker. (1989). Allelopathic

effects of wheat straw on cotton germination, emergence, and yield. *Crop Science* 29: 1057-1061.

Hill, M., H. Witsenboer, M. Zabeau, P. Vos, R. Kesseli, and R. Michelmore. (1996). PCR-based fingerprinting using AFLPs as a tool for studying genetic relationships in *Lactuca* spp. *Theoretical and Applied Genetics* 93: 1202-1210.

Hinen, J.A. and A.D. Worsham. (1991). Evaluation of rye varieties for weed suppression in no-till corn, soybeans and sorghum. *Proceedings of the Southern Weed Science Society* 44: 339.

Hoffman, M.L., L.A Weston, J.C. Snyder, and E.E. Regnier. (1996). Separating the effects of sorghum (*Sorghum bicolor*) and rye (*Secale cereale*) root and shoot residues on weed development. *Weed Science* 44: 402-407.

Hoult, A.H.C. and J.V. Lovett. (1993). Biologically active secondary metabolites of barley. III. A method for identification and quantification of hordenine and gramine in barley by high-performance liquid chromatography. *Journal of Chemical Ecology* 19: 2245-2254.

Jansen, R.C. (1996). Complex plant traits: Time for polygenic analysis. *Trends in Plant Science* 1: 89-94.

Johnson, G.A., M.S. Defelice, and Z.R. Helsel. (1993). Cover crop management and weed control in corn (*Zea mays*). *Weed Technology* 7: 425-430.

Jordan, N. (1993). Prospects for weed control through crop interference. *Ecological Applications* 3: 84-91.

Leather, G.R. (1983). Sunflowers (*Helianthus annuus*) are allelopathic to weeds. *Weed Science* 31: 37-42.

Lemerle, D., B. Verbeek, R.D. Cousens, and N.E. Coombes. (1996). The potential for selecting wheat varieties strongly competitive against weeds. *Weed Research* 36: 505-513.

Liebl, R., F.W. Simmons, L.M. Wax, and E.W. Stoller. (1992). Effect of rye (*Secale cereale*) mulch on weed control and soil moisture in soybeans (*Glycine max*). *Weed Technology* 6: 838-846.

Loarce, Y., G. Hueros, and E. Ferrer. (1996). A molecular linkage map of rye. *Theoretical and Applied Genetics* 93: 1112-1118.

Lovett, M. (1994). Fishing for complements: Finding genes by direct selection. *Trends in Genetics* 10: 352-357.

Mangan, F., R. DeGregorio, M. Schonbeck, S. Herbert, K. Guillard, R. Hazzard, E. Sideman, and G. Litchfield. (1995). Cover cropping systems for brassicas in the Northeastern United States: 2. Weed, insect and slug incidence. *Journal of Sustainable Agriculture* 5: 15-36.

Martin, G.B., S.H. Brommonschenkel, J. Chunwongse, A. Frary, M.W. Ganal, R. Spivey, T. Wu, E.D. Earle, and S.D. Tanksley. (1993). Map-based cloning of a protein kinase gene conferring disease resistance in tomato. *Science* 262: 1432-1436.

Masiunas, J.B., L.A. Weston, and S.C. Weller. (1995). The impact of rye cover crops on weed populations in a tomato cropping system. *Weed Science* 43: 318-323.

Massantini, F., F. Caporali, and G. Zellini. (1977). Evidence for allelopathic control of weed in lines of soybeans. *Proceedings of the EWRS Symposium Methods of Weed Control and Their Integration* 1: 20-32.

McCloskey, W.B. and J.S. Holt. (1990). Triazine resistance in *Senecio vulgaris* parental and nearly isonuclear backcrossed biotypes is correlated with reduced productivity. *Plant Physiology* 92: 954-962.

Meyerowitz, E.M. and C.R. Somerville, eds. (1994). *Arabidopsis*. Plainview, NY: Cold Springs Harbor Laboratory Press.

Michelmore, R. (1996). Big news for plant transformation. *Nature Biotechnology* 14: 1653-1654.

Michelmore, R.W., I. Paran, and R.V. Kesseli. (1991). Identification of markers linked to disease-resistance genes by bulked segregant analysis: A rapid method to detect markers in specific genomic regions by using segregating populations. *Proceedings of the National Academy of Sciences USA* 88: 9828-9832.

Mickleson, J.A. and R.G. Harvey. (1997). Control of giant foxtail (*Setaria faberi*), wild-proso millet (*Panicum millaceum*), and wooly cupgrass (*Eriochloa villosa*) in glufosinate-resistant field corn. *Proceedings of the Weed Science Society of America* 37: 87.

Mohler, C.L. (1991). Effects of tillage and mulch on weed biomass and sweet corn yield. *Weed Technology* 5: 545-552.

Molisch, H. (1937). *Der Einfluss einer Pflanze auf die andere: Allelopathie*. Jena, Germany: Gustave Fischer Verlag.

Moore, M.J., T.J. Gillespie, and C.J. Swanton. (1994). Effect of cover crop mulches on weed emergence, weed biomass, and soybean (*Glycine max*) development. *Weed Technology* 8: 512-518.

Muller, C.H. (1969). Allelopathy as a factor in ecological process. *Vegetatio* 18: 348-357.

Mwaja, V.N., J.B. Masiunas, and L.A. Weston. (1995). Effects of fertility on biomass, phytotoxicity, and allelochemical content of cereal rye. *Journal of Chemical Ecology* 21: 81-96.

Nicol, D., S.V. Copaja, S.D. Wratten, and H.M. Niemeyer. (1992). A screen of worldwide wheat cultivars for hydroxamic acid levels and aphid antixenosis. *Annals of Applied Biology* 121: 11-18.

Niemeyer, H.M. (1988a). Hydroxamic acids (4-hydroxy-1,4-benzoxazin-3-ones), defense chemicals in the Gramineae. *Phytochemistry* 27: 3349-3358.

Niemeyer, H.M. (1988b). Hydroxamic acid content of *Triticum* species. *Euphytica* 37: 289-293.

Nimbal, C.I., J.F. Pedersen, C.N. Yerkes, L.A. Weston, and S.C. Weller. (1996). Phytotoxicity and distribution of sorgoleone in grain sorghum germplasm. *Journal of Agricultural and Food Chemistry* 44: 1343-1347.

Olofsdotter, M., D. Navarez, and K. Moody. (1995). Allelopathic potential in rice (*Oryza sativa* L) germplasm. *Annals of Applied Biology* 127: 543-560.

Overland, L. (1966). The role of allelopathic substances in the "smother crop" barley. *American Journal of Botany* 53: 423-432.

Padgette, S.R., K.H. Kolacz, X. Delannay, D.B. Re, B.J. Lavallee, C.N. Tinius, W.K. Rhodes, Y.I. Otero, G.F. Barry, D.A. Eichholtz, V.M. Peschke, D.L. Nida, N.B. Taylor, and G.M. Kishore. (1995). Development, identification, and characterization of a glyphosate-tolerant soybean line. *Crop Science* 35: 1451-1461.

Paterson, A.H., T-H. Lan, K.P. Reischmann, C. Chang, Y-R. Lin, S-C. Liu, M.D.

Burow, S.P. Kowalski, C.S. Katsar, T.A. DelMonte, K.A. Feldmann, K.F. Schertz, and J.F. Wendel. (1996). Toward a unified genetic map of higher plants, transcending the monocot-dicot divergence. *Nature Genetics* 14: 380-382.

Perez, F.J. and J. Ormeno-Nunez. (1991). Difference in hydroxamic acid content in roots and root exudates of wheat (*Triticum aestivum* L.) and rye (*Secale cereale* L.): Possible role in allelopathy. *Journal of Chemical Ecology* 17: 1037-1043.

Phillips, U., P. Wehling, and G. Wricke. (1994). A linkage map of rye. *Theoretical and Applied Genetics* 88: 243-248.

Purvis, C.E., R.S. Jessop, and J.V. Lovett. (1985). Selective regulation of germination and growth of annual weeds by crop residues. *Weed Research* 25: 415-421.

Putnam, A.R. (1986). Allelopathy: Can it be managed to benefit horticulture? *HortScience* 21: 411-413.

Putnam, A.R. and J. DeFrank. (1983). Use of phytotoxic plant residues for selective weed control. *Crop Protection* 2: 173-181.

Putnam, A.R. and W.B. Duke. (1974). Biological suppression of weeds: Evidence for allelopathy in accessions of cucumber. *Science* 185: 370-372.

Raimbault, B.A., T.J. Vyn, and M. Tollenaar. (1990). Corn response to rye cover crop management and spring tillage systems. *Agronomy Journal* 82: 1088-1093.

Rice, E.L. (1974). *Allelopathy*. New York, NY: Academic Press, Inc.

Rice, E.L. (1984). *Allelopathy*. Orlando, FL: Academic Press, Inc.

Rice, E.L. (1995). *Biological Control of Weeds and Plant Diseases: Advances in Applied Allelopathy*. Norman, OK: University of Oklahoma Press.

Robbins, W.W., A.S. Crafts, and R.N. Raynor. (1942). *Weed Control–A Text Book and Manual*. New York, NY: McGraw-Hill Book Company, Inc.

Sankula, S., M.P. Braverman, F. Jodari, S.D. Linscombe, and J.H. Oard. (1997). Evaluation of glufosinate on rice (*Oryza sativa*) transformed with the BAR gene and red rice (*Oryza sativa*). *Weed Technology* 11:70-75.

Schonbeck, M., S. Herbert, R. DeGregorio, F. Mangan, K. Guillard, E. Sideman, J. Herbst, and R. Jaye. (1993). Cover cropping systems for brassicas in the Northeastern United States: 1. Cover crop and vegetable yields, nutrients and soil conditions. *Journal of Sustainable Agriculture* 3: 105-132.

Schreiber, M.M. (1992). Influence of tillage, crop rotation, and weed management on giant foxtail (*Setaria faberi*) population dynamics and corn yield. *Weed Science* 40: 645-653.

Schuler, M.A. (1996). The role of cytochrome P450 monooxygenase in plant-insect interactions. *Plant Physiology* 112: 1411-1419.

Sebastian, S.A., G.M. Fader, J.F. Ulrich, D.R. Forney, and R.S. Chaleff. (1989). Semidominant soybean mutation for resistance to sulfonylurea herbicides. *Crop Science* 29: 1403-1408.

Shankle, M.W., D.R. Shaw, and C.R. Medlin. (1997). Weed management using glufosinate-resistant (Liberty-Link®) soybean technology. *Proceedings of the Weed Science Society of America* 37: 88.

Shilling, D.G., R.A. Liebl, and A.D. Worsham. (1985). Rye (*Secale cereale* L.) and wheat (*Triticum aestivum* L.) mulch: The suppression of certain broadleaf weeds and the isolation and identification of phytotoxins. In: *The Chemistry of Allelopa-*

*thy-Biochemical Interactions Among Plants*, ed. A.C. Thompson. Washington, DC: American Chemical Society, pp. 243-271.

Simcox, K.D. and D.F. Weber. (1985). Location of the benzoxaxinless (*bx*) locus in maize by monosomic and B-A translocation analyses. *Crop Science* 25: 827-830.

Smeda, R.J. and S.C. Weller. (1996). Potential of rye (*Secale cereale*) for weed management in transplant tomatoes (*Lycopersicon esculentum*). *Weed Science* 44: 596-602.

Strauch, E., W. Wohlleben, and A. Puhler. (1988). Cloning of the phosphinothricin-*N*-acetyl-transferase gene from *Streptomyces viridochromogenes* Tu 494 and its expression in *Streptomyces lividans* and *Escherichia coli. Gene* 63: 65-74.

Swanton, C.J. and S.D. Murphy. (1996). Weed Science beyond the weeds: The role of integrated weed management (IWM) in agroecosystem health. *Weed Science* 44: 437-445.

Swanton, C.J. and S.F. Weise. (1991). Integrated weed management: The rationale and approach. *Weed Technology* 5: 657-663.

Tanksley, S.D., M.W. Ganal, and G.B. Martin. (1995). Chromosome landing: A paradigm for map-based gene cloning in plants with large genomes. *Trends in Genetics* 11: 63-68.

Teasdale, J.R., C E. Beste, and W.E. Potts. (1991). Response of weeds to tillage and cover crop residue. *Weed Science* 39:195-199.

Teasdale, J. R. (1996). Contribution of cover crops to weed management in sustainable agricultural systems. *Journal of Production Agriculture* 9:475-479.

Vidal, R.A. and T.T. Bauman. (1996). Surface wheat (*Triticum aestivum*) residues, giant foxtail (*Setaria faberi*) and soybean (*Glycine max*) yield. *Weed Science* 44: 939-943.

Vos, P., R. Hogers, M. Bleeker, M. Reijans, T. van de Lee, M. Hornes, A. Frijters, J. Pot, J. Peleman, M. Kuiper, and M. Zabeau. (1995). AFLP: A new technique for DNA fingerprinting. *Nucleic Acids Research* 23: 4407-4414.

Waddington, J. (1978). Growth of barley, bromegrass and alfalfa in the greenhouse in soil containing rapeseed and wheat residues. *Canadian Journal of Plant Science* 58: 241-248.

Waddington, J. and K.E. Bowren. (1978). Effects of crop residues on production of barley, bromegrass and alfalfa in the greenhouse and of barley in the field. *Canadian Journal of Plant Science* 58: 249-255.

Wahle, E.A. (1996). Optimization of rye (*Secale cereale*) cover cropping systems for vegetable production. M.S. thesis. Urbana-Champaign, IL: University of Illinois.

Weidenhamer, J.D. (1996). Distinguishing resource competition and chemical interference: Overcoming the methodological impasse. *Agronomy Journal* 88: 866-875.

Weston, L.A. (1996). Utilization of allelopathy for weed management in agroecosystems. *Agronomy Journal* 88: 860-866.

Williams, J.G.K., A.R. Kubelik, J. Livak, J.A. Rafalski, and S.V. Tingey. (1990). DNA polymorphisms amplified by arbitrary primers are useful as genetic markers. *Nucleic Acids Research* 18: 6531-6535.

Wyse, D.L. (1994). New technologies and approaches for weed management in sustainable agriculture systems. *Weed Technology* 8: 403-407.

Yenish, J.P., A.D. Worsham, and A.C. York. (1996). Cover crops for herbicide replacement in no-tillage corn (*Zea mays*). *Weed Technology* 10: 815-821.

Young, F.L., A.G. Ogg, Jr., R.I. Papendick, D.C. Thill, and J.R. Alldredge. (1994). Tillage and weed management affects winter wheat yield in an integrated pest management system. *Agronomy Journal* 86: 147-154.

RECEIVED: 03/25/97
ACCEPTED: 08/01/97

# Improving Soil Quality:
# Implications for Weed Management

Eric R. Gallandt
Matt Liebman
David R. Huggins

**SUMMARY.** Future advancements in crop production will rely on increased understanding of ecological principles that control interactions among cropping system components. Our interest in linking soil quality and weed management derives from the belief that greater understanding of key processes and properties that define soil-weed relationships will lead to the design of agroecosystems with greater capacity and opportunity to suppress weeds. We identified seed-bank persistence, seedling establishment, and interspecific interference as key processes that affect annual weed population dynamics. We then examined how soil processes and properties can affect each of these factors and how, in turn, soil-improving management practices and cropping system design may affect weed dynamics. We established weed-related soil management objectives as: (i) reducing the persistence of seeds in the soil;

---

Eric R. Gallandt, Assistant Professor, Department of Crop and Soil Sciences, Washington State University, P.O. Box 646420, Pullman, WA 99164-6420. Matt Liebman, Associate Professor, Department of Applied Ecology and Environmental Sciences, University of Maine, 5722 Deering Hall, Orono, ME 04469. David R. Huggins, Research Agronomist, USDA-ARS, Department of Crop and Soil Sciences, Washington State University, P.O. Box 646420, Pullman, WA 99164-6420.

Address correspondence to: Eric R. Gallandt, Department of Crop and Soil Sciences, Washington State University, P.O. Box 646420, Pullman, WA 99164-6420 (E-mail: gallandt@wsu.edu).

Manuscript 9803-17, Washington State University Department of Crop and Soil Sciences.

[Haworth co-indexing entry note]: "Improving Soil Quality: Implications for Weed Management." Gallandt, Eric R., Matt Liebman, and David R. Huggins. Co-published simultaneously in *Journal of Crop Production* (Food Products Press, an imprint of The Haworth Press, Inc.) Vol. 2, No. 1 (#3), 1999, pp. 95-121; and: *Expanding the Context of Weed Management* (ed: Douglas D. Buhler) Food Products Press, an imprint of The Haworth Press, Inc., 1999, pp. 95-121. Single or multiple copies of this article are available for a fee from The Haworth Document Delivery Service [1-800-342-9678, 9:00 a.m. - 5:00 p.m. (EST). E-mail address: getinfo@haworthpressinc.com].

© 1999 by The Haworth Press, Inc. All rights reserved.

(ii) reducing the abundance of safe-sites for weed establishment and the filling of available sites; and (iii) reducing crop yield loss caused by a given density of weeds. Soil factors that can be managed to achieve these goals include: (i) chemical, physical, and biological conditions that affect resources required for weed seed germination, establishment and growth; (ii) habitat for herbivores and pathogens that attack weed seeds and seedlings; and (iii) phytotoxin production. We concluded that many as yet unexplored opportunities exist to manipulate the soil environment and to design cropping systems that create multiple weed suppressive conditions at critical junctures of weed seed-bank persistence, establishment, and interference. *[Article copies available for a fee from The Haworth Document Delivery Service: 1-800-342-9678. E-mail address: getinfo@haworthpressinc.com]*

**KEYWORDS.** Agroecosystems, crop ecology, weed biology

## *INTRODUCTION*

Soil quality has been defined as the capacity of a specific kind of soil to function, within natural or managed ecosystem boundaries, to sustain plant and animal productivity, maintain or enhance water and air quality, and support human health and habitation (Karlen et al., 1997). Soils function within the framework of major controlling variables of climate, biota, parent material, and topography that drive key processes and operate over time to generate soil properties (Jenny, 1941). Key processes that effect soil properties include radiation interception, plant production, decomposition of organic materials, nutrient cycling, erosion, leaching, and gas exchange (Coleman, Reid, and Cole, 1983). In agroecosystems, these processes interact with management practices and can lead to soil damage and decreased function resulting from soil erosion, acidification, compaction, salinization, water-logging, organic matter decline and nutrient depletion (Arshad and Coen, 1992). Recognition among agriculturalists for the need to establish a balance between degradative and restorative soil processes has raised interest in the concept of soil quality and management practices that restore or improve soil function (Lal and Stewart, 1995; Parr et al., 1992; Warkentin, 1995).

Holistic or ecological perspectives of soil and agroecosystem management have contributed to developing concepts of soil quality and agricultural sustainability. The merging of ecology and agriculture can lead to a greater understanding of nature and the development of management practices that use ecological principles to optimize agroecosystems (Coleman and Hendrix, 1988). For example, the successful implementation of integrated pest management relies on understanding ecological principles and processes such as population dynamics, predator-prey relationships, community structure, and resource allocation. Our interest in linking soil quality and weed management derives from the belief that greater understanding of key processes and properties that define soil-weed relationships will lead to the design of agroecosystems with greater capacity and opportunity to suppress weeds. In particular, we are interested in what effects soil-improving practices may have on weed dynamics.

Practices useful for improving soil quality may be regionally-or cropping system-specific, but are likely to include additions of organic amendments (Parr and Hornick, 1992), use of cover crops and green manures in rotations (Cannell and Hawes, 1994; Wardle, 1995), and management of fertilizer inputs. These soil-improving practices can affect crop production by altering physical properties, nutrient dynamics, insect pests, pathogens, and weeds (Drinkwater et al., 1995; Patriquin, Baines and Abboud, 1995; van Bruggen, 1995; Liebman and Ohno, 1998). We begin with a discussion of critical processes affecting the population dynamics of annual weeds, with particular reference to points that have been shown to be, or could be, modulated by the soil environment. We then discuss tillage, fertility, the use of cover crops or green manures, and resultant modifications of the soil environment that affect weed dynamics. Lastly we discuss how weeds may be affected by the integration of multiple soil-improving practices at the level of the cropping system.

## CRITICAL PROCESSES AFFECTING POPULATION DYNAMICS OF ANNUAL WEEDS

Seedbank persistence, seedling establishment, and interspecific interference are key processes that affect annual weed population dynamics (Figure 1). Soil processes and properties interact with weed characteristics to determine the availability of many essential plant resources, and the community structure and activity of soil fauna and flora including saprophytes, pathogens, and predators. At the species level these processes have direct or indirect effects on seedbank persistence, seedling establishment, and interspecific interference which ultimately affect weed community dynamics (Swanton, Clements, and Derksen, 1993).

### Seedbank Persistence

Establishment of a persistent seedbank contributes to the success of annual weeds by dispersing a population over time (Cavers and Benoit, 1989; Cook, 1980). Lacking input of new seeds, seed banks decline at rates that may differ among species (Cook, 1980), locations (Burnside, Wicks and Fenster, 1977;

FIGURE 1. Key processes controlling annual weed dynamics that may be modulated through changes in soil quality. The soil environment may be manipulated to affect seedbank persistence (A), the abundance of safe-sites and thus seedling establishment (B), and interspecific interference (C).

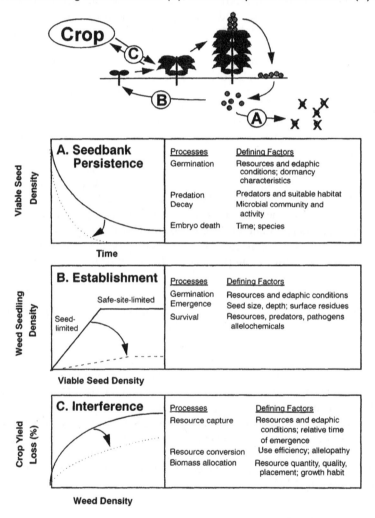

Donald and Zimdahl, 1987; Robocker et al., 1969), and seed sources within a single species (Donald, 1993). Mechanistic explanations of variations in location- and species-specific decay rates are generally not available, but the considerable degree of variation that has been observed suggests that opportunities exist to manipulate seedbanks (Figure 1A).

Losses from the seedbank occur as a result of germination, predation, microbial invasion and decay, and age-related embryo death. While these processes include species-specific characteristics, they are inextricably related to the soil environment. Germination, for example, could vary as a result of tillage and residue effects on soil water potential, temperature, and gas exchange (Currie, 1972). Likewise, the absence of tillage and the presence of surface residues could encourage the activity of numerous seed predators (e.g., Brust and House, 1988). Circumstantial evidence also suggests that enhanced biological activity resulting from improved soil quality could affect the persistence of weed seeds in soil.

Although microorganisms may be isolated from rotting seeds, only rarely have studies distinguished saprophytic colonization from pathogenic attack (Kremer, 1993). A notable exception is the work of Lonsdale (1993), who examined the effects of fungicide treatment on losses from the dormant seedbank of *Mimosa pigra* L., a shrub that has invaded northern Australia. Germination was responsible for most of the seed losses but fungicide treatment reduced loss of *M. pigra* seeds by 10 to 16%. Similarly, Fellows and Roeth (1992) found that winter survival of shattercane [*Sorghum bicolor* (L.) Moench.] was increased by fungicide treatment. While it is clear that soil microorganisms affect weed seed mortality, additional research to determine the relationship between soil environmental factors and weed seed pathogens is necessary before soils may be managed for depletion of the weed seedbank (Kremer, 1993).

Simulation models show strong effects of seed mortality on future weed population size (Gonzalez-Andujar and Fernandez-Quintanilla, 1991; Jordan, 1993; Jordan et al., 1995) and support the premise that seedbanks are important targets for management efforts. The models show strong effects of seed mortality on future weed population dynamics. Thus, a first goal in the use of soil management to manipulate weed dynamics is to reduce the persistence of seeds in the soil (Figure 1A).

### Seedling Establishment

A relatively small proportion (< 20%) of the germinable seedbank actually produces seedlings in a given year (Forcella, 1992; Roberts and Ricketts, 1979). Successful establishment requires that seeds reside within "safe-sites" in which they are protected from hazards and experience conditions that promote germination, emergence, and growth (Harper, 1977). Safe-site occurrence is a function of seed depth, light, alternating temperatures, water, crop residues, and freedom from predators and diseases, which effectively act as an "environmental sieve" to determine final seedling densities (Naylor, 1985).

The potential for weed establishment increases with increasing density of germinable seeds, reaching a maximum where the available safe-sites are filled (Figure 1B). The abundance of safe-sites will therefore influence: (i) the slope of the density dependent, seed-limited portion of the graph; and (ii) the point at which establishment is independent of seed density, i.e., where safe-sites become limiting. Soil-improving practices and resultant changes in seed and residue distribution, bulk density, water relations, and soil solution chemistry could affect both the abundance and filling of safe-sites. Therefore, a second goal in the use of soil management to manipulate weed dynamics is to reduce the abundance and/or filling of safe-sites (Figure 1B).

### Interspecific Interference

Weed/crop interference ultimately determines crop yield loss as well as weed biomass and seed production. Individual species performance within a mixed stand is dependent on species-specific resource capture and conversion, biomass allocation, and in certain instances, response to allelochemicals (Liebman and Gallandt, 1997). The relationship between crop yield loss and weed density has been described by a hyperbolic model (Cousens, 1985), the parameters of which are subject to high variation among years and locations (Lindquist et al., 1996). This variation indicates that a given weed density may result in different levels of crop yield loss depending on genetic, environmental, and management factors. With regard to the latter, soil-improving practices may contribute to differential species performance primarily through changes in the temporal or spatial resource supply and residue-mediated effects.

Weed growth and interspecific interference may also be affected by changes in soil microorganisms. Deleterious rhizobacteria (Schippers, Bakker, and Bakker, 1987) have been shown to selectively inhibit weed seedling growth without harming crop plants (Kremer et al., 1990; Skipper, Ogg, and Kennedy, 1996). Soil-improving farming systems have, for example, increased survival of mycorrhizal fungal spores (Douds, Janke, and Peters, 1993) and the abundance of actinomycetes (Drinkwater et al., 1995), bacterial, and fungal populations (Sivapalan, Morgan, and Franz, 1993). It may therefore be possible to manage soils to encourage the activity of deleterious rhizobacteria (Boyetchko, 1996). A final goal in the use of soil management to manipulate weed dynamics is therefore to reduce yield loss per individual weed (Figure 1C).

### STRATEGIES TO IMPROVE SOIL QUALITY AND SUPPRESS WEEDS: A SYSTEMS APPROACH

Harper (1974) noted that "While a study of ecosystems requires the description of its parts, it is the analysis of the interactions between the parts that marks the difference between ecological study and less dynamic scientific endeavors." Cropping system studies offer a valuable context within which to examine component interactions. Most weed ecology and management research could be considered "component research," i.e., it examines relatively few parameters isolated from the system in which they would eventually operate. Component studies are critical for identifying driving forces within a system and are the best approach to elucidate mechanisms underlying main and interactive effects. A criticism, however, is that because component-based research requires "field-truthing" by growers, it is difficult to predict how individual components will perform within a cropping system (Vereijken, 1989). Cropping system experiments, on the other hand, provide some of the field-truthing experience usually reserved for growers.

## Soil Management Components

Key soil management components that affect weed dynamics include: disturbance, primarily due to tillage; fertility management, including use of both synthetic and organic nutrient sources; and crop rotation including the use of cover crops and green manures. Each of these soil management components has major effects on soil processes, properties, and weed ecology.

### Tillage

The reduced disturbance in conservation tillage systems creates a soil environment that is unique in many physical, chemical, and biological properties that affect weed seed dormancy, species adaptation, and occurrence as well as behavior of weed attacking herbivores and pathogens (Cardina, Regnier, and Harrison, 1991). The most dramatic changes in the environment surrounding surface-residing weed seeds occurs when no-till planting is adopted. As compared to conventional inversion tillage, no-till systems are characterized by greater amounts of surface residues, microbial activity (Doran, 1980), and a greater proportion of weed seeds near the soil surface (Yenish, Doll, and Buhler, 1992). It is possible that soil microbial agents could be more effectively used to suppress weeds in no-till systems. Pitty, Staniforth, and Tiffany (1987), for example, found that fungal colonization of *Setaria* spp. caryopses was related to crop residue placement in soil. In no-till plots, colonization occurred predominately on seeds near the soil surface, whereas moldboard plowing and deeper residue incorporation resulted in greater colonization of seeds at lower depths. Because surface applications of microbial inoculants and soil amendments can be placed close to the majority of weed seeds, and soil moisture is conserved under surface residues (Cardina, Regnier, and Harrison, 1991), no-till conditions offer potential to augment the survival of applied microbial biocontrol agents.

The physical environment created by surface residues in no-till systems may also encourage the activity of weed seed predators (House and Brust, 1989). In the southeastern U.S., ground beetles (*Harpulus* spp., Carabidae: Coleoptera), field crickets (*Gryllus* spp., Orthoptera), ants (Formicidae), and field mice (*Peromyscus maniculatus*) were important weed-seed predators within a low-input, no-till cropping system in which soybean (*Glycine max* L.) was grown in a surface mulch of wheat (*Triticum aestivum* L.) straw (Brust and House, 1988). Over a five-week period in the fall, 2.3 times more seeds were removed from this no-till system compared to an adjacent conventional tillage system.

Earthworms (Lumbricidae) are generally more abundant in no-till relative to conventionally-tilled fields and can improve soil quality by increasing soil porosity, organic matter turnover, and nitrogen mineralization (Brussaard, 1994; Logsdon and Linden, 1992; Marinissen and deRuiter, 1993). Earthworms may affect weed dynamics by consuming seeds (Grant, 1983; McRill and Sagar, 1973), by redistributing seeds vertically in the soil profile (van der Reest and Rogaar, 1988), and by grazing on germinated seedlings within underground chambers (Shumway and Koide, 1994). Because seeds of certain species are preferred over others (Shumway and Koide, 1994), earthworms could ultimately affect weed community dynamics.

Tillage has a major impact on soil bulk density, pore size distribution, aggregation, and surface roughness. Within the stratum where weed seeds germinate and successfully establish, the environment may be highly heterogeneous. Modifications affecting tilth could therefore determine the abundance of safe-sites (Harper, 1977) which could, in turn, affect recruitment of weed populations (Buhler, 1995). For instance, soil aggregates are known to differ in their physical and chemical properties (Braunack and Dexter, 1989); would an increase in the size or abundance of water stable aggregates or larger clods affect the number of safe-sites available for a given weed species?

Roberts and Hewson (1971) used mechanical disturbance to create seedbeds with three degrees of surface fineness: a relatively coarse seedbed adequate for large-seeded crops, a medium seedbed, and a fine, fluffy seedbed. One half of each plot was subsequently rolled to increase compaction. Over three years of field experiments weed densities in the coarse, medium, and fine seedbeds mean density were 103, 134, and 206 weed seedlings m$^{-2}$, respectively. Rolling increased weed density 29%. Because data were reported as total weeds, it is not known whether species responded differently to the treatments.

Fine seed beds may not benefit establishment of more deeply buried seeds. Cussans et al. (1996) examined establishment of catchweed bedstraw (*Galium aparine* L.), blackgrass (*Alopecurus myosuroides* Hudson), and common chickweed (*Stellaria media* L.) when buried under 2 to 11 cm of soil sieved into four clod size classes: fine (< 6 mm), small (6-13 mm), medium (14-25 mm), and large (26-50 mm). The time to 50% emergence was, for each species, unaffected by clod size at depths less than approximately 5 cm. Emergence from a depth of 6 cm, however, was 2 to 5 days sooner where seeds were covered with large clods relative to those covered with fine clods.

Although species responded similarly to manipulation of safe-sites in the previous examples, other studies indicate that safe-sites may be species-specific. Harper, Williams, and Sagar (1965) found a higher proportion of *Bromus rigidus* established on soil with low micro-topographic variation, whereas a higher proportion of *B. madritensis* became established on rougher soil. Apparently the straight awn of *B. rigidus* encouraged the seed to lie flat and establish close seed-soil contact in the finer soils whereas the curved awn of *B. madritensis* wrapped around soil clods and increased the probability of favorable seed-soil contact on the rougher soils. In related experiments conducted with common lambsquarters (*Chenopodium album* L.) and *Brassica oleracea* L., the former species had the poorest emergence from the roughest seedbed, whereas the latter had poorest emergence from the smoothest soil surface (Harper, Williams, and Sagar, 1965). These results supported Harper's original concept of species-specific safe-sites and suggested that, for many species, maximum population size is not density dependent but is a function of the number of suitable micro-sites in the soil (Figure 1B).

In addition to seed-soil contact necessary for adequate water absorption (Harper, Williams, and Sagar, 1965), safe-sites for successful germination and establishment may be determined by light conditions. Large aggregates, i.e., clods, may affect light penetration into soil and dictate whether conditions are suitable for germination. Cussans et al. (1996) examined light penetration below 2 to 11 cm of a clay soil that had been sieved to four size classes (< 6 mm, 6-13 mm, 14-25 mm, and 26-50 mm). Measured at a depth of 2.5 cm, light penetration was <1% with the three smallest clod classes compared to about 12.5% under the largest clods. Complete prevention of light penetration in the largest clods was possible only at a depth of 10 cm of soil.

Improvements in soil tilth will not change the requirement for good crop seed-soil contact at planting. It may, however, affect the availability of safe-sites for weed establishment (Figure 1B). Following disturbance, soils with improved secondary structure may have a coarser soil surface due to greater proportions of water-stable aggregates. An Iowa grower offers the following planting suggestions: "pack seed with a small, one-inch press wheel, and cover seed with 2 to 3 inches of loose soil," because " . . . the loose soil will dry out and the weed seed will not germinate" (Thompson, Thompson, and Thompson, 1995). In this example, the coarser loose soil could result in a net increase in safe-sites for species with germination requirements similar to those of *Bromus madritensis*, but a decrease in sites for species with requirements similar to those of common lambsquarters (Harper, Williams, and Sagar, 1965). Therefore, modifications of the soil physical environment may or may not be useful for desired manipulations of weed populations. Because of the potential for important effects on other weed management strategies such as cultivation, relationships between soil physical characteristics and weed establishment clearly warrant further study.

*Fertility Management*

Chemical indicators of soil quality include salinity, pH, cation exchange capacity, and the availability of plant nutrients (Karlen, Eash, and Unger, 1992). Improving the chemical quality of soils generally aims to minimize salinity, maintain pH suitable for the crops being grown, and optimize nutrient use effiency. Management of soil chemistry commonly involves additions of synthetic fertilizers and lime, but may include incorporation of legume residues and other organic materials as nutrient sources (Donald, 1993; Stute and Posner, 1995; Varco et al., 1993; Westcott and Mikkelsen, 1987).

*Synthetic nutrient sources:* Broadcast application of fertilizer, especially nitrogen (N), can promote germination of common lambsquarters (Henson, 1970; Williams, and Harper, 1965), wild oat (*Avena fatua* L.) (Sexsmith and Pittman, 1963), and certain other weed species (Fenner, 1985). Following establishment, the effects of fertilization on weed/crop mixtures are varied. Fertilization, for example, may ameliorate weed interference caused by competition for nutrients. Staniforth (1957) found that corn (*Zea mays* L.) yield loss caused by yellow foxtail [*Setaria lutescens* (Weigel) Hubb] and green foxtail [*S. viridis* (L.) Beauv.] was 20%, 10%, and 5% when N fertilizer was applied at 0, 78, and 157 kg N ha$^{-1}$, respectively (see also Nieto and Staniforth, 1961; Vengris, Colby, and Drake, 1955). Tollenaar et al. (1994) examined the effects of a mixed weed population on corn yield and also found that yield loss decreased as N fertilizer rate increased. Alternatively, if species respond similarly to an added resource, or if the available supplies of resources do not limit growth, there may be no effect of fertilization on interspecific interference (Pfeiffer and Holmes, 1961; Wells, 1979). If, however, species respond differentially to alleviation of competition for a limited resource, competition for another resource may be exacerbated, causing one species to dominate (Tilman, 1990). Consequently, if application of inorganic fertilizers is of greater relative benefit to weeds than to the crop, yields may actually be reduced by fertilization. For example, Carlson and Hill (1985) found that N fertilizer application increased interference from wild oat and reduced spring wheat yields. When pre-plant N was not applied, 20 wild oat plants m$^{-2}$ reduced spring wheat yield 26%, but with 168 kg N ha$^{-1}$ of pre-plant N, the same wild oat density reduced yield 43%. Relative to most crop plants, annual weeds have higher relative growth rates (Seibert and

Pearce, 1993) and are able to accumulate higher concentrations of nitrogen, phosphorus, potassium, calcium, and magnesium (Alkamper, 1976; DiTomaso, 1995).

Liebman (1989) examined the effects of N fertilization with ammonium sulfate and irrigation on competition between the weed white mustard (*Brassica hirta* Moench), and either a long-vined pea (*Pisium sativum* L. 'Century'), or a short-vined ('Alaska') pea, each intercropped with barley (*Hordeum vulgare* L.). Nitrogen fertilizer increased white mustard biomass 472% compared to the control (no added N), averaged across all treatment combinations. By comparison, crop seed yield increase from applied N was 31%. Under irrigated conditions, applied N increased the weed:crop ratio 5-fold in the long-vined 'Century' pea treatments but 11-fold in the short-vined 'Alaska' pea treatments. With added N, white mustard overtopped and shaded 'Alaska' pea.

Species may also differ in their requirements for, and responses to, phosphorus (P) and potassium (K). Bhaskar and Vyas (1988) reported that common lambsquarters interfered with wheat, grown in pots, by competing primarily for P and secondarily for N. In a replacement series, changing the wheat:common lambsquarters ratio from 1:1 to 1:3 resulted in a 153-fold decrease in wheat P uptake but only a 2-fold decrease in N uptake (on a per plant basis). Common lambsquarters has been reported by other investigators to have a high requirement for K. Two separate pot studies showed that common lambsquarters growing in a high K environment was more competitive with tomato (*Lycopersicon esculentum* Miller) (Minotti, 1977), and the weed common groundsel (*Senecio vulgaris* L.) (Qasem and Hill, 1995), than in low K treatments.

*Organic amendments:* Animal manures and composts are important sources of nutrients on many farms, and can increase soil biological activity, improve soil structure and water holding capacity, and provide organic compounds that stimulate plant nutrient uptake and growth (Parr, Papendick and Colacicco, 1986; Russell, 1973; Valdrighi et al., 1996). Despite these beneficial characteristics, manure and compost are often sources of weed introduction and dispersal in cropland (Mt. Pleasant and Schlather, 1994). Consequently, management of manure and compost has great importance both to soil quality and weed dynamics.

The density of viable weed seeds in manure is a function of the types of animals generating the manure and the types of plant materials fed to them. Poultry manure typically contains few viable weed seeds because they are broken down within the animal's gizzard (Cudney et al., 1992). Dairy manure may contain tens of thousands of viable seeds per ton (Cudney et al., 1992; Mt. Pleasant and Schlather, 1994), because seeds of certain species survive passage through the animal's digestive system (Sarapatka, Holub, and Lhotska, 1993). The manure of milking cows, which are fed mostly concentrate feeds, can contain many fewer weed seeds than that of dry cows, which are fed coarser (and weedier) forages (Cudney et al., 1992).

Storage and handling of amendments can also affect weed seeds. Composting is thought to kill weed seeds contained in manure and plant materials through seed exposure to elevated temperatures, organic acids, methane, and carbon dioxide produced during the fermentation process (Sarapatka, Holub, and Lhotska, 1993). A combination of moist conditions and high temperature is especially effective in killing weed seeds (Bloemhard et al., 1992). Cudney et al. (1992) found that composts made aerobically from dairy manure contained significantly fewer viable weed seeds than uncomposted manure. Aerobic composting did not entirely eliminate viable weed seeds suggesting that longer composting periods, deeper piles, supplemental watering, and turning of piles should be tested as ways to further reduce weed seed density. An additional alternative is anaerobic composting of manure, which can be used to conserve nutrients more effectively than aerobic systems and to generate methane for fuel. Anaerobic composting of a mixture of straw and dairy manure for 30 days completely killed seeds of seven weed species and, at a depth of 40 cm in the pile, reduced the viability of four other species by 64 to 96% (Sarapatka et al., 1993). At a depth of 180 cm in the compost pile, seeds of all eleven weed species tested were killed. These results show that considerable opportunities exist to develop composting technologies and manure handling methods that greatly reduce densities of viable weed seeds.

## Cover Cropping and Green Manuring

Cover crops are grown between periods of regular cropping, primarily to reduce soil erosion. Green manure crops are grown specifically for soil improvement through additions of organic matter and, in the case of legume green manures, nitrogen. Other functions of cover crops and green manures include capturing and recycling of nutrients, breaking of plow pans, producing mulch for water retention, and supplying emergency forage (Sarrantonio, 1994). Cover cropping and green manuring practices may also play an important and desirable role in weed management. While growing, cover and green manure crops may suppress weed growth (McLenaghen et al., 1996). Furthermore, the surface residues of these crops can directly affect weed population dynamics by the action of leached allelochemicals (Putnam, 1994), increased moisture, lower temperature, and reduced light quantity (Teasdale, 1993). Teasdale (1996) noted the following salient features regarding weed control by the surface residues of winter annual cover crops: (i) weed control increased with increasing residue biomass; (ii) weed control was species-specific; and (iii) complete weed control was not achieved.

Incorporated crop residues and organic amendments can supply nutrients to a subsequent crop thereby decreasing the amount of synthetic fertilizer required to attain desired yields. Legume green manures, for example, may mineralize N in a temporal pattern coincident with crop demand (Stute and Posner, 1995), but it is also possible that supplemental fertilization is required to compensate for sub-optimal early season N release (Varco et al., 1993; Westcott and Mikkelsen, 1987). Given that fertilization practices can affect weed germination, growth, and interference with crops (DiTomaso, 1995), strategies that combine organic and synthetic nutrient sources may be useful to provide optimal delivery of nutrients to the crop while avoiding excessive nutrient levels that could enhance a weed's competitive advantage. Alternatively, residue effects on weeds may be explained by nutrient reductions. Residues with high carbon:nitrogen residues, such as those from cereals, can affect small-seeded weeds by immobilizing N in soil (Samson, Drudy, and Omielan, 1992).

There is also evidence that incorporation of cover and green manure crop residues into soil may suppress weed growth in subsequent cash crops compared to areas with no cover or green manure crops. Weed suppressive effects of a cruciferous green manure crop were reported by Boydston and Hang (1995), who measured weed biomass and potato (*Solanum tuberosum* L.) tuber yields following bare fallow or rapeseed (*Brassica napus* var. *napus*) green manure treatments. Incorporation of rapeseed green manure reduced weed biomass in a subsequent potato crop 50 to 96% below levels measured following the bare fallow treatment. The rapeseed green manure increased potato yield 10 to 18% compared to fallow, whether or not weeds were present. Enhancement of potato growth was attributed to improved crop nutrition and suppression of soil-borne pathogens. Weed suppression was attributed to phytotoxic isothiocyanate compounds derived from rapeseed tissues.

An example of how a leguminous green manure may contribute to weed management is provided by the results of Dyck, Liebman, and Erich (1995), who conducted a field study comparing two N fertility regimes for sweet corn production: crimson clover (*Trifolium incarnatum* L.) green manure and ammonium nitrate fertilizer. Interference between sweet corn and common lambsquarters was examined following soil management treatments that consisted of clover grown for two months before incorporation into the soil and bare fallow maintained for two months before being amended with different rates of synthetic N fertilizer. Following the clover and bare fallow regimes, sweet corn was grown alone and in combination with a fixed density of common lambsquarters. Crop biomass was 20% greater and weed biomass was approximately 40% lower when N fertility was supplied to the crop-weed mixtures by crimson clover green manure rather than ammonium nitrate fertilizer (Figure 2). Comparisons of corn growth under weed-free and weed-infested conditions showed that weed competition reduced corn bio-

FIGURE 2. Effects of ammonium nitrate fertilizer (45 kg N ha$^{-1}$) and soil-incorporated crimson clover green manure on above-ground biomass production by sweet corn and common lambsquarters grown in mixture in field plots at Stillwater, ME, measured 77 days after planting in 1989 (A), and 86 days after planting in 1990 (B). Estimated N fertilizer equivalence value of the clover green manure was 52 kg N ha$^{-1}$ in 1989 and 58 kg N ha$^{-1}$ in 1990. Asterisks indicate significant differences (p < 0.05). Adapted from Dyck, Liebman, and Erich (1995).

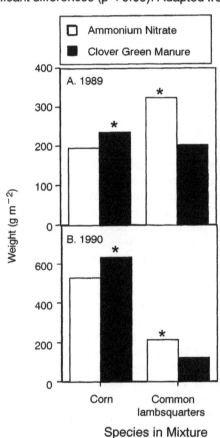

mass by up to 14% when clover green manure was used, but as much as 36% when synthetic fertilizer was used. Results of two companion studies, in which common lambsquarters was grown alone in soils managed with the same treatments as in the sweet corn plus common lambsquarters study, indicated that the weed-suppressive effect of crimson clover residue occurred when corn was not present and when different rates of ammonium nitrate were applied (Dyck and Liebman, 1994; Dyck, Liebman, and Erich, 1995). Dyck and Liebman (1994) suggested that allelopathic interactions may have been responsible for the suppressive effect of crimson clover residue on common lambsquarters, a hypothesis consistent with the results of White, Worsham, and Blum (1989).

Recently Gallandt and Liebman (*in preparation*), compared the effects of soil-incorporated red clover (*Trifolium pratense* L.) residue and ammonium nitrate fertilizer on the performance of dry bean (*Phaseolus vulgaris* L.) and wild mustard (*Sinapis arvensis* L.). Although bean is a legume, the short season, determinate varieties grown in Maine often exhibit a positive yield response to increases in soil N (Liebman et al., 1995). Use of red clover green manure in place of ammonium nitrate significantly increased seed yield of bean grown in mixture with wild mustard in one out of three years, although it had no effect on bean yield in two other years (Figure 3). Use of red clover green manure in place of ammonium nitrate also reduced weed biomass in two out of three years; a neutral effect was observed in a third year (Figure 3). Overall, use of N from clover green manure rather than synthetic sources increased bean yield by an average of 7% and decreased wild mustard biomass by an average of 37%.

Doolan (1997) indicated that allelopathic interactions may be responsible for the suppressive effect of red clover residue on wild mustard. Red clover residue was inoculated with soil microbes and allowed to decompose for 0 to 5 weeks and aqueous extracts of the residue were then filtered to remove microbes and applied to newly germinated seedlings of wild mustard. Significant reductions in root growth of the weed seedlings were observed with concentrations of red clover similar to those used in the field experiment with bean and wild mustard. Variations between years in the weed-suppressive effects of red clover residue under field conditions may have been caused by variations in rainfall and soil moisture (Doolan, 1997).

Why may a selective effect of allelopathic plant residues occur between crop and weed species, such that weeds are suppressed and crops are unaffected or enhanced by residue incorporation? Westoby, Leishman, and Lord (1996) noted that seedlings of larger-seeded species may be better able than small-seeded species to survive various forms of stress. Large-seeded species have greater amounts of seed reserves that remain uncommitted to seedling structure during early growth. These reserves allow maintenance, growth, and repair functions to continue even when stress factors limit photosynthetic carbon gain. It is thus conceivable that allelochemicals from cover crops and green manures may have a greater effect on small-seeded weeds than on large-seeded or vegetatively reproduced crops or weeds (e.g., corn, bean, potato, or velvetleaf). If allelochemicals were localized in upper soil profile, crops with large seed reserves and access deeper layers of the soil profile,

FIGURE 3. Effects of ammonium nitrate fertilizer (84 kg N ha$^{-1}$) and soil-incorporated red clover green manure on seed yield of 'Marafax' bean and total (above- plus below-ground) biomass of wild mustard. Measurements were made on plants in two-species mixtures in field plots at Stillwater, ME, 13 to 14 weeks after planting in 1994 (A), 1995 (B), and 1996 (C). Asterisks indicate significant differences (p < 0.05); "ns" indicates no significant difference (from Gallandt and Liebman, *in preparation*).

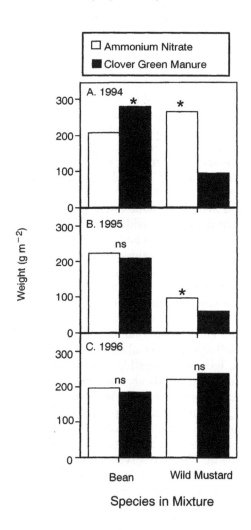

could establish largely unaffected by allelochemicals. Crops would therefore have a competitive advantage over weeds conferred by earlier canopy establishment. If this hypothesis was true, then long-term use of cover crops could cause weed species to shift toward larger-seeded or allelochemically-tolerant taxa.

Despite their potential to suppress weeds and improve soil quality, cover crops and green manures also have the potential to interfere with crop production. Crops planted soon after incorporation of cover crops and green manures into soil may be stressed due to allelochemicals (Miller, 1996; Weston, 1996) or pathogens (Dabney et al., 1996) associated with the incorporated residues. Problems with certain insect pests, such as seed corn maggot (*Delia paltura*), may be exacerbated following incorporation of cover crops and green manures (Hammond, 1990). In drier areas, excessive depletion of soil moisture by cover crops and green manures may compromise the growth of subsequent cash crops (Zentner et al., 1996). These types of effects need to be addressed when cover cropping and green manuring are considered as weed management tactics.

## Cropping Systems Study: The Maine Potato Ecosystem Project

In the Maine Potato Ecosystem Project, contrasting soil management systems were established to test the hypothesis that improved soil quality will lead to greater crop productivity because of physical and nutritional benefits to the crop, and altered pest dynamics. A factorial design of two soil management systems with three contrasting pest management systems allowed for explicit tests of possible interactions of soil quality and pest dynamics (Alford et al., 1996).

### Treatments

The project was established in 1991 to contrast amended *versus* unamended soil management strategies and conventional *versus* reduced-input *versus* bio-intensive pest management strategies. Conventional pest management relied on application of pesticides recommended by University of Maine Cooperative Extension; the reduced-input system made use of the same pesticides but attempted to reduce doses or increase thresholds on which application decisions were based; and the biointensive pest management system made use of mechanical or cultural practices, natural chemicals, or biological controls such as insect predators. The soil management systems were designed to compare a typical fertilizer-based system (*unamended*) to reduced rates of fertilizer plus legume green manure, animal manure, and compost (*amended*).

The unamended soil management system used a two year rotation of barley-potato; barley was grown for grain and intercropped with red clover. Potatoes were fertilized with 134 kg N ha$^{-1}$, 134 kg P$_2$O$_5$ ha$^{-1}$, and 134 kg K$_2$O ha$^{-1}$ at planting, with an additional side-dressed application of urea-ammonium nitrate at 62 kg N ha$^{-1}$, supplied at a later date. The amended system used a two-year green manure-potato rotation: the green manure consisted of a mixture of oat (*Avena sativa* L.), pea and hairy vetch (*Vicia villosa* L.). The amended system received 45 Mg ha$^{-1}$ beef manure plus 22 Mg ha$^{-1}$ compost in the spring, and potato was fertilized with 67 kg N ha$^{-1}$, 67 kg P$_2$O$_5$ ha$^{-1}$, and 67 kg K$_2$O ha$^{-1}$ at planting. Annually, the compost, manure and green manure contributed an average of 528 kg N ha$^{-1}$, 180 kg P$_2$O$_5$ ha$^{-1}$, and 126 kg K$_2$O ha$^{-1}$.

*Soil Quality*

Compared to the unamended soil management system, the amended soil management system increased soil organic matter content, water-stable aggregate content, effective cation exchange capacity, pH, potassium, magnesium, calcium, and soluble inorganic phosphorus. Furthermore, the amended system reduced requirements for synthetic fertilizers, enhanced late-season crop vigor and canopy duration, increased tuber quality, and increased yields 12% to 30% relative to the unamended system (Porter and McBurnie, 1996).

*Weed Biomass*

In the pest management system that relied exclusively on cultivation for direct weed control, weed biomass in potato was relatively high overall, but 75% lower in 1994, and 73% lower in 1995, in the amended soil management system as compared to the unamended system (Gallandt et al., 1998) (Figure 4). In contrast, weed biomass in the reduced-input and conventional systems (which included herbicides) was low with no significant difference between the unamended and amended soil management treatments. Although weeds did not cause significant yield loss in 1994, yield loss due to weeds in 1995 was 37% in the unamended soil management system but only 12% in the amended system (Table 1).

Weed management benefits observed in the amended soil management system could be due to direct or indirect effects of the potato fertilization strategy, the pea/oat/hairy vetch rotation crop preceding potatoes, or some combined effect of the two. Direct effects, for example, could include weed mortality caused by rotation crop interference or fertility-induced alterations in weed germination, emergence, or establishment. If the competitive ability of the potato crop were affected, soil management would affect weeds through crop interference. The latter is a probable mechanism because weed biomass was equivalent in the barley/red clover and the pea/oat/hairy vetch

FIGURE 4. Effects of three levels of weed management and two soil management systems on above-ground weed biomass in potatoes. Weed management systems were conventional (CONV; metribuzin and paraquat at full rates + cultivation), reduced input (RI; metribuzin and paraquat at 1/2 × rates + cultivation), and biointensive (BIO; cultivation only). Measurements were made mid-August in 1994 and 1995. Within a year, bars not topped by the same letter are significantly different (p < 0.05). Adapted from Gallandt et al. (1998).

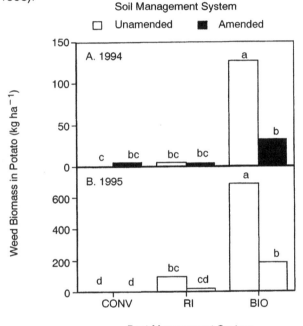

rotation crops, and compared to the unamended crop, amended potato had greater or equivalent leaf area and consistently greater yields (Gallandt et al., 1998). Also, the relative yield increase due to compost application was greater during dry growing seasons than in relatively moist seasons suggesting that a major component of the potato yield response was due to improvements in soil physical properties and/or other, possibly biotic, changes in the bulk soil, or rhizosphere (G. A. Porter, personal communication).

## CONCLUSIONS

The goals of integrated weed management are to reduce weed densities, minimize the damage caused per surviving individual, reduce extreme popu-

TABLE 1. Tuber yield in potato managed without herbicides, with hand weeding, "Weed free," and without hand weeding, "Ambient weeds," in contrasting soil management systems. Within a year, means followed by the same letter are not significantly different (p < 0.05). Adapted from Gallandt et al. (1998).

| | Tuber yield | | |
|---|---|---|---|
| Soil management system | Weed free | Ambient weeds | Yield loss |
| | (kg m$^{-2}$) | | (%) |
| 1994 | | | |
| Unamended | 1.95 b | 1.95 b | 0 |
| Amended | 2.57 a | 2.29 ab | 11 |
| 1995 | | | |
| Unamended | 2.02 b | 1.28 c | 37 |
| Amended | 2.42 a | 2.14 ab | 12 |

lation fluctuations to avoid epidemics, and slow the development of resistance to deployed tactics. This approach to weed management requires an ecologically-based understanding of the many factors that determine weed performance including relevant interactions between soil quality and weeds. Specifically, our interest in linking weed and soil ecology led to three management objectives: (i) reducing the persistence of seeds in the soil; (ii) reducing the abundance of safe-sites and the filling of available sites; and (iii) reducing crop yield loss per individual weed. Accomplishment of these objectives involves understanding management effects on soil processes critical to weed dynamics such as the quantity and availability of soil resources including water, light, and nutrients; the species composition and function of biotic communities, including pathogens and herbivores that attack and kill weeds; and allelopathic affects on weed germination, emergence, and survival. Major soil management practices that affect soil processes related to weed dynamics are tillage, water and nutrient management, and crop rotation, including the use of cover crops and green manures. Combined in a cropping system these practices can be used to reduce soil degradative processes, improve soil quality, and provide the opportunity to suppress weeds. However, a better understanding of how soil improving practices affect weed dynamics will be needed to design cropping systems that simultaneously address soil quality and weed management objectives. Areas of research include evaluating soil management impacts on seed bank density over time; the formation, quantity, and availability of safe-sites; and weed density effects on crop yield. These component research efforts need to be incorporated into long-term studies to improve understanding of weed community and population dynamics within cropping systems that include soil improving management strategies. We conclude that many as yet unexplored opportunities exist to manipulate the soil environment and to design cropping systems that improve soil quality and create multiple weed suppressive conditions. While this strategy will not eliminate weeds, it should slow the development of weed resistance to a given tactic and provide broader safe-guards to weed epidemics.

## REFERENCES

Alford, A. R., F. A. Drummond, E. R. Gallandt, E. Groden, D. A. Lambert, M. Liebman, M. C. Marra, J. C. McBurnie, G. A. Porter, and B. Salas. (1996). *The Ecology, Economics, and Management of Potato Cropping Systems: A Report of the First Four Years of the Maine Potato Ecosystem Project*, Orono, ME: Maine Agricultural and Forest Experiment Station, University of Maine, Bulletin 843.

Alkamper, J. (1976). Influence of weed infestation on effect of fertilizer dressings. *Pflanzenschutz-Nachrichten* 29: 191-235.

Arshad, M. A. and G. M. Coen. (1992). Characterization of soil quality: physical and chemical criteria. *American Journal of Alternative Agriculture* 7: 25-31.

Bhaskar, A. and K. G. Vyas. (1988). Studies on competition between wheat and *Chenopodium album* L. *Weed Research* 28: 53-58.

Bloemhard, C. M. J., M. W. M. F. Arts, P. C. Scheepens, and A. G. Elema. (1992). Thermal inactivation of weed seeds and tubers during the drying of pig manure. *Netherlands Journal of Agricultural Science* 40:11-19.

Boydston, R. A. and A. Hang. (1995). Rapeseed (*Brassica napus*) green manure crop suppresses weeds in potato (*Solanum tuberosum*). *Weed Technology* 9: 669-675.

Boyetchko, S. M. (1996). Impact of soil microorganisms on weed biology and ecology. *Phytoprotection* 77: 41-56.

Braunack, M. V. and A. R. Dexter. (1989). Soil aggregation in the seedbed: a review. I. Properties of aggregates and beds of aggregates. *Soil & Tillage Research* 14: 259-279.

Brussaard, L. (1994). Interrelationships between biological activities, soil properties and soil management. In *Soil Resilience and Sustainable Land Use*, ed. D. J. Greenland and I. Szabolcseds. Wallingford, UK: CAB International, pp. 309-329.

Brust, G. E. and G. J. House. (1988). Weed seed destruction by arthropods and rodents in low-input soybean agroecosystems. *American Journal of Alternative Agriculture* 3: 19-25.

Buhler, D. D. (1995). Influence of tillage systems on weed population dynamics and management in corn and soybean in the central USA. *Crop Science* 35: 1247-1258.

Burnside, O. C., G. A. Wicks, and C. R. Fenster. (1977). Longevity of shattercane seed in soil across Nebraska. *Weed Research* 17: 139-143.

Cannell, R. Q. and J. D. Hawes. (1994). Trends in tillage practices in relation to sustainable crop production with special reference to temperate climates. *Soil & Tillage Research* 30: 245-282.

Cardina, J., E. Regnier, and K. Harrison. (1991). Long-term tillage effects on seed banks in three Ohio soils. *Weed Science* 39: 186-194.

Carlson, H. L. and J. E. Hill. (1985). Wild oat (*Avena fatua*) competition with spring wheat: plant density effects. *Weed Science* 33: 176-181.

Cavers, P. B. and D. L. Benoit. (1989). Seed banks in arable land. In *Ecology of Soil Seed Banks*, ed. M. A. Leck, V. T. Parker and R. L. Simpsoneds. San Diego, CA: Academic Press, Inc., pp. 309-328.

Coleman, D. C., C. P. P. Reid, and C. V. Cole. (1983). Biological strategies of nutrient cycling in soil systems. *Adv. Ecol. Res.* 13:1-55.

Coleman, D. C. and P. F. Hendrix. (1988). Agroecosystem processes. In *Concepts of Ecosystem Ecology*, ed. L. R. Pomeroy and J. J. Alberts. New York: Springer-Verlag, pp. 149-170.

Cook, R. E. (1980). The biology of seeds in soil. In *Demography and Evolution in Plant Populations*, ed. O. T. Solbrig. Berkeley, CA: University of California Press, pp. 107-129.

Cousens, R. (1985). A simple model relating yield loss to weed density. *Advances in Applied Biology* 107: 239-252.

Cudney, D. W., S. D. Wright, T. A. Shultz, and J. S. Reints. (1992). Weed seed in dairy manure depends on collection site. *California Agriculture* 46(3): 31-32.

Currie, J. A. (1972). The seed-soil system. In *Seed Ecology*, ed. W. Heydecker. University Park, PA: The Pennsylvania University Press, pp. 463-480.

Cussans, G. W., S. Raudonius, P. Brian, and S. Cumberworth. (1996). Effects of depth of seed burial and soil aggregate size on seedling emergence of *Alopecurus myosuroides*, *Galium aparine*, *Stellaria media* and wheat. *Weed Research* 36: 133-141.

Dabney, S. M., J. D. Schreiber, C. S. Rothrock, and J. R. Johnson. (1996). Cover crops affect sorghum seedling growth. *Agronomy Journal* 88: 961-970.

DiTomaso, J. M. (1995). Approaches for improving crop competitiveness through the manipulation of fertilization strategies. *Weed Science* 43: 491-497.

Donald, W. W. (1993). Models and sampling for studying weed seed survival with wild mustard (*Sinapis arvensis*) as a case study. *Canadian Journal of Plant Science* 73: 637-645.

Donald, W. W. and R. L. Zimdahl. (1987). Persistence, germinability, and distribution of jointed goatgrass (*Aegilops cylindrica*) seed in soil. *Weed Science* 35: 149-154.

Doolan, K. L. (1997). The Allelopathic Potential of Red Clover Residue. M.S., University of Maine, Orono, ME.

Doran, J.W. (1980). Soil microbial and biochemical changes associated with reduced tillage. *Soil Sci. Soc. Am. J.* 44: 765-771.

Douds, D. D. J., R. R. Janke, and S. E. Peters. (1993). VAM fungus spore populations and colonization of roots of maize and soybean under conventional and low-input sustainable agriculture. *Agriculture, Ecosystems and Environment* 43: 325-335.

Drinkwater, L. E., D. K. Letourneau, F. Workneh, A. H. C. van Bruggen, and C. Shennan. (1995). Fundamental differences between conventional and organic tomato agroecosystems in California. *Ecological Applications* 5: 1098-1112.

Dyck, E. and M. Liebman. (1994). Soil fertility management as a factor in weed control: the effect of crimson clover residue, synthetic nitrogen fertilizer, and their interaction on emergence and early growth of lambsquarters and sweet corn. *Plant and Soil* 167: 227-237.

Dyck, E., M. Liebman, and M. S. Erich. (1995). Crop-weed interference as influenced by a leguminous or synthetic fertilizer nitrogen source: I. double cropping experiments with crimson clover, sweet corn, and lambsquarters. *Agriculture, Ecosystems and Environment* 56: 93-108.

Fellows, G. M. and F. W. Roeth. (1992). Factors influencing shattercane (*Sorghum bicolor*) seed survival. *Weed Science* 40: 434-440.

Fenner, M. (1985). *Seed Ecology*, New York: Chapman and Hall.

Forcella, F. (1992). Prediction of weed seedling densities from buried seed reserves. *Weed Research* 32: 29-38.

Gallandt, E. R., M. Liebman, S. Corson, G. A. Porter, and S. D. Ullrich. (1998). Effects of pest and soil management systems on weed dynamics in potato. *Weed Science* 46:238-248.

Gonzalez-Andujar, J. L. and C. Fernandez-Quintanilla. (1991). Modelling the population dynamics of *Avena sterilis* under dry-land cereal cropping systems. *Journal of Applied Ecology* 28: 16-27.

Grant, J. D. (1983). The activities of earthworms and the fates of seeds. In *Earthworm Ecology, From Darwin to Vermiculture*, ed. J. E. Satchell. London: Chapman and Hall Ltd., pp. 107-122.

Hammond, R. B. (1990). Influence of cover crops and tillage on seedcorn maggots (Diptera:Anthomyiidae) populations in soybeans. *Environmental Entomology* 19: 510-514.

Harper, J. L. (1974). Agricultural ecosystems. *Agro-Ecosystems* 1: 1-6.

Harper, J. L. (1977). *Population Biology of Plants*, San Diego, CA: Academic Press, Inc.

Harper, J. L., J. T. Williams, and G. R. Sagar. (1965). The behaviour of seeds in soil. I. The heterogeneity of soil surfaces and its role in determining the establishment of plants from seed. *Journal of Ecology* 53: 273-286.

Henson, I. E. (1970). The effects of light, potassium, nitrate and temperature on the germination of *Chenopodium album* L. *Weed Research* 10: 27-39.

House, G. J. and G. E. Brust. (1989). Ecology of low-input, no-tillage agroecosystems. *Agriculture, Ecosystems and Environment* 27: 331-345.

Jenny, H. (1941). *Factors of Soil Formation*. New York: McGraw-Hill.

Jordan, N. (1993). Simulation analysis of weed population dynamics in ridge-tilled fields. *Weed Science* 41: 468-474.

Jordan, N., D. A. Mortensen, D. M. Prentflow, and K. C. Cox. (1995). Simulation analysis of crop rotation effects on weed seedbanks. *American Journal of Botany* 82: 390-398.

Karlen, D. L., N. S. Eash, and P. W. Unger. (1992). Soil and crop management effects on soil quality indicators. *American Journal of Alternative Agriculture* 7: 48-55.

Karlen, D. L., M. J. Mausbach, J. W. Doran, R. G. Cline, R. F. Harris, and G. E. Schuman. (1997). Soil quality: a concept, definition, and framework for evaluation (a guest editorial). *Soil Science Society of America Journal* 61: 4-10.

Kremer, R. J. (1993). Management of weed seed banks with microorganisms. *Ecological Applications* 3: 42-52.

Kremer, R. J., M. F. T. Begonia, L. Stanley, and E. T. Lanham. (1990). Characterization of rhizobacteria associated with weed seedlings. *Applied Environmental Microbiology* 56: 1649-1655.

Lal, R. and B. A. Stewart. (1995). Need for long-term experiments in sustainable use of soil resources. In *Soil Management, Experimental Basis for Sustainability and Environmental Quality*, eds. R. Lal and B. A. Stewart. Boca Raton, FL: CRC Press, Inc., Lewis Publishers, pp. 537-545.

Liebman, M. (1989). Effects of nitrogen fertilizer, irrigation, and crop genotype on canopy relations and yields of an intercrop/weed mixture. *Field Crops Research* 22: 83-100.

Liebman, M., S. Corson, R. J. Rowe, and W. A. Halteman. (1995). Dry bean response to nitrogen fertilizer in two tillage and residue management systems. *Agronomy Journal* 87: 538-546.

Liebman, M. and E. R. Gallandt. (1997). Many little hammers: ecological approaches for management of crop-weed interactions. In *Ecology in Agriculture*, ed. L. E. Jackson. San Diego, CA: Academic Press, Inc., pp. 291-343.

Liebman, M. and T. Ohno. (1998). Crop rotation and legume residue effects on weed emergence and growth: implications for weed management. In *Integrated Weed and Soil Management*, eds. J.L. Hatfield, D.D. Buhler, and B.A. Stewart. Chelsea, MI: Ann Arbor Press, pp. 181-221.

Lindquist, J. L., D. A. Mortensen, S. A. Clay, R. Schmenk, J. J. Kells, K. Howatt, and P. Westra. (1996). Stability of coefficients in the corn yield loss-velvetleaf density relationship across the North Central U.S. *Weed Science* 44: 309-313.

Logsdon, S. D. and D. R. Linden. (1992). Interactions of earthworms with soil physical conditions influending plant growth. *Soil Science* 154: 330-337.

Lonsdale, W. M. (1993). Losses from the seed bank of *Mimosa pigra*: soil micro-organisms vs. temperature fluctuations. *Journal of Applied Ecology* 30: 654-660.

Marinissen, J. C. Y. and P. C. deRuiter. (1993). Contribution of earthworms to carbon and nitrogen cycling in agro-ecosystems. *Agriculture, Ecosystems and Environment* 47: 59-74.

McLenaghen, R. D., K. C. Cameron, N. H. Lampkin, M. L. Daly, and B. Deo. (1996). Nitrate leaching from plowed pasture and the effectiveness of winter catch crops in reducing leaching losses. *New Zealand Journal of Agricultural Research* 39: 413-420.

McRill, M. and G. R. Sagar. (1973). Earthworms and seeds. *Nature* 243: 482.

Miller, D. A. (1996). Allelopathy in forage crop systems. *Agronomy Journal* 88: 854-859.

Minotti, P. L. (1977). Differential response of tomato and lambsquarters seedlings to potassium level. *Journal of the American Society of Horticultural Science* 102: 646-648.

Mt. Pleasant, J. and K. J. Schlather. (1994). Incidence of weed seed in cow (*Bos* sp.) manure and its importance as a weed source for cropland. *Weed Technology* 8: 304-310.

Naylor, R. E. L. (1985). Establishment and peri-establishment mortality. In *Studies* on *Plant Demography: A Festschrift for John L. Harper*, ed. J. White. London, UK: Academic Press, Inc., pp. 95-109.

Nieto, J. H. and D. W. Staniforth. (1961). Corn-foxtail competition under various production conditions. *Agronomy Journal* 53: 1-5.

Parr, J. F. and S. B. Hornick. (1992). Agricultural use of organic amendments: a historical perspective. *American Journal of Alternative Agriculture* 7: 181-189.

Parr, J. F., R. I. Papendick, and D. Colacicco. (1986). Recycling of organic wastes for a sustainable agriculture. *Biological Agriculture and Horticulture* 3: 115-130.

Parr, J. F., R. I. Papendick, S. B. Hornick, and R. E. Meyer. (1992). Soil quality: attributes and relationship to alternative and sustainable agriculture. *American Journal of Alternative Agriculture* 7: 5-11.

Patriquin, D. G., D. Baines, and A. Abboud. (1995). Soil fertility effects on pests and diseases. In *Soil Management in Sustainable Agriculture*, eds. H. F. Cook and H. C. Leeeds. Wye, UK: Wye College Press, pp. 161-174.

Pfeiffer, R. K. and H. M. Holmes. (1961). A study of the competition between barley and oats as influenced by barley seed rate, nitrogen level and barban treatment. *Weed Research* 1: 5-18.

Pitty, A., D. W. Staniforth, and L. H. Tiffany. (1987). Fungi associated with caryopses of *Setaria* species from field-harvested seeds and from soil under two tillage systems. *Weed Science* 35: 319-323.

Porter, G. A. and J. C. McBurnie. (1996). Crop and soil research. In *The Ecology, Economics, and Management of Potato Cropping Systems: A Report of the First Four Years of the Maine Potato Ecosystem Project*, ed. M. C. Marra. Orono, ME: Maine Agricultural and Forest Experiment Station, University of Maine, Bulletin 843, pp. 8-62.

Putnam, A. R. (1994). Phytotoxicity of plant residues. In *Managing Agricultural Residues*, ed. P. W. Unger. Boca Raton, Florida: Lewis Publishers, pp. 285-314.

Qasem, J. R. and T. A. Hill. (1995). Growth, development and nutrient accumulation in *Senecio vulgaris* L. and *Chenopodium album* L. *Weed Research* 35: 187-196.

Roberts, H. A. and R. T. Hewson. (1971). Herbicide performance and soil surface conditions. *Weed Research* 11: 69-73.

Roberts, H. A. and M. E. Ricketts. (1979). Quantitative relationships between the weed flora after cultivation and the seed population in the soil. *Weed Research* 19: 269-275.

Robocker, W. C., M. C. Williams, R. A. Evans, and P. J. Torell. (1969). Effects of age, burial, and region on germination and viability of halogeton seed. *Weed Science* 17: 63-65.

Russell, E. W. (1973). *Soil Conditions and Plant Growth*, 10th/Ed., London, UK: Longman.

Samson, R., C. Drury, and J. Omielan. (1992). Effect of winter rye mulches and fertilizer amendments on nutrient and weed dynamics in no-till soybeans, Rep. No. #57B. Soil and Environmental Enhancement Program (SWEEP), Agriculture Canada, Guelph, Ontario.

Sarapatka, B., M. Holub, and M. Lhotska. (1993). The effect of farmyard manure anaerobic treatment on weed seed viability. *Biological Agriculture and Horticulture* 10: 1-8.

Sarrantonio, M. (1994). *Northeast Cover Crop Handbook*, Emmaus, PA: Rodale Institute.

Schippers, B., A. W. Bakker, and P. A. H. M. Bakker. (1987). Interactions of deleterious and beneficial rhizosphere microorganisms and the effect of cropping practices. *Annual Reviews of Phytopathology* 25: 339-358.

Seibert, A. C. and R. B. Pearce. (1993). Growth analysis of weed and crop species with reference to seed weight. *Weed Science* 41: 52-56.

Sexsmith, J. J. and U. J. Pittman. (1963). Effect of nitrogen fertilizers on germination and stand of wild oats. *Weeds* 11: 99-101.

Shumway, D. L. and R. T. Koide. (1994). Seed preferences of *Lumbricus terrestris* L. *Applied Soil Ecology* 1: 11-15.

Sivapalan, A., W. C. Morgan, and P. R. Franz. (1993). Monitoring populations of soil microorganisms during a conversion from a conventional to an organic system of vegetable growing. *Biological Agriculture and Horticulture* 10: 9-27.

Skipper, H. D., A. G. J. Ogg, and A. C. Kennedy. (1996). Root biology of grasses and ecology of rhizobacteria for biological control. *Weed Technology* 10: 610-620.

Staniforth, D. W. (1957). Effects of annual grass weeds on the yield of corn. *Agronomy Journal* 49: 551-555.

Stute, J. K. and J. L. Posner. (1995). Synchrony between legume nitrogen release and corn demand in the upper midwest. *Agronomy Journal* 87: 1063-1069.

Swanton, C. J., D. R. Clements, and D. A. Derksen. (1993). Weed succession under conservation tillage: A hierarchial framework for research and management. *Weed Technology* 7: 286-297.

Teasdale, J. R. (1993). Interaction of light, soil moisture, and temperature with weed suppression by hairy vetch residue. *Weed Science* 41: 46-51.

Teasdale, J. R. (1996). Contribution of cover crops to weed management in sustainable systems. *Journal of Production Agriculture* 9: 475-479.

Thompson, D., S. Thompson, and R. Thompson. (1995). *Alternatives in Agriculture*, Thompson On-Farm Research, Boone, IA.

Tilman, D. (1990). Mechanisms of plant competition for nutrients: the elements of a predictive theory of competition. In *Perspectives on Plant Competition*, eds. J. B. Grace and D. Tilmaneds. San Diego, CA: Academic Press, Inc., pp. 117-141.

Tollenaar, M., S. P. Nissanka, A. Aguilera, S. F. Weise, and C. J. Swanton. (1994). Effect of weed interference and soil nitrogen on four maize hybrids. *Agronomy Journal* 86: 596-601.

Valdrighi, M. M., A. Pera, M. Agnolucci, S. Frassinetti, D. Lunardi, and G. Vallini. (1996). Effects of compost-derived humic acids on vegetable biomass production and microbial growth with a plant (*Cichorium intybus*)-soil system: a comparative study. *Agriculture, Ecosystems and Environment* 58: 133-144.

van Bruggen, A. H. C. (1995). Plant disease severity in high-input compared to reduced-input and organic farming systems. *Plant Disease* 79: 976-984.

van der Reest, P. J. and H. Rogaar. (1988). The effect of earthworm activity on the vertical distribution of plant seeds in newly reclaimed polder soils in the Netherlands. *Pedobiologia* 31: 211-218.

Varco, J. J., W. W. Frye, M. S. Smith, and C. T. MacKown. (1993). Tillage effects on legume decomposition and transformation of legume and fertilizer nitrogen-15. *Soil Science Society of America Journal* 57: 750-756.

Vengris, J., W. G. Colby, and M. Drake. (1955). Plant nutrient competition between weeds and corn. *Agronomy Journal* 47: 213-216.

Vereijken, P. (1989). The DFS farming systems experiment. In *Development of Farming Systems, Evaluation of the Five-Year Period 1980-1984*, ed. J. C. Zadoks. Wageningen, the Netherlands: Pudoc, Centre for Agricultural Publishing and Documentation.

Wardle, D. A. (1995). Impacts of disturbance on detritus food weebs in agro-ecosystems of contrasting tillage and weed management practices. *Advances in Ecological Research* 26: 105-185.

Warkentin, B. P. (1995). The changing concept of soil quality. *Journal of Soil and Water Conservation* 50: 226-228.

Wells, G. J. (1979). Annual weed competition in wheat crops: the effect of weed density and applied nitrogen. *Weed Research* 19: 185-191.

Westcott, M. P. and D. S. Mikkelsen. (1987). Comparison of organic and inorganic nitrogen sources for rice. *Agronomy Journal* 79: 937-943.

Westoby, M., M. Leishman, and J. Lord. (1996). Comparative ecology of seed size and dispersal. *Phil. Trans. R. Soc. Lond.* 351: 1309-1318.

Weston, L. A. (1996). Utilization of allelopathy for weed management in agroecosystems. *Agronomy Journal* 88: 860-866.

White, R. H., A. D. Worsham, and U. Blum. (1989). Allelopathic potential of legume debris and aqueous extracts. *Weed Science* 37: 674-679.

Williams, J. T. and J. L. Harper. (1965). Seed polymorphism and germination I. The influence of nitrates and low temperatures on the germination of *Chenopodium album. Weed Research* 5: 141-150.

Yenish, J. P., J. D. Doll, and D. D. Buhler. (1992). Effects of tillage on vertical distribution and viability of weed seed in soil. *Weed Science* 40: 429-433.

Zentner, R. P., C. A. Campbell, V. O. Biederbeck, and F. Selles. (1996). Indianhead black lentil as green manure for wheat rotations in the Brown soil zone. *Canadian Journal of Plant Science* 76: 417-422.

RECEIVED: 07/21/97
ACCEPTED: 04/30/98

# Soil Microorganisms
# for Weed Management

## A. C. Kennedy

**SUMMARY.** Traditional methods of weed management have not considered the microbial or other biological factors that influence plant growth; however, incorporating this knowledge may expand weed management possibilities to develop weed-suppressive soils. Alternative weed management strategies are needed to expand the capability of weed control as weed pressures continue to limit optimum yield and the use of synthetic chemical herbicides for weed control becomes more restricted. Biotic factors can influence the distribution, abundance, and competitive abilities of plant species. It has been shown that soil microorganisms are capable of suppressing weeds in the field, and seed decay phenomena are most likely microbial. It is imperative that an understanding of soil microorganisms and their ecology be developed, so that they may be used to benefit agriculture, especially weed management. Further study is required so that the ecological and biological effects of the resident soil microbial population on weed growth can be used effectively in weed management strategies to assist in reducing inputs. *[Article copies available for a fee from The Haworth Document Delivery Service: 1-800-342-9678. E-mail address: getinfo@haworthpressinc.com]*

**KEYWORDS.** Weed-suppressive soils, biocontrol

## INTRODUCTION

Herbicide efficiency, as well as herbicide use, has been increasing in the U.S. (Gianessi and Puffer, 1991; USDA, 1993). Public awareness and con-

A. C. Kennedy, Soil Scientist, USDA-ARS, 221 Johnson Hall, Washington State University, Pullman, WA 99164-6421 (E-mail: akennedy@wsu.edu).

[Haworth co-indexing entry note]: "Soil Microorganisms for Weed Management." Kennedy, A. C. Co-published simultaneously in *Journal of Crop Production* (Food Products Press, an imprint of The Haworth Press, Inc.) Vol. 2, No. 1 (#3), 1999, pp. 123-138; and: *Expanding the Context of Weed Management* (ed: Douglas D. Buhler) Food Products Press, an imprint of The Haworth Press, Inc., 1999, pp. 123-138. Single or multiple copies of this article are available for a fee from The Haworth Document Delivery Service [1-800-342-9678, 9:00 a.m. - 5:00 p.m. (EST). E-mail address: getinfo@haworthpressinc.com].

cern, has also increased and the restriction of synthetic chemical herbicide use is limiting options for weed control (Bridges, 1994). Ecologically oriented weed management within viable, integrated systems is gaining popularity as cultivation and excess herbicide use for weed control is less favored (Wyse, 1994; Liebman and Gallandt, 1997).

Microbial means of weed management need to be explored to add to the total package of successful weed management systems. Microorganisms can influence plant growth both positively and negatively, and thus should be considered in weed management strategies (Tarr, 1972; Christensen, 1989; TeBeest, 1991; Kremer and Kennedy, 1996). The classical and inundative means of biological control have been investigated and in some cases are successful (DeBach, 1964; TeBeest, 1991). These two approaches involve the importation or addition of pathogenic agents for release, dissemination and self-perpetuation, or to saturate the system. The potential for microbially based weed management has been demonstrated (Mortensen, 1986; TeBeest, 1991; Kennedy, Ogg, and Young, 1992); however, the economic benefit is yet to be seen. The role of soil microorganisms in modern weed management is often not realized (TeBeest, 1991) and a greater awareness of the soil microbial community and its impact on plant growth and chemical plant interactions is critical. Microorganisms in weed management could increase crop efficiency, decrease the need for tillage, and decrease the use of synthetic chemical herbicides. The future demands that integrative approaches be used that involve management of the whole system for weed suppression, including practices to conserve or enhance native weed pathogens.

Weed management that takes advantage of multiple stresses may be more successful than relying on a single means of control (Kennedy and Kremer, 1996). Use of microbes or biotic forms of management, however, demands an understanding of the ecology of the system, its participants, and their plant specificity. Biologically based weed management practices take advantage of natural microbe/plant interactions and plant/plant competition (TeBeest, 1991). The use of microbial dynamics in management systems, however, may require special management considerations to enhance their activity and achieve a reduction in weed growth. Ecologically-based approaches, which consider not only the weed, but also the microbial community and the environment, will result in the greatest success for weed management (Kennedy and Kremer, 1996, Liebman and Gallant, 1997).

## MICROBIAL DIVERSITY AND ECOLOGY

The number of microbial species that reside on earth is unknown; however, it has been estimated that less than 1% of all microorganisms on earth have been classified and named (Torsvik, Goksoy, and Daae, 1990; Hawksworth,

1991; Ward et al., 1992). Presently about 5,000 species of microorganisms have been described and identified (Amann, Ludwig, and Schleifer, 1995), but estimates have suggested that up to 1 million microorganisms may exist (ASM, 1994). Species identification may result in phenotypic or genetic characterizations that have little relevance to the functions occurring in soil systems and a microorganism's influence on weed growth (Kennedy, 1995). Understanding the full extent of the weed/plant ecology in ecosystems requires knowledge of the functions of microbial species and the effect of the microbial community on plant health and vigor.

The role of microorganisms in agroecosystems is often understated (Kennedy and Smith, 1995). The survival of all organisms depends on the beneficial functions of microorganisms. Microorganisms are responsible for many different functions ranging from nutrient cycling and residue decomposition to maintenance of soil structure and plant health (Metting, 1993). Microorganisms may also have detrimental effects on plants by producing plant-suppressive compounds, decreasing plant available nutrients, and causing plant diseases. These impacts on plant health should be considered in pest management strategies, as they influence both crop competitiveness and weed control. Development of management approaches to take advantage of naturally occurring communities of microorganisms in soil is needed.

The diversity of soil microorganisms is more extensive than in any other group of organisms in the world (Kennedy and Gewin, 1997); however, our knowledge base must expand to gain more useful information on these organisms. The functions of whole microbial communities is of greater consequence in management systems than that of individuals. The composition of microbial communities can vary in response to the type of disturbance or environmental perturbation. Understanding microbial diversity, as well as function, will advance our knowledge of soil microbial communities as they relate to (1) management of microbial action for weed suppression in agroecosystems and the development of weed-suppressive soils, (2) the application of biocontrol agents, and (3) the effective use of genetically-engineered microorganisms for weed management.

## THE SOIL ENVIRONMENT

The microbial diversity and function of a given soil system are the result of many different factors, such as water, temperature and climate. The mineralogy and texture of soils are used to classify soils and can influence the development of microbial communities in those soils. The physical position of the microbes can also affect the level of diversity and function. Differences in function have been found due to their proximity to rhizosphere (Metting, 1993), position in landscape (Turco and Bezdicek, 1987), depth in soil profile

(Fredrickson et al., 1991), and arrangement in macro- and micropore spaces (Lee and Foster, 1991). Plant communities and their aboveground ecology can impact not only nutrient availability, but also microbial function (Christensen, 1989). We cannot assume that the plant and microbial ecology of one system will function the same as in another system; therefore, more information is needed on each type of interaction. Plant root systems release different types of root exudates which support different microflora (Rovira, 1965; Neal, Larson, and Atkinson, 1973; Miller, Henken, and Van Veen, 1989). These differences are responsible for differences in microbial community structure (DeLeij and Whipps, 1993; Garland, 1996). The availability of substrates may have a greater impact on diversity and makeup of the microbial community than the composition of exudates. To effectively utilize biocontrol measures in weed management practices, it will be necessary to develop environments that will sustain the biocontrol agents (Boyetchko, 1996) or communities (Kennedy and Gewin, 1997).

## INTERACTIONS

### Weed-Microbe

Spatial and temporal differences in plant development alter the susceptibility to pathogen stress and can be used to benefit weed management strategies (Figure 1). Many diseases are caused by saprophytic and parasitic microorganisms and have potential to be used selectively in weed control. The stages of plant growth that can be affected include the seed, the seedling and the established plant throughout its various growth stages. The microorganisms can be categorized into the following groups as to their affect on plant growth: (1) seed decay, (2) pre-emergence seedling blight or rot, (3) root rots caused by cortical parasites, (4) vascular wilt diseases, (5) hypertrophy diseases, and (6) non-parasitic root pathogens (Tarr, 1972; Heydecker, 1973).

Seed decay is promoted by viruses, bacteria, and fungi (Barton, 1961; Neergaard, 1977). Bacteria and fungi can cause (1) seed abortion, (2) seed rot, and (3) seed discoloration. Fungal infection also can result in (1) shrunken seeds, (2) sclerotization or stromatization of seed, (3) seed necrosis, (4) reduction or elimination of germination capacity, and (5) physiological alterations in seed. The water content of the seed dramatically affects species composition of fungal flora colonizing the seed. Mechanical damage to seeds, especially in the seed coat, makes seeds more susceptible to invasion by microbes (Kremer and Spencer, 1989). While the method of attack on seeds may vary, many microorganisms affect seeds through the production of suppressive metabolites (Halloin, 1986). Field or storage fungi associated with crop seed are specific examples of these microorganisms.

FIGURE 1. Stages of potential impact on plant growth by microorganisms

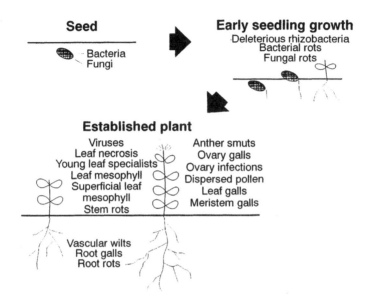

**Seed**
- Bacteria
- Fungi

**Early seedling growth**
Deleterious rhizobacteria
Bacterial rots
Fungal rots

**Established plant**

| Viruses | Anther smuts |
| Leaf necrosis | Ovary galls |
| Young leaf specialists | Ovary infections |
| Leaf mesophyll | Dispersed pollen |
| Superficial leaf | Leaf galls |
| mesophyll | Meristem galls |
| Stem rots | |

Vascular wilts
Root galls
Root rots

The majority of biological control agents presently being studied attack the developing plant rather than the seed. Fungal biological weed control agents formulated as mycoherbicides have been commercially developed and are generally foliarly applied. Several biocontrol agents have been registered under the trade names of DeVine®, Collego™, and BioMal™. Devine® (*Phytophthora palmivora*) has activity against stranglervine (*Morrenia odorata* (H. & A.) Lindl.) by attacking basal stems and roots of the seedling (Ridings, 1986). Collego™ promotes anthracnose, a plant disease involving lesions, necrosis, and hypoplasia. The causal agent, *Colletotrichum gloeosporioides* f. sp. *Aeschynomene* (Daniel et al., 1973), is used for the control of northern jointvetch [*Aeschynomene virginica* (L.)] by foliar application (Templeton, 1982) in rice and soybean. BioMal™ is registered for the control of round leaf mallow (*Malva pusilla*; Mortensen, 1988) in small grains and lentil. *Alternaria cassiaeis,* a blight inducing pathogen (CASST), is a mycoherbicide for control against sickle pod (*Cassia obtusifolia* L.) and can be used in the control of weeds in soybean (*Glycine max* L.) and peanut (*Arachis hypogaea*) (Walker and Boyette, 1985).

Many other individual organisms have been investigated for control of specific weeds. *Colletotrichum coccodes* has been used for the control of eastern black nightshade (*Solanum ptycanthum* Dun.) (Andersen and Walker,

1985) by attacking fresh plant shoots. *Phomopsis convolvulus* has been investigated for the control of field bindweed (*Convolvulus arvensis* L.; Ormeno-Nunez, Reeleder, and Watson, 1988) and *Bipolaris sorghicola* has been studied for johnsongrass (*Sorghum halepense* L.Pers.; Winder and Van Dyke, 1989) control. *Alternaria crasse* is also being investigated for control of jimsonweed (*Datura stramonium* L.) and *Sclerotinia sclerotiorum* (Lib.) de Bary is being studied for control of Canada thistle (*Cirsium arvense* (L.) Scop.), spotted knapweed (*Centauria maculosas* Lam.), and dandelion (*Taraxacum officinale* Weber; Brosten and Sands, 1986; Miller, Henken, and Van Veen, 1989; Riddle, Burpee, and Boland, 1991). All indicate the potential impact that the microorganisms may have on weed populations.

Several bacterial biological control agents are soil applied and rely on suppression of the weed seedling for control. Examples of this type of biological control include deleterious rhizobacteria (Kremer, 1986; Kremer et al., 1990; Kennedy, 1997). Soil bacterial isolates inhibited velvetleaf, morning glory (*Ipomoea* sp.), cocklebur (*Xanthium canadense* Mill.), pigweed (*Amaranthus* sp.), lambsquarters (*Chenopodium* spp.), smartweed (*Polygonum* sp.) and jimsonweed (Kremer, 1986). The grass weeds downy brome (*Bromus tectorum* L.; Kennedy et al., 1991), japanese brome (*Bromus japonicus* L.; Harris and Stahlman, 1995) and jointed goatgrass (*Aegilops cylindrica* Host.; Kennedy, Ogg, and Young, 1992) have also been shown to be inhibited by bacteria in laboratory greenhouse and field (Kennedy et al., 1991; Kennedy, Ogg, and Young, 1992).

### *Weed-Microbe-Insect*

Plant pathogens combined with insects have been shown to achieve greater weed control than when used alone. Insects can act as disseminators of pathogens and frequently as vectors of the pathogen, thus spreading the pathogen from plant to plant. Many pathogenic viruses and some bacteria and fungi are transmitted in this way (Tarr, 1972). Saprophytic fungi, which survive on live tissue, can enter through injuries caused by insects. The effect can be indirect as the presence of fungi can reduce the photosynthetic capability of the plant, thus adversely affecting it. Pathogens can enter through wounds caused by insects. Scentless plant bugs (*Niesthrea louisianica*) were found to attack velvetleaf seeds, resulting in increased *Fusarium* infection, decreased seed survival, and greater weed control than either the insect or fungus alone would provide (Kremer and Spencer, 1989).

### *Weed-Microbe-Herbicide*

Although biocontrol organisms and/or products have been used successfully, biocontrol agents used in concert with synthetic herbicides can produce a synergistic effect resulting in greater weed control than either agent could achieve alone (Figure 2). The synergistic weed control observed by this approach is brought about by a multiple stress phenomenon that is created when reduced levels of chemical herbicides and microbes are used together, which has broadened the range of control (Charudattan and De Loach, 1988). For example, Kremer and Schulte (1989) saw enhanced control of velvetleaf using the fungal pathogen *Fusarium oxysporum* in conjuction with chemical herbicides. The stress imparted to plants by the herbicide resulted in increased colonization of a *Pseudomonas* spp. biocontrol agent (Greaves and Sargent, 1986). Multiple stresses through the use of microbe/herbicide or microbe/chemical amendment may impact weed growth and competitiveness and aid existing weed management technology.

### SEED BANK DECLINE

The greatest source of weed infestation in cropland is the weed seed bank (Cavers and Benoit, 1989). Seed mortality in soil can be due to fluctuating physical characteristics of temperature (Lonsdale, 1993) and moisture (Simpson, 1990). The biological characteristics of the seed, such as seed age and

FIGURE 2. Potential interactions among herbicides, microorganisms, the crop and weeds

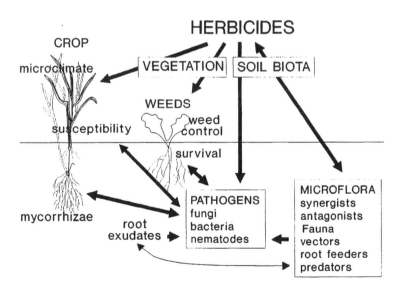

viability of seed (Priestly, 1986; Copeland and McDonald, 1995), the biological characteristics of the soil (Kennedy and Kremer, 1996), macrofaunal predation (Brust and House, 1988), and soil microbial community degradation or inhibition of the germinating seed (Kremer, 1993) all can affect seed viability. These stresses can work in concert to reduce seed viability and vigor. Weed management in the past generally focused on reducing weed populations by use of herbicides, with little or no attention paid to weed control possibilities using microorganisms (Kremer, 1993). Soil microorganisms may have a direct or indirect impact on the plant composition of a community, specifically on the weed and crop populations (Boyetchko, 1996; Westover, Kennedy, and Kelley, 1998). Most research involving microorganisms involves the microbial role in seed-borne disease, rather than their role in seed deterioration (Halloin, 1986). Recent research indicated the possible role of microorganisms in "unexplained" losses of seeds in soil, especially since microbial activity has been found to contribute to nonviability of seed (Harper, 1977; Egley and Duke, 1985). Kremer (1993) suggests the use of phytopathogenic microbes to accelerate the depletion of the weed seed bank.

Microbial activity has been proposed as a possible mechanism of seed decay in several studies when significant weed seed losses occurred (e.g., Roberts and Feast, 1973; Froud-Williams, Chancellor, and Drennan, 1983; Zorner, Zimdahl, and Schweizer, 1984; Colosi, Cavers, and Bough, 1988). *Fusarium* has been found to be responsible for weed seed decay and often causes the greatest seed deterioration in velvetleaf (Kremer and Schulte, 1989). Although microorganisms are recognized as having negative effects on weed seeds, very few studies have been directed at characterizing microorganisms important to weed decay (Kremer, 1993). Kirkpatrick and Bazzaz (1979) identified seedborne fungi associated with different weed species and found that the fungi affected germination and seedling development. Likewise, Mishra and Srivastava (1977) and Roberts, Robertson, and Hanlin (1986) found that weed seed-associated fungi had deleterious effects on weed seeds. Microbial decay was also suggested as the cause of losses of wild oat seed viability (Zorner, Zimdahl and Schweizer, 1984). In other studies, wild oat germination was reduced when seeds were planted into untreated soil compared to steam-treated soil (Kiewnick, 1964).

The microbial affect on seed germination was studied by treating the seed and soil with antimicrobial agents. For example, Fellows and Roeth (1992) found that shattercane seed (*Sorghum bicolor* L. Moench) treated with carboxin and thiram had dramatic survival increases. Survival of *Mimosa pigra* seed treated with benomyl increased from 7 to 10% depending on the site (Lonsdale, 1993). Temperature fluctuations at the soil surface was a larger factor controlling seed mortality than any seed treatment. Fungal colonization of *Setaria* spp. caryopses was related to crop residue placement (Pitty,

Staniforth, and Tiffany, 1987). Greater fungal colonization of seed occurred nearer the soil surface in no-till plots while with intensive tillage, fungal invasion of seed was increased with depth.

Greater weed mortality occurs during the seed stage of the life cycle than at any other phase; therefore, factors affecting seed survival rate are important to management decisions (Yenish, Doll, and Buhler, 1992). Since the seed coat protects the seed from microbial attack (Halloin, 1986), weed control approaches utilizing microorganisms should aim to circumvent or overcome seed resistance (Kremer, 1993). Allelopathic effects of microbial metabolites can be toxic to weed seeds and do not require invasion of the seed or seed coat by the pathogen (Harman, 1983).

Research has revealed that some chemicals have functions other than originally intended, such as promoting fungal infection of weed seeds and affecting weed seed decline. Velvetleaf seedling emergence was reduced 15% compared to controls when a fungus (*Fusarium oxysporum*) was applied with ethephon (a germination stimulant) or carbofuran (an insecticide). An increase in nonviable seeds, 23% with ethephon and 17% with carbofuran, also resulted (Kremer and Schulte, 1989). Insects can also promote microbial action on seeds as well as to the plant. For example, insect feeding on velvetleaf seeds increased fungal infection from 8% to 98% (Kremer and Spencer, 1989).

Differences exist between the seed banks of grassland and cultivated land (Wilson, 1988). The composition and density of weed seeds is closely linked to the cropping history of the land. Seed density has been found to be four times greater in cropped land of 5 years than in long-term grassland systems (Kelley and Bruns, 1975). Differences may stem from the location of the seeds in these two different systems; in grasslands, most seeds are located in the top 2 cm of the soil surface (Hayashi and Numato, 1971), whereas in the cultivated soil the majority is in the upper 15 cm of the soil surface and can go as deep as cultivation (Idris and Beshir, 1979). Seed mortality of wild oats (*Avena fatua* L.) was highest at the surface and decreased with increasing depth of seed placement in soil (Zorner, Zimdahl, and Schweizer, 1984). The seed bank moves closer to the soil surface as the intensity of tillage declines (Wicks and Somerhalder, 1971), indicating the potential of management of the seed bank through soil microorganisms.

## INTEGRATED MANAGEMENT

Within cropping systems, management practices will affect the composition and density of weed seeds and diversity of germinating weed species (Streibig, 1988; Wilson, 1988). For example, conservation tillage has resulted in shifts towards perennials, grasses and wind-disseminated weeds

(Swanton, Clements, and Derksen, 1993). In addition to the differences found in weed seed number and composition, weed seed distribution is also affected by management practice. In no-till systems, the weed seeds were concentrated in the surface soil whereas in conventional tillage weed seeds were more evenly distributed throughout plow depth (Yenish, Doll, and Buhler, 1992). Differences in soil structure may also affect weed seed germination. Weed seed germination in tilled soils was less in the large peds or clods of soil than in the smaller sizes (Roberts and Feast, 1972).

The shift to predominantly surface-residing weed seeds in no-till could be advantageous in controlling weeds, from both chemical and biological control standpoints. Surface distribution promotes greater germination initially, but if seed production is minimized, long term weed control would be enhanced (Froud-Williams, Chancellor, and Drennan, 1981). From the biological perspective, no-till systems result in greater microbial activity, especially in the top few inches of soil. No-till resulted in higher microbial biomass (Carter and Rennie, 1982; Carter, 1986; Doran, 1987; Simard, Angers, and Lapierre, 1994) mineralizable carbon (Franzluebbers, Hons, and Zuberer, 1994; Simard, Angers, and LaPierre, 1994), and dehydrogenase and phosphatase activity (Doran, 1980) than in conventionally-managed comparisons. Cardinia, Reginier, and Harrison (1991) found that the increased soil moisture in surface-managed residue management systems resulted in greater success of biological control agents and predisposed the system to greater seed decomposition. Given greater microbial activity (Doran, 1980) and a high proportion of weed seeds in the surface layer of soil (Yenish, Doll, and Buhler, 1992), it is possible that microbial agents could be especially effective in no-till systems.

Changes in microbial communities are readily observed with changes in management (Kennedy and Smith, 1995). In biologically-based weed management, weed-suppressive soils may be an important component of weed management (Kirkpatrick and Bazzaz, 1979; Kremer, 1986; Roberts, Robertson, and Hanlin, 1986; Kennedy and Kremer, 1996). Microbial activity of indigenous organisms contributes to the depletion of the weed seed bank and can be managed for greater weed seed decay. Native biocontrol agents can be utilized and enhanced for weed control. This is no easy task, since the soil system is complex. The diversity of biocontrol agents residing within the soil, a diverse population of target weeds, and environmental variables all contribute to the development of microbial populations that could, in various ways, suppress weed growth (Kennedy and Kremer, 1996). The number of weed-suppressive bacteria in soil varies with management, which may influence the competitive ability of the weed, and thus may be a useful consideration in developing biological control agents (Kennedy, unpublished). The horizonation of residue and microbial activity in some systems, such as no-till (Doran,

1980), establishes areas of increased seed decay potential within the residue zone. Assuming weed-suppressive populations follow the same trends as the general populations in this system, the potential for weed-suppression increases in surface-managed residues relative to other soil microsites.

## CONCLUSIONS

Microbially based weed management takes advantage of biotic factors that influence the distribution, abundance, and competitive abilities of plant species. Biological measures offer additional practices to assist herbicides in suppressing weed growth and establishment. Biocontrol is successful when the weed, the microbial communities, and environment interact in such a manner that weed control or suppression occurs. The challenge ahead is to develop management systems that favor microbial activity in weed control by increasing the interaction of these three components. If we can understand biological and environmental influences control weed seed mortality, strategies can be developed to manage for weed-suppressive soils. Researchers need to develop an understanding of microorganisms and the soil and phyllosphere ecology of microorganisms, so that they may be utilized to benefit weed management and agriculture.

## REFERENCES

Amann, R.I., W. Ludwig and K.H. Schleifer. (1995). Phylogenetic identification and *in situ* detection of individual microbial cells without cultivation. *Microbiol. Rev.* 59:143-169.

American Society for Microbiology. (1994). *Microbial diversity research priorities.* American Society for Microbiology, Washington DC. 7 p.

Andersen, R.N. and H.L. Walker. (1985). *Colletotrichum coccodes*: A biocontrol agent of east black nightshade (*Solanum ptycanthum*). *Weed Sci.* 33: 902-905.

Barton, L.V. (1961). *Seed Preservation and Longevity.* Leonard Hill, London.

Boyetchko, S.M. (1996). Impact of soil microorganisms on weed biology and ecology. *Phytoprotection* 77:41-56.

Bridges, D.C. (1994). Impact of weeds on human endeavors. *Weed Technol.* 8:392-395.

Brosten, B.S. and D.C. Sands. (1986). Field trials of *Sclerotinia sclerotiorum* to control Canada thistle (*Cirsium arvense*). *Weed Sci.* 34:377-380.

Brust, G.E. and G.J. House. (1988). Weed seed destruction by arthropods and rodents in low-input soybean agroecosystems. *Am J. Alt. Agri.* 3:19-25.

Cardinia, J., E. Reginier and K. Harrison. (1991). Long-term tillage effects on seed banks in three Ohio soils. *Weed Sci.* 39:186-194.

Carter, M.R. (1986). Microbial biomass as an index for tillage-induced changes in soil biological properties. *Soil Till. Res.* 7:29-40.

Carter, M.R. and D.A. Rennie. (1982). Changes in soil quality under zero tillage farming systems: distribution of microbial biomass and mineralizable C and N potentials. *Can. J. Soil Sci.* 62:587-597.

Cavers, P.B. and D.L. Benoit. (1989). Seed banks in arable land. In *Ecology of Soil Seed Banks*, eds. M.A. Leck, V.T. Parker and R.L. Simpson. San Diego, CA: Academic Press, pp. 309-328.

Charudattan, R. and C.J. De Loach, Jr. (1988). Management of pathogens and insects for weed control in agroecosystems. In *Weed Management in Agroecosystems: Ecological Approaches*, eds. M.A. Altieri and M. Liebman. Boca Raton, FL: CRC Press, pp. 245-264.

Christensen, M. (1989). A view of fungal ecology. *Mycologia* 81:1-19.

Colosi, J.C., P.B. Cavers and M.A. Bough. (1988). Dormancy and survival in buried seeds of proso millet *(Panicum miliaceum)*. *Can. J. Bot.* 66:161-168.

Copeland, L.O. and M.B. McDonald. (1995). *Seed Science and Technology*, Third Edition, New York, NY: Chapman and Hall, 409 p.

Daniel, J.T., G.E. Templeton, R.J. Smith, Jr., and W.T. Fox. (1973). Biological control of northern jointvetch in rice with an endemic fungal disease. *Weed Sci.* 21:303-307.

DeBach, P. (1964). *Biological Control of Insects, Pests and Weeds.* New York, New York: Reinhold, 844 p.

DeLeij, F.A.A.M. and J.M. Whipps. (1993). The use of colony development for the characterization of bacterial communities in soil and on roots. *Microb. Ecol.* 27:81-97.

Doran, J.W. (1980). Soil microbial and biochemical changes associated with reduced tillage. *Soil Sci. Soc. Am. J.* 44:765-771.

Doran, J.W. (1987). Microbial biomass and mineralizable nitrogen distributions in no-tillage and plowed soils. *Biol. Fert. Soils* 5:68-75.

Egley, G.H. and S.O. Duke. (1985). Physiology of weed seed dormancy and germination. In *Weed Physiology: Reproduction and Ecophysiology*, ed. S.O. Duke. Boca Raton, Florida: CRC Press, pp. 27-64.

Fellows, G.M. and F.W. Roeth. (1992). Factors influencing shattercane *(Sorghum bicolor)* seed survival. *Weed Sci.* 40:434-440.

Franzluebbers, A.J., F.M. Hons and D.A. Zuberer. (1994). Long-term changes in soil carbon and nitrogen pools in wheat management systems. *Soil Sci. Soc. Am. J.* 58:1639-1645.

Fredrickson, J.K., D.L. Balkwill, J.M. Zachara, S.M.W. Li, F.J. Brockman and M.A. Simmons. (1991). Physiological diversity and distributions of heterotrophic bacteria in deep cretaceous sediments of the Atlantic coastal plain. *Appl. Environ. Microbiol.* 57:402-411.

Froud-Williams, R.J., R.J. Chancellor and D.S.H. Drennan. (1981). Potential changes in weed floras associated with reduced-cultivation systems for cereal production in temperate regions. *Weed Res.* 21:99-109.

Froud-Williams, R.J., R.J. Chancellor and D.S.H. Drennan. (1983). Influence of cultivation regime upon buried weed seeds in arable cropping systems. *J. Appl. Ecol.* 20:199-208.

Garland, J.L. (1996). Patterns of potential C source utilization by rhizosphere communities. *Soil Biol. Biochem.* 28:223-230.

Gianessi, L.P. and C. Puffer. (1991). *Herbicide Use in the United States: National Summary Report.* Washington, DC: Resources for the Future.

Greaves, M.P. and J.A. Sargent. (1986). Herbicide-induced microbial invasion of plant roots. *Weed Sci.* 34:50-53.

Halloin, J.M. (1986). Microorganisms and seed deterioration. In *Physiology of Seed Deterioration*, eds. M.B. McDonald, Jr. and C.J. Nelson. Madison, WI: CSSA, pp. 89-99.

Hawksworth, D.L. (1991). *The Biodiversity of Microorganisms and Invertebrates: Its Role in Sustainable Agriculture.* Melksham, UK: CAB International, Redwood Press Ltd., 302 p.

Harman, G.E. (1983). Mechanisms of seed infection and pathogenesis. *Phytopath.* 73: 326-329.

Harper, J.L. (1977). *Population Biology of Plants.* London, England: Academic Press.

Harris, P.A. and P. Stahlman. (1995). Soil bacteria as a selective biological control agent of winter annual grass weeds in winter wheat. *Appl. Soil Ecol.* 3:275-281.

Hayashi, I. and M. Numato. (1971). Viable buried-seed population in the Miscanthus-and Zoysia type grasslands in Japan-ecological studies on the buried-seed population in the soil related to plant succession. VI. *Japanese Journal of Ecology* 20:243-245.

Heydecker, W. (1973). *Seed Ecology.* University Park, PA : Pennsylvania State University Press.

Idris, M. and M.E. Beshir. (1979). On the distribution and dynamics of weed populations in Sudan Gezira. *Proceedings of Symposium for Weed Research*, Sudan 18:15-17.

Kelley, A.D. and V.F. Bruns. (1975). Dissemination of weed seeds by irrigation water. *Weed Sci.* 23:486-490.

Kennedy, A.C. (1995). Soil microbial diversity in agricultural systems. In *Exploring the Role of Diversity in Sustainable Agriculture*, eds. R.K. Olson, C.A. Francis and S. Kaffka. Madison, WI: Am. Soc. Agron., pp. 35-54.

Kennedy, A.C. (1997). Deleterious rhizobacteria and weed control. In *Ecological Interactions and Biological Control*, eds. D.A. Andow, D.W. Ragsdale and R.F. Nyvall. Boulder, CO: Westview Press. pp. 164-177.

Kennedy, A.C. and V.L. Gewin. (1997). Soil microbial diversity: Present and future considerations. *Soil Sci.* 162:607-617.

Kennedy, A.C., L.F. Elliott, F.L. Young and C.L. Douglas. (1991). Rhizobacteria suppressive to the weed downy brome. *Soil Sci. Soc. Amer. J.* 55:722-727.

Kennedy, A.C. and R.J. Kremer. (1996). Microorganisms in weed control strategies. *J. Prod. Ag.* 9:480-485.

Kennedy, A.C., A.G. Ogg, Jr. and F.L. Young. (1992). Biocontrol of jointed goatgrass. Patent number 07/597,150. November 17, 1992.

Kennedy, A.C. and K.L. Smith. (1995). Soil microbial diversity and the sustainability of agricultural soils. *Plant Soil* 170:75-86.

Kiewnick, L. (1963). Experiments on the influence of seedborne and soilborne microflora on the viability of wild oats seeds (*Avena fatua* L.) II. Experiments on the influence of microflora on the viability of seed in the soil. *Weed Res.* 4:31-43.

Kirkpatrick B.L. and F.A. Bazzaz. (1979). Influence of certain fungi on seed germination and seedling survival of four colonizing annuals. *J. Appl. Ecol.* 16:515-527.

Kremer, R.J. (1986). Antimicrobial activity of velvetleaf (*Abutilon theophrasti*) seeds. *Weed Sci.* 34:617-622.

Kremer, R.J. (1993). Management of weed seed banks with microorganisms. *Ecol. Appl.* 3:42-52.

Kremer, R.J. and A.C. Kennedy. (1996). Rhizobacteria as biocontrol agents of weeds. *Weed Tech.* 10:601-609.

Kremer, R.J. and L.K. Schulte. (1989). Influence of chemical treatment and *Fusarium oxysporum* on velvetleaf. *Weed Technol.* 3:369-374.

Kremer, R.J. and N.R. Spencer. (1989). Impact of seed-feeding insects and microorganisms on velvetleaf (*Abutilon theophrasti*) seed viability. *Weed Sci.* 37: 211-216.

Kremer, R.J., M.F.T. Begonia, L. Stanley and E.T. Lanham. (1990). Characterization of rhizobacteria associated with weed seedlings. *Appl. Environ. Microbiol.* 56: 1649-1655.

Lee, K.E. and R.C. Foster. (1991). Soil fauna and soil structure. *Aust. J. Soil Res.* 29:745-775.

Liebman, M. and E.R. Gallandt. (1997). Many little hammers: Ecological approaches for management of crop-weed interactions. In *Ecology in Agriculture*, ed. L.E. Jackson. San Diego CA: Academic Press. pp. 291-343.

Lonsdale, W.M. 1993. Losses from the seed bank of *Mimosa pigra*: Soil microorganisms vs. temperature fluctuations. *J. Appl. Ecol.* 30:654-660.

Metting, F.B., Jr., ed. (1993). *Soil Microbial Ecology: Applications in Agricultural and Environmental Management.* New York, NY: Marcel Dekker, Inc. 646 pp.

Miller, H., J.G. Henken and J.A. Van Veen. (1989). Variations and composition of bacterial populations in the rhizosphere of maize, wheat and grass cultivars. *Can. J. Microb.* 35:656-660.

Mishra, R.R. and W.B. Srivastava. (1977). Comparison of mycoflora associated with certain crop and weed seeds. *Acta Mycol.* 13:145-149.

Mortensen, K. (1986). Biological control of weeds with plant pathogens. *Can. J. Plant Pathol.* 8:229-231.

Mortensen, K. (1988). The potential of an endemic fungus, *Colletotrichum gloeosporiodes* f.sp. *malvae*, for biological control of round-leaved mallow (*Malva pusilla*) and velvetleaf (*Abutilon theophrasti*). *Weed Sci.* 36:473-478.

Neal, J.L., R.I. Larson and T.G. Atkinson. (1973). Changes in rhizosphere populations of selected physiological groups of bacteria related to substitution of specific pairs of chromosomes in spring wheat. *Plant Soil* 39:209-212.

Neergaard, P. (1977). *Seed Pathology.* Wiley, New York, NY.

Ormeno-Nunez, J., R.D. Reeleder and A.K. Watson. (1988). A foliar disease of field bindweed (*Convolvulus arvensis* L.) caused by *Phomopsis convolvulus*. *Plant Dis.* 72:338-342.

Pitty, A., D.W. Staniforth and L.H. Tiffany. 1987. Fungi associated with caryopses of

*Setaria* species from field-harvested seeds and from soil under two tillage systems. *Weed Sci.* 35:319-323.

Priestly, D.A. (1986). *Seed Aging: Implications for Seed Storage and Persistence in the Soil.* Ithaca, NY. Comstock Publ. Assoc., 304 p.

Riddle, G.E., L.L. Burpee and G.H. Boland. (1991). Virulence of *Sclerotinia sclerotiorum* and *S.minor* on dandelion (*Taraxacum officinale*). *Weed Sci.* 39:109-118.

Ridings, W.H. (1986). Biological control of stranglervine (*Morrenia odorata* Lindl.). in citrus-a researcher's view. *Weed Sci.* Suppl. 34: Supp. 1.

Roberts, H.A. and P.M. Feast. (1972). Fate of seeds of some annual weeds in different depths of cultivated and undisturbed soil. *Weed Res.* 12:316-324.

Roberts, H.A. and P.M. Feast. (1973). Changes in the numbers of viable weed seeds in soil under different regimes. *Weed Res.* 13:298-303.

Roberts, R.G., J.A. Robertson and R.T. Hanlin. (1986). Fungi occurring in the achenes of sunflower (*Helianthus annuus*). *Can. J. Bot.* 64:1964-1971.

Rovira, A.D. (1965). Plant root exudates and their influence upon soil microorganisms. In *Ecology of Soil Borne Pathogens–Prelude to Biological Control*, eds. K.F. Baker and W.C. Snyder. Berkeley, CA: University of California Press, pp. 170-186.

Simard, R.R., D.A. Angers and C. Lapierre. (1994). Soil organic matter quality as influenced by tillage, lime and phosphorous. *Biol. Fert. Soils* 18:13-18.

Simpson, G.M. and C.J. Schollenberger. (1990). *Seed Dormancy in Grasses.* Cambridge, UK: Cambridge Univ. Press. 297 p.

Streibig, J.C. (1988). Weeds-the pioneer flora of arable land. *Ecological Bulletins* 39:59-62.

Swanton, C.J., D.R. Clements and D.A. Derksen. (1993). Weed succession under conservation tillage: A hierarchical framework for research and management. *Weed Technol.* 7:286-297.

Tarr, S.A.J. (1972). *Principles of Plant Pathology.* Winchester, New York.

TeBeest, D.O. (1991). *Microbial Control of Weeds*, New York, NY: Chapman and Hall, 284 p.

Templeton, G.E. (1982). Status of weed control with plant pathogens. In *Biological Control of Weeds with Plant Pathogens*, eds. R. Charudattan and H.L. Walker. New York, NY: Wiley Press, pp. 29-44.

Torsvik, V., J. Goksoy and F.L. Daae. (1990). High diversity in DNA of soil bacteria. *Appl. Environ. Microbiol.* 56:782-787.

Turco, R.F. and D.F. Bezdicek. (1987). Diversity within two serogroups of *Rhizobium leguminosarum* native to soils in the Palouse of eastern Washington. *Ann. Appl. Biol.* 111:103-114.

USDA Economic Research Service. (1993). Economic indicators of the farm sector, *National finance summary.* January. 1301 New York Ave. NW. Washington, DC.

Walker, H.L. and C.D. Boyette. (1985). Biological control of sicklepod in soybeans (*Glycine max*) with *Alternaria cassiae*. *Weed Sci.* 33:212-215.

Ward, D., M. Bateson, R. Weller and A. Ruff-Roberts. (1992). Ribosomal RNA analysis of microorganisms as they occur in nature. *Adv. Microbial Ecology* 12:219-286.

Westover, K.M., A.C. Kennedy and S. Kelley. (1997). Patterns of rhizosphere microbial community structure associated with co-occurring plant pairs. *J. Ecology* (In press).

Wicks, G.A. and B.R. Somerhalder. (1971). Effect of seedbed preparation for corn on distribution of weed seed. *Weed Sci.* 19:666-669.

Wilson, R.G. (1988). Biology of weed seeds in the soil. In *Weed Management in Agroecosystems: Ecological Approaches*, eds. M.A. Altieri and M. Liebman. Boca Raton, FL: CRC Press, pp. 25-39.

Winder, R.S. and C.G. Van Dyke. (1989). The pathogenicity, virulence and biocontrol potential of two *Bipolaris* species on johnsongrass *(Sorghum halepense)*. *Weed Sci.* 38:89-94.

Wyse, D.L. (1994). New technologies and approaches for weed management in sustainable agricultural systems. *Weed Technol.* 8:403-407.

Yenish, J.P., J.D. Doll and D.D. Buhler. (1992). Effects of tillage on vertical distribution and viability of weed seed in soil. *Weed Sci.* 40:429-433.

Zorner, P.S., R.L. Zimdahl and E.E. Schweizer. (1984). Sources of viable seed loss in buried dormant and non-dormant populations of wild oat *(Avena fatua* L.) seed in Colorado. *Weed Res.* 24:143-150.

RECEIVED: 09/18/97
ACCEPTED: 01/15/98

# Soil Weed Seed Banks
# and Weed Management

## Jack Dekker

**SUMMARY.** Studies of soil weed seed banks are of relatively recent origin considering their importance as sources of diversity and continued occupation of many types of habitats, including agroecosystems. The management of weed seed banks is based on knowledge and modification of the behavior of seeds within the soil seed bank matrix. The behavior of seeds defines the phenotypic composition of the floral community of a field. Selection and adaptation over time have led to the highly successful weed populations that exploit resources unused by crops. The weed species infesting agricultural seed banks are those populations that have found successful trait compromises within and between the five roles of seeds: dispersal and colonization, persistence, embryonic food supply, display of genetic diversity, and as a means of species multiplication. Diverse weed seed populations provide seed banks the opportunity to exploit any change in conditions to ensure their enduring survival and spread. The soil seed bank matrix is the spatial arrangement of environmental and physical factors over time. The behavior of soil seed banks at any level of biological, spatial, or temporal organization is a consequence of the accumulated, emergent behavior at lower levels of organization. Weed seed behavior arises from their sensitivity to environmental conditions within the physical structure of the soil seed bank. This sensitivity is reflected in changes of short duration (e.g., germination), during the annual life cycle, over multiple years (e.g., population shifts), and over evolutionary time. Understanding the

---

Jack Dekker, Associate Professor, Department of Agronomy, Iowa State University, Ames, IA 50011 (E-mail: jdekker@iastate.edu).

The author would like to thank Milton Haar, Kari Jovaag, Ed Luschei, and Randy Thornhill for their insights about seeds, the seed bank, and the seed bank matrix.

[Haworth co-indexing entry note]: "Soil Weed Seed Banks and Weed Management." Dekker, Jack. Co-published simultaneously in *Journal of Crop Production* (Food Products Press, an imprint of The Haworth Press, Inc.) Vol. 2, No. 1 (#3), 1999, pp. 139-166; and: *Expanding the Context of Weed Management* (ed: Douglas D. Buhler) Food Products Press, an imprint of The Haworth Press, Inc., 1999, pp. 139-166. Single or multiple copies of this article are available for a fee from The Haworth Document Delivery Service [1-800-342-9678, 9:00 a.m. - 5:00 p.m. (EST). E-mail address: getinfo@haworthpressinc.com].

© 1999 by The Haworth Press, Inc. All rights reserved.

processes that drive and control seed behavior will allow us to manipulate and manage weed seed banks in an economic and sustainable manner. This knowledge will allow us to implement improved, more informed, weed management systems and strategies. Important weed bank management strategies include prevention of seed introduction on farm, acquisition of weed biology information (including predictive tools), decision making about weed seed infestation levels and their implementation (eradication, reduction, tolerance), weed seed population shifts (within the seed bank, between species, increased diversity), and manipulations encouraging beneficial weed species. Environmental modification and changes in cropping systems can also be of considerable strategic importance in weed management. *[Article copies available for a fee from The Haworth Document Delivery Service: 1-800-342-9678. E-mail address: getinfo@haworthpressinc.com]*

**KEYWORDS.** Population shifts, seed germination, weed dispersal, weed diversity

## *INTRODUCTION*

Soil seed banks are reserves of viable seeds present in the soil and on its surface. Seed banks consist of both recent and older seed shed in, and dispersed into, a locality. This reserve of propagules is the source of local diversity, and is essential for the continuing existence of the flora in that locality. The first studies of soil seed banks date to little over a century ago (Putensen, 1882; Peter, 1893; Brenchley, 1918). There has been an increase in interest in agricultural soil seed banks in the last several decades due to questions of the nature of weedy adaptation, the spatial demography of weeds, the life cycle dynamics of weed populations, and the critical importance of these areas to weed management (Grime, 1979; Leck, Parker and Simpson, 1989; Miles, 1979; Rabotnov, 1978; Roberts, 1981; Sagar and Mortimer, 1976; Thompson and Grime, 1979).

Large numbers of viable seeds in the soil is a common attribute of a wide range of plant communities, including both undisturbed and disturbed habitats: grasslands (e.g., Major and Pyott, 1966), arable land (e.g., Cavers and Benoit, 1989), forests (e.g., Strickler and Edgerton, 1978), and aquatic ecosystems (van der Valk and Davis, 1976, 1978, 1979). Additionally, crop seeds can carryover in the soil seed bank and can become problems themselves (Hughes, 1974). The exception to these seed banks occur in temperate and tropical forests, where reserves of persistent seedlings perform this function (Grime, 1979).

Seeds as used herein include seeds (fertilized ripened ovule of a flowering plant) and fruits, but not vegetative propagules or spores. The focus herein will be primarily on agricultural seed banks and seeds of annual weed species, since the vast majority of seeds in the seed bank are often from annual weed species (Kropac, 1966; Roberts, 1981).

Management of agricultural weed seed banks is based on the behavior of seed within the seed bank matrix. The seed bank matrix consists of the spatial arrangement of environmental and physical factors over time. The behavior of seed in the seed bank defines the weed phenotype, the sum of the observable properties of the plant. Phenotype polymorphism results in a broad exploitation of the seed bank by an individual species as well as by the community of weed species. The niches available in a soil seed bank are the ecological roles played by a species in the community, and are loosely equivalent to the microhabitats.

The objectives of this paper are to review information of weed seed behavior in soil seed banks, and how weed seed interact with the soil seed bank matrix (environment-space-time) to define this behavior. The second objective is to suggest ways that this information could be used to manage agricultural soil seed banks. The concluding goal is to provide a discussion of experimental considerations limiting our understanding of weed seed banks.

## WEED SEED IN THE SOIL SEED BANK

The diversity of weed seeds are a consequence of different evolutionary compromises within an individual species caused by selection and adaptation in the soil seed bank matrix. The results of these compromises arise from the several roles performed by the weed seed in the continuation of the species. Harper (1977) has described five roles of weed seeds: dispersal, persistence, embryo food supply, display of genetic diversity, and a means of species multiplication. These different seed roles are not necessarily compatible with each other. Individual species have apportioned resources differentially to these roles, an indication of the inevitable tradeoffs involved in the survival and enduring occupation of a locality by a species.

### Dispersal and Colonization

Dispersal of seed includes both the processes by which an established seed bank maintains itself in a locality, as well as the process by which it expands its range and population size into new localities (Ehrendorfer, 1965; van der Pijl, 1969). Dispersal of seed occurs horizontally, in soil depth and vertical movement through the air or water, and in time. The distribution of the seed rain within and between localities is usually very uneven. Agricultural activity also results in seed movement within the seedbank (e.g., tillage).

Colonization is both the occupation of a soil site by a seedling, as well as the successful invasion of a new seed bank by a weed seed. Colonizing species often possess similar traits. Life form traits associated with colonizing ability include plastic growth, seed dormancy, rapid growth, annual life cycle, quick flowering, neutral photoperiodism, and economy of pollen (Baker, 1965; Barrett and Shore, 1989). Genetic traits associated with colonizing ability include polyploidy, self-pollination (Allard, 1965; Baker, 1965), marked interpopulation differentiation (Wang, Wendel and Dekker, 1995a, b), high degree of multi-locus association, and an optimum balance between flexibility (high rate of variation output to be tested by selection) and stability (fixation and multiplication of successful biotypes) (Ehrendorfer, 1965; Barrett and Richardson, 1986; Brown and Marshall, 1981).

## Persistence

Possibly the most important trait responsible for weed colonization success is the accumulation of dormant seed in the soil seed bank, ensuring their persistence in that locality (Grime, 1981). Seeds are the morphological form in which the plant endures the most hostile conditions in its life cycle, and for the longest portion of that life cycle. Seeds persist in the soil by means of several different types of adaptations to these hostile soil conditions (Roberts, 1972): resistance to predation and decay (Crist and Friese, 1993); prolific seed production; and seed dormancy. The ability of a seed to germinate at a favorable time in the season which results in a plant that reproduces is as important as seed dormancy. The mechanisms by which seeds regulate dormancy and germinability are numerous and complex. These mechanisms can derive from developmental regulation in the embryo itself as well as from envelope tissues surrounding the embryo (Dekker et al., 1996).

## Embryonic Food Supply

The naked seed embryo cannot exist on its own in the seed bank. Food reserves are part of the seed structure and perform an essential role in seed survival (Harper, Lovell, and Moore, 1970). When seeds germinate they face harsh competition from the environment as well as competitors. The amount of food reserves the seed has to rely on until it can begin independent photosynthetic food production is critical in this early struggle. Allocations of limited energy by the parent plant to an individual seed involves compromises between food reserves and seed numbers (Harper, 1977). Large numbers of seed with smaller seed food reserves (e.g., pigweed species, *Amaranthus* spp.) compared to smaller seed numbers per plant with larger food reserves (e.g., velvetleaf, *Abutilon theophrasti*) have important implications for a weed seeds colonizing ability in agroecosystems (Grime and Jeffrey, 1965; Harper and Obeid, 1967; Twamley, 1967).

## Display of Genetic Diversity

As conditions in soil seed banks change over long periods of time, the conditions favorable for persistence and survival also change. A species that consists of many different genotypes, each with potentially different combinations of adaptive traits, is more likely to survive and thrive with these long-term changes in habitat. Seeds are the form in which a species first displays this genetic diversity to the rigors of selection in the soil seed bank. They are the time in a species life cycle when new genetic combinations first appear and are released for dispersal. This display of genetic diversity in the seed bank takes the form of both intra- and inter-species biodiversity.

## Means of Multiplication of Species

The seed is the means by which the plant, and the species, multiplies. Seed number and fecundity are an important component of the fitness of a weed species. Because of the primary importance of yield, agriculturalists often equate the fitness of a weed species entirely with yield, or seed number produced (Dekker, 1993). The survivors selected over evolutionary time in the soil seed bank are not necessarily the ones that produce the most seed. Fisher's "Fundamental Theorem of Natural Selection" (Fisher, 1929) is: "The rate of increase in fitness of any organism at any time is equal to its genetic variance in fitness at that time." Harper has stated (1977): "The abundance of a species has more to do with the abundance of habitable sites and the genotypic and phenotypic plasticity that permit a wide range of sites to be occupied–it has nothing to do directly with the reproductive capacity of plants." The success of a weed species is only indirectly related to its fecundity; it is directly related to the genetic variance in fitness of the soil seed bank.

## Interactions Among the Seed Roles

The allocation of resources to a seed by a parent plant between these several roles is revealed in the successful phenotypes present in soil seed banks. The consequences of this selection and adaptation in weed seeds are also observable in complex interactions of these different roles. These interactions reveal the importance of diverse seed populations and the consequences of aggregated behavior in dynamic systems.

## Biodiversity and Heterogeneous Seed Populations

The sources of weed biodiversity and heterogeneous seed populations arises from several sources. These sources include somatic variability and phenotypic plasticity, as well as genetic diversity (Dekker, 1997; Pigliucci, 1997; Schlichting, 1986; Silvertown, 1984). All these sources act within a plant, a population (intra-specificity), and among species of a community (inter-specificity). The heterogeneity that arises from compromises among the several seed roles involve fecundity (size, number), dormancy and germination. A species that produces seed populations with individuals that each have different germination requirements is more likely to succeed (Dekker, 1997). Hedgebetting allows heterogeneous seed populations to take advantage of whatever conditions may exist during seasonal germination periods have a distinct selective advantage over those with one germination phenotype.

### Aggregate Behavior in Dynamic Soil Seed Bank Systems

Soil seed bank behavior at any level of biological system organization is a consequent, emergent, property of accumulated behaviors at lower levels of system organization. The seed behavior, like all dynamic systems, is organized and consists of many levels of organization. Seed phenotype heterogeneity is a consequence of its constituent biological groups. The behavior of grassy weed seeds like those of the foxtails (*Setaria* spp.) results from the interaction of all the parts of the plant system (Dekker et al., 1996). A soil seed bank in a corn field in Iowa may contain five different foxtail species (*S. faberi, S. geniculata, S. glauca, S. verticillata, S. viridis*) acting together as a species-group to exploit the available niches of that agroecosystem. Green foxtail (*S. viridis*) in that seed bank may consist of numerous genotypes (Wang, Wendel, and Dekker, 1995a, b), each with different adaptive traits allowing it to exploit opportunities left available by the particular cropping system (Wang and Dekker, 1995). The seed of giant foxtail in the soil will possess a very wide range of germination requirements when shed from the parent plant. This heterogeneity will arise from many sources in an individual plant, including the type of tiller the panicle appears on, the position of the seed on the panicle, and the position of the seed on the fascicle in the panicle (Dekker et al., 1996). An individual giant foxtail seed's behavior is the consequence of interactions among the several compartments and envelopes (embryo, caryopsis, seed hull) of the seed (Dekker et al., 1996). The behavior of individual axes within the giant foxtail embryo can also be independently controlled. The cumulative behavior of a population of cells may result in activity by the embryo shoot in which they exist. In each of these instances of foxtail seed bank behavior, the consequences of activity at lower levels of organization define the behavior at higher levels.

## WEED SEED BEHAVIOR: THE INTERACTION OF WEED SEED AND THE SOIL SEED BANK MATRIX

The weed seed behavior is the result of the weed interacting with the soil seed bank matrix. The soil seed bank matrix consists of the spatial and temporal arrangement of environmental factors. Seed banks are the evolutionary consequence of selection among heterogeneous genotypes and phenotypes within the soil seed bank matrix (environment-space-time).

## *Environment*

The environmental factors that interact and define the soil seed bank matrix are numerous and complex. Prediction of these effects is usually problematic. The environmental factors include rainfall and moisture, temperature and solar radiation, and gases. The environment also includes other physical factors of the habitat, including the soil, subsoil and underlying layers; slope, aspect, topography; longitude, latitude, hemisphere; or continental, maritime, or aquatic. The biotic environment influences the weed seed bank profoundly and includes the floral, animal, and microbial environments. Each of these interacting biotic components can result in interference, competition, symbiosis, or neutrality in relation to the weed seed. Finally, the most important component of environment from a human perspective is crop and weed management. The practices used by farmers that influence the fate of weed seed banks include tillage and other soil disturbances, planting and crop rotation, herbicide use and other weed control measures, fertilizer use and manure management, irrigation and drainage, governmental public policy and legal restrictions on land use, as well as many other human controlled activities associated with farming.

These factors of the soil seed bank matrix are received and sensed by weed seeds in several ways. They are sensed as dosages (quantity × time), and these dosages can result in differential seed responses depending on the sensitivity of the seed at the time of interception. The quality of the dosages can also vary considerably. The dosages of these factors can be readily available, in excess or act as limitations to seed behavior. The dosages can be received continuously or discretely, as constant signals or as alternating signals of different frequencies. As such these factors are modifiers of behavior as well as stimulators and/or inhibitors of activity (Thompson, Grime, and Mason, 1977; Wesson and Waring, 1967). Often the same factor can act as either stimulator or inhibitor depending on the context and sensitivity of the receiving seed. Combinations of factors can often result in responses not predictable by the responses of individual factors alone, and are often quite complex and difficult to interpret.

## *Space*

The soil seed bank matrix also consists of the spatial distribution of seed and environment over time. Seed and environment are dispersed and distributed unevenly over the landscapes of the world. On the spatial scale of the seed, resources and inhibitors interact to define which microsites are favorable for persistence, germination, and seedling establishment (Harper, 1977) and which sites are not. This uneven distribution of seed and environment over time results in patchiness and spatial patterns of weed seed and plants in agroecosystems (Cardina, Johnson, and Sparrow, 1997; Cavers, 1983; Johnson, Mortensen, and Gotway, 1996). Understanding the causative reasons for safe and hostile microsite distribution in agricultural soil seed banks is fundamental to planning sustainable weed management systems (Oriade et al., 1996). Soil seed bank patchiness and aggregate distribution of seed is emergent property accumulating over many spatial scales.

## Time

Spatially distributed weed seed interact with the environment over time. The changes that occur in seed banks over time differ with different time scales: short term changes, the seasonal and annual life cycle of the seed bank, multiple year changes, and evolutionary changes. Soil seed bank behavior at any of these levels of temporal organization is a consequence of accumulated behaviors at smaller time scales.

## Short Term Changes in the Weed Seed Bank

At relatively short time scales many important seed phenomena occur. Some of these are continuous processes and some are discrete events. The ability to germinate changes continuously in a seed or embryo from shortly after fertilization until germination (Dekker et al., 1996; Le Page-Degivry, Barthe, and Garello, 1990). Trewavas (1987) has made an elegant analogy of this situation to animal or human memory: the germination state of a seed at any one instant is a reflection of its previous experiences. Discrete phenomena also occur over the entire life cycle of the seed, including fertilization, abscission, mortality, germination and emergence (Dekker and Meggitt, 1986). These discrete events provide reproducible time markers allowing comparison to be made between different experiments.

## The Annual Weed Seed Bank Life Cycle

The temporal rhythm most relevant to agriculture is the annual cycle of weed emergence, growth, reproduction, and seed dispersal. Seed of many species germinate in the spring, establish reproductive plants, and shed dormant seed to the soil seed bank. The changes in the composition of the weed seed bank include both inputs and losses (Figure 1). Understanding the dy-

FIGURE 1. The annual life cycle of weed seed in the soil seed bank, and changes in the seed bank with time.

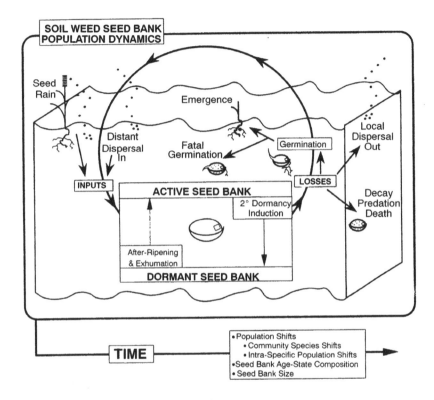

namic processes that regulate these changes is the challenge of weed seed bank management. When we understand the processes driving the fate of seed we will have achieved the ability to manipulate and manage weed populations at their most fundamental level.

*Inputs to the seed bank.* The heterogeneous seed rain provides additions to the active and dormant seed bank (Harper, 1977). Large numbers of seed shed from local plants join those of previous years to determine the new seed bank composition. To these are joined seeds that disperse into the local seed bank from distant sources.

*Seed fate: losses from the seed bank.* The subsequent fates of the seed entering the soil seed bank are numerous (Cavers, 1983; Zorner, Zimdahl, and Schweizer, 1984). Seed losses occur due to mortality: predation, pathogenic decay, physiological seed death, and cropping system mortality. Seed can be lost by dispersal out of a locality to distant locations. The seed bank is depleted every year by losses due to germination. Germination can result in emergence and seedling establishment as well as mortality.

*Changes over the annual cycle.* Within the seed bank many changes occur over the annual cycle. Many weed seeds are shed dormant in the seed rain, and over time lose those restraints to germination. Others remain germinable awaiting suitable conditions conducive to germination. Other appear to reacquire dormancy after having once lost it (secondary dormancy; Forcella et al., 1997; Karssen, 1980; Taylorson, 1982). An annual pattern, or rhythm, or changes in germinability have been reported for many species in the soil seed bank (Baskin and Baskin, 1985). Seasonal periods of species-specific germination and emergence also occur (Buhler et al., 1997b; Roberts and Lockett, 1978). Dormant seeds may undergo cyclic physiological changes in response to both endogenous and environmental conditions. This may be an important adaptation for weed survival in terms of regulation of germination and emergence (Montegut, 1975) to appropriate times of the annual cycle, as well as increasing soil longevity by avoiding fatal germination events and times. Also of importance are the changes that occur in the seed bank composition with the annual seed rain. Each year new seed are added, changing the size and character of the seed bank as the new seed mix with the surviving older seeds from previous years.

*Shifts between the active and dormant seed bank.* The soil seed bank can be described by two groups of seeds that have a large overlap in terms of their physiological status, the active and dormant seed banks (Figure 1). Dormant seed entering the seed bank can after-ripen and become germinable, entering the active seed bank. Germinable seed can have dormancy reinduced in them and pass from the active to the dormant seed bank. This later transition is common when seed are buried deeper in the soil, below the layer of favorable germination and emergence, by tillage and soil disturbances. Reentry into the active seed bank can occur as they are exhumed from depth by the same movement processes.

### Multiple Year Changes

As time passes other changes in the soil seed bank occur. The floral composition of the seed bank can change in terms of size, species present, and age of seed (Warr, Thompson, and Kent, 1993). Agricultural activities and changes in the environment are the two most important factors influencing these multiple year changes in the soil seed bank.

*Population shifts over time.* The floral composition of seed banks change with time and changing conditions (Thompson and Grime, 1979). Species composition of a site is a reflection of the history of seed production, as well as the different adaptive strategies of those species (Grime, 1979). Species present can be replaced, or numerically reduced over time, as well as populations within a species can be displaced by other intra-specific variants. The most dramatic examples of these population shifts in the last 50 years have been caused by herbicide usage (Le Baron and Gressel, 1982). Population shifts have also occurred in response to other crop and weed management practices.

*Seed bank longevity.* Seed banks consist of seed produced in several years (Kalisz, 1991; Kalisz and McPeek, 1992). The seeds of some weed species are able to survive in the soil longer than others (Burnside et al., 1996; Kivilaan and Bandurski, 1981; Toole and Brown, 1946). Seed age may be a factor in the numbers of seeds that germinate and emerge from the seed bank relative to those that remain in the soil (Naylor, 1972; Wilson and Cussans, 1975).

*Seed bank size.* The numbers of seed in the soil change with time. The relative balance of seed inputs and losses reflect seed bank dynamics (Rothrock, Squires, and Sheeley, 1993). There may be a trend towards smaller seed bank size with increasing latitude (Johnson, 1975), with increasing altitude, and in the later stages of plant succession (Thompson, 1978). These trends may be attributable to disturbance and stress resulting in lower rates of biomass production inhibiting reproduction.

## WEED SEED BANK MANAGEMENT SYSTEMS

Understanding the dynamic processes that control seed bank behavior is the challenge of weed seed bank management (Buhler, Hartzler, and Forcella, 1997; Dekker, 1997; Forcella, Durgan, and Buhler, 1996). When we understand the processes driving the fate of seed we will have achieved the ability to manipulate and manage weed populations at a fundamental level. There are several ways we can utilize our knowledge of soil seed banks to manage weeds in cropping systems. The first step in the development of a weed management system occurs with the defining, planning, and integrating of several strategies. Modeling seed banks, and developing predictive tools of their behavior, is and will continue to be an important improvement in the management of weeds. Finally, understanding the forces that drive the appearance of weed populations, shifts in weed populations, and mortality of weed populations will allow environmental and cropping system modifications leading to improved weed seed management techniques and systems.

### Weed Seed Bank Management Strategies

Weed management planning involves consideration of several types of strategic approaches. These strategies include prevention of weed introduction, acquisition of seed bank information, decisions about weed control goals and their implementation, shifting of weed seed populations, and manipulation of seed banks to encourage beneficial weed populations.

### Strategy: Prevention of Weed Seed Introduction on Farm

The first strategy of any weed seed management system must be to prevent the introduction of new weed seed onto the farm. This preventative approach includes any source of new weed seed to the farm: new equipment, human traffic, crop seed, forages and silage, manure, and dispersal into the farm from adjacent land. Related to these preventative tactics are support of governmental actions including seed purity laws, noxious weed laws, public policy, and local and federal quarantines.

### Strategy: Acquire More Weed Biology Information to Guide Management

### Strategy: Access Sources of Information

The more informed an individual manager is about weed seed biology and seed banks the more capable that person will be to devise and integrate weed management strategies and implement control tactics. Sources of information include institutional education, short courses, conferences, books, local extension information and experts, cooperative and seed company agronomists, herbicide and pesticide technical support personnel and literature, as well as information and wisdom from other growers.

### Strategy: Modeling Seed Banks and Predicting Seed Behavior

Modeling seed banks and development of research and extension predictive tools are an important source of weed management information (Buhler, et al., 1997b; Lybecker, Schweizer, and Westra, 1994; Oriade et al., 1996; Swinton and King, 1994; see also discussion in Experimental Considerations). These tools allow exploitation of vulnerable times in the weed's life cycle by prediction of timing of key life cycle events including dormancy, germination, emergence, interference periods in the crop, and the seed rain. Research model development has helped in understanding community, species, population, and individual seed behavior. Prediction and timing of postemergence herbicide applications, and of spring weed emergence (Forcella et al., 1992, 1997) have been the focus of on-farm extension computer tools. Economic as well as production information can be of benefit to growers in conjunction with their other ongoing planning activities. These tools are of benefit to growers in many ways, including timing of preemergence and postemergence herbicides, timing of the duration of season emergence by species (Buhler et al., 1997b), timing of induction of summer dormancy in some weed species, timing of tillage operations for maximum weed control (e.g., stale seed bed practices), and management of the seed rain. Continuing efforts to develop research models of seed bank behavior may provide new insights and increase the robustness of on-farm tools.

### Strategy: Decisions and Actions About Weed Seed Infestation Levels

Each weed manager makes a conscious or unconscious decision about the level of weed control acceptable to them in their fields. These decisions are based on many interacting factors, many of which are subjective. Weed control expectations are based on past experience, local community and neighbor standards, personal pride in management, economic considerations, yield expectations, labor availability, risk assumption thresholds, time availability in critical farming periods, and long-term goals. The decisions of weed control expectations in individual farmers range from total eradication to high levels of tolerance to weed infestations. Considerable improvements could be made in many individual farm management systems with a more critical evaluation of reasons behind these weed control goals.

### Strategy: Eradicate Weed Seed from the Seed Bank

For some growers the presence of any weed infestations is unacceptable. High value crops such as horticultural and ornamental production often demand this level of weed control. Crops that are hand-harvested often require complete elimination of interfering weeds. Poisonous weeds, as well as weeds interfering and preventing mechanical harvesting, can also necessitate total weed control. Finally, some weed species are perceived as so undesirable that a zero tolerance approach is required if crop production is to continue in the future [e.g., parasitic weeds such as *Orobanche* sp.; problem weeds such as velvetleaf and cocklebur (*Xanthium strumarium*)].

### Strategy: Decrease Weed Seed from the Seed Bank

The most common weed control expectation of growers is to decrease the amount of weed seed from the seed bank. These reductions can be from small to near complete elimination, and often are species specific with greater reductions of more undesirable species.

Elimination of seed production in an agricultural field results in an inevitable decrease in the seed bank size. Roberts (1962) has shown that the decrease in viable seed follows an exponential rate of decline in both disturbed and undisturbed soils, but varied by weed species and tillage practices (Forbes,

1963; Roberts and Feast, 1973b; Thurston, 1966). Control of all weed infestations in a field is usually very difficult, and only a few escaping weeds can
produce large numbers of seeds, replenishing the seed bank, and making elimination impossible. Information about the patterns of loss of seeds, and the
variability of losses under different conditions, from seed bank is mostly lacking.
Studies of artificial seed banks have provided some information about these loss
patterns (Roberts, 1962; Roberts and Feast, 1973a). Depletion of shallow seed
banks is more rapid than in deeper ones, possibly due to the predation and
widely changing environmental conditions (Roberts and Feast, 1972, 1973a). It
is more difficult to kill seed buried in deeper soil layers, where conditions are
more constant, and longevity may be more closely related to adaptations by the
seed itself (Burnside, Wicks, and Fenster, 1977; Harper, 1957).

Weed control tactics can take advantage of the high mortality of soil seed
banks with increased awareness of the forces that drive seed death. Tactics
that enhance mortality include those that exacerbate seed decay, predation,
and fatal germination. Practices that enhance dispersal of seed out of the
locality can also increase on-farm losses (e.g., capture in harvesting operations). Cropping practices are also a major source of weed seed mortality,
including herbicides, tillage and mechanical weed control, as well as several
types of fallowing (Kropac, 1966). Tactics and management practices that
allow the exploitation of vulnerable times in the life cycle of individual
weeds require more information availability about weed biology, and more
highly motivated and educated managers.

*Strategy: Tolerate Weed Seed in the Seed Bank*

Some growers tolerate relatively greater amounts of weed infestations
than others. Tolerance to high weed infestations may stem from two opposite
attitudes about weed management: a lack of concern about weed interference,
or a well-informed decision based on weed thresholds and yield losses. The
development and utilization of predictive tools of weed behavior has allowed
many growers to accurately predict how much weed interference, and the
duration of that competition in the crop, they can tolerate in their production
fields. Accurate and timely information, coupled with an analysis of maximum economic return per unit of land (as opposed to maximum yield), have
led many managers to profitably tolerate higher levels of weed infestations.
Also, accurate knowledge of site-specific weed infestations in individual's
fields by well-informed growers aids in this higher level of tolerance.

*Strategy: Shift Weed Seed Populations*

Closely related to weed infestation tolerance are strategies that change the
weed seed floral composition of individual farm fields. Population shifts occur

every time management practices are changed, and with the changing environment of the growing seasons. Understanding the forces that lead to these weed population shifts is the challenge offered by these management strategies.

*Strategy: Shift Weed Seed from Active to Dormant Part of Seed Bank*

Burial of weed seed out of harms way is an important strategy in weed seed management. Weed seeds often live longer at deeper soil depths, but deep burial of seed can greatly increase the chances that they will not emerge and thus interfere with crop production. Sequestration by deep burial for long enough periods will also result in the eventual mortality of that seed. Careful use of tillage can both effectively bury weed seed, and prevent or reduce its subsequent exhumation to shallower depths (Buhler and Mester, 1991; Yenish, Doll, and Buhler, 1992).

*Strategy: Shift Weed Seed from Dormant to Active Part of Seed Bank*

Keeping weed seed on the soil surface or shallow layers, or by exhuming buried seed to those shallow layers, usually increases the mortality of weed seed (Roberts and Feast, 1972, 1973a). Exposure to harsh environmental conditions, as well as easier access to predation, may be the reason for this enhanced surface mortality. Shallow placement of seed often leads to accelerated after-ripening of the dormant seed rain, speeding the entry of those seeds into the active seed bank. Careful management of these shallow seeds is important as they are more likely to germinate, emerge, and interfere with crops when they are in that shallow active seed bank. Conversely, stimulating seed to germinate in this shallow zone can lead to rapid depletion of the seed bank if carefully timed to coincide with weed control tactics. No-till and reduced tillage systems encourage shallow soil distribution of seed (Buhler and Mester, 1991; Yenish, Doll, and Buhler, 1992).

*Strategy: Shift Weed Seed from Undesirable to Desirable Species*

The appearance of new, and often more troublesome, weed species results from changes in our crop and weed management systems. The forces that drive these population shifts, if understood, can be used strategically to shift weed seed banks from those undesirable weed species to more desirable ones. An excellent example of the use of this strategy relates to herbicide resistant weed management. Continuous use of the same herbicide(s), the same crop, and a lack of tillage weed control over time will lead to the selection for herbicide resistant weeds (LeBaron and Gressel, 1982). Management of herbicide resistant weeds includes the rotation of herbicides with different modes of action, rotation of crops, and the use of several weed control tactics to ensure weed populations are confronted with many different tactics. These management practices prevent shifts to herbicide resistant weeds and maintain the susceptible biotypes in local populations.

This same approach can be utilized if the forces that drive the infestation of more desirable weed species are known and can be implemented in a weed management system. Practices that favor less aggressive and less competitive weeds (e.g., purslane (*Portulaca oleracea*) and other prostrate habit weeds) at the expense of very troublesome weeds (e.g., woolly cupgrass (*Eriochloa villosa*), wild proso millet (*Panicum miliaceum*), or weeds with no effective control tactics) will help implement this strategy. Weeds have also been purposefully used as cover crops within crops to smother and interfere with other less desirable weeds (Dekker, 1986; McCormick and Dekker, 1986). Finally, this strategy could potentially be used to shift weed populations that are not only easier to control but also economically less expensive to control.

### Strategy: Encourage Weed Diversity

The benefits conferred to agroecosystems by enhanced biodiversity are numerous (Dekker, 1997). Seed banks are the source of floristic diversity and plant population stability at a particular site (Baskin and Baskin, 1978; Levin and Wilson, 1978). Encouragement of weed seed diversity in seed banks may result in seed banks not only easier to manage than ones dominated by a few problematic weeds, but also crop management may be enhanced by the soil microflora, soil insect and soil macro-fauna diversity that thrive in more diverse agroecosystems. Diversity in crops and cropping systems, as well as diversity in weeds and weed management systems, result in more complete exploitation of available resources (Harper, 1977). The shift of weed seed in the soil seed bank to a few dominant species is an indication that the cropping systems are leaving a large opportunity for those species adapted to exploit it. The shift of weed seed to a diverse array of many different species is an indication that smaller opportunities are available to the weed flora, forcing small numbers of plants of many species to take advantage of the paucity of resources unused by crops. Crop and weed management systems that leave few unused resources available are more likely to have this smaller, more diverse, more easily managed weed seed bank flora.

### Strategy: Manipulate Seed Banks to Encourage
### Beneficial Weed Species

Some weed species may be of direct benefit to crop production. Manipulation of soil weed seed banks to encourage these species could result in improved production as well as weed management. Soil tillage timing to particular times of the year has been suggested as a means to manipulate weed communities to increase the populations of beneficial insects (Altieri and Whitcomb, 1979). Seed banks are beneficial for the restoration of contaminated sites including areas of chemical spills and strip-mined sites (Beauchamp, Lang, and May, 1975; Johnson and Bradshaw, 1979).

### Strategy: Environmental Modification of the Seed Bank Matrix Strategies

Strategies that manage the forces driving the appearance of weed populations, shifts in weed populations, and mortality of weed populations can be based on environmental and cropping system modifications. Some of these driving forces are uncontrollable, while modification of the environment and cropping systems may be manageable.

### Strategy: Environmental Modification of Soil Seed Banks

The soil seed bank environment can be modified in several ways. Choices of crop and tillage systems can modify the conditions under the crop canopy and affect seed germination and emergence. Modifications to the crop canopy, to the soil surface residue and soil itself, to soil fertility, herbicide use and herbicide residues persisting in the soil, all can have significant effects on soil temperature; light; soil gases and their permeability within the soil; soil moisture and drainage; soil tilth, structure, aggregation; and the spatial heterogeneity of seed in a field.

### Strategy: Cropping System Modification of Soil Seed Banks

Changes in crop and weed management systems and tactics can modify soil seed banks (see discussion in Experimental Considerations). Seed bank size reductions, as well as species and population shifts with can occur with changes in crop rotations, tillage, fertility, and most commonly with the use of herbicides (Chancellor, 1981; Dyer, 1995).

## EXPERIMENTAL CONSIDERATIONS

Our knowledge of agricultural soil seed banks is limited by several important experimental and theoretical considerations. Understanding the dynamic processes that regulate the changes in seed bank behavior, and then exploiting them in weed and crop management, are the challenges of weed seed bank research. When we understand the processes driving the fate of seed in the soil we will have achieved the ability to manipulate and manage weed populations at a fundamental level. If we are to manage seed banks at this level we need new information, we need to utilize reproducible experimental methodologies, and we must strive to elucidate a theoretical basis to seed bank behavior to guide this search.

### Knowledge of Seed Banks

Studies of soil weed seed banks did not begin until relatively recently considering their importance to agriculture. Early research has been descriptive, and there exists a need to continue efforts in quantifying seed bank dynamics as well as the mechanisms driving seed inputs and losses (Roberts, 1981). There are gaps in our knowledge of seed banks, and much of what we do know is still rudimentary. There also exists little information about seed banks below the level of the seed. Also lacking is information about the fate of seed in seed banks, estimation of viable seed in seed banks, seed germination microsites, seed bank spatial distribution, the effects of crop and weed management on seed bank dynamics, the need for robust predictive models of seed bank behavior, and seed biology.

### Fate of Seeds in Seed Banks

Accounting for the inputs and losses of seed in the soil seed bank is a very difficult experimental and methodological challenge. Despite this, some studies have attempted to quantify the inputs and losses in seed banks (Chambers and MacMahon, 1994; Sagar and Mortimer, 1976; Teo-Sherrell, Mortensen, and Keaton, 1996; Van der Vegte, 1978; Wilson and Cussans, 1975). Lacking are more complete input-loss accounting by category (e.g., dispersal in and out, germination and emergence, mortality, predation) of seed in agricultural seed banks under a wide range of geographic, cropping system, weed management system, and environmental conditions. There is a need for more information at the intra-specific level to determine how populations of the same species might interact to exploit a site more efficiently and for enduring occupation (van der Vegte, 1978). Understanding the fate of seeds will provide insights guiding weed and crop management practices to manipulate seed banks.

### Estimation of Viable Seed in Seed Banks

Estimation of the presence of viable seed in seed banks is crucial to understanding seed bank dynamics (Ambrosio, Dorado, and Del Monte, 1997; Benoit, Kenkel, and Cavers, 1989; Cardina and Sparrow, 1996; Roberts, 1981). Viable seed in the soil seed bank have been estimated using seedling emergence and direct extraction from the soil. Estimates of seedling emergence have been made (e.g., Forcella et al., 1992, 1997; Grundy, Mead, and Bond, 1996) but are incomplete because unknown quantities of dormant seed remain in the soil. Viable seed in the soil can also be estimated by extraction of soil from the soil seed bank and either directly separating seed from the soil (e.g., Fay and Olson, 1978; Gross, 1990; Thorsen and Crabtree, 1977) or germinating seed under uniform environmental conditions (e.g., Feast and Roberts, 1973). But, there is often a disparity between the species present in exhumed seed at any one time and what species emerged seedlings are present above the surface at that time (Ball and Miller, 1989; Chancellor, 1965; Roberts and Feast, 1973a; Roberts and Ricketts, 1979; Sagar and Mortimer, 1976; Williams and Egley, 1977).

*Seed Germination Microsites*

Understanding the causes for safe and hostile microsite distribution in agricultural soil seed banks is fundamental to weed management systems. There is a lack of understanding of the influences of environmental factors on weed seed behavior, on seed bank characteristics in different soils and on geographic and climatic regions.

*Seed Bank Spatial Distribution*

Seeds and the seed bank matrix are not uniformly distributed in agricultural fields. Studies of the spatial distribution of seed banks are very difficult but have been done (Cardina, Johnson, and Sparrow, 1997; Cardina, Sparrow, and McCoy, 1995; Kellman, 1978). We also lack information to determine what the relevant spatial unit of study is for seed banks. Determinations of seed number per unit area by species, and by soil depth have been made (Kropac, 1966; Pareja, Staniforth, and Pareja, 1985). Errors of heterogeneity can occur on several spatial scales with this technique. Large numbers of samples are needed to obtain accurate estimates (Rabotnov, 1978). Often it is better to take large numbers of small samples rather than relatively smaller numbers of large samples (Kropac, 1966; Roberts, 1970). Improvements in these estimations have been made by sampling over time: sampling the same time of year in successive years (Roberts and Dawkins, 1967; Warington, 1958); and by taking more frequent samples over the annual cycle (Knipe and Springfield, 1972).

*Effects of Crop and Weed Management on Seed Bank Dynamics*

We lack information on interactions and effects of crop and weed management systems on seed bank characteristics. The seed bank of an agricultural field is a tangible history of the successes and failures of cropping and weed management practices. The failures are indicated by both the overall seed bank size, and by the relatively larger numbers of seeds of individual species. Many, if not all, cultural practices in farming are potentially capable of reducing (or increasing) the seed bank size, either as a whole or of individual species. The influence of crop rotations (e.g., Jordan et al., 1995), tillage systems (e.g., Egley and Williams, 1990; Fay and Olson, 1978; Forcella and Gill, 1986; Forcella, Buhler, and McGiffen, 1994; Lueschen and Andersen, 1978; Mohler and Galford, 1997; Roberts and Dawkins, 1967; Roberts and Feast, 1973b; Soriano et al., 1968), fertility programs (Banks, Santelmann, and Tucker, 1976; Brenchley and Warington, 1930; Dotzenko, Ozkan, and Storer, 1969), herbicides (e.g., Burnside, 1978; Gressel, 1978; Schweizer and Zimdahl, 1979) and other weed control tactics all have important, and poorly understood, impacts on seed banks.

## Predictive Models of Seed Bank Behavior

One of the needs of farmers in managing soil seed banks is for accurate prediction of weed emergence and weed development as an aid in weed management decisions. Emergence prediction aids have been developed, but lack robustness for a wide variety of agricultural conditions (e.g., Buhler et al., 1997a, b; Forcella et al., 1992; Lybecker, Schweizer, and Westra, 1994; Naylor, 1970; Roberts and Ricketts, 1979; Swinton and King, 1994).

## Seed Biology

There exists a considerable lack information about some of the primary biochemistry and physiology within the seed driving its behavior in the soil. These include information about mechanisms regulating seed dormancy and germination, and about weed seed and population genetics (Levin and Wilson, 1978).

## Experimental Reproducibility

Seed bank studies are inherently difficult to reproduce because of many factors associated with comparing samples from diverse populations in diverse seed bank matrices (see matrix discussion above). Other sources of error can also occur, including inadequate phenotype characterization and population identification of sampled seed (Wang, Wendel, and Dekker, 1995a, b), seed sampling and storage information, and complete germination assay descriptions. Some of these errors could be eliminated by consistent use of reproducible time markers allowing comparisons to be made between different experiments including use of discrete life cycle events such as fertilization, abscission, mortality, germination, and emergence; as well as sampling at the same time each year in multi-year experiments.

## Theoretical Framework

Finally, we lack a general theory of seed bank behavior and seed bank dynamics. Several researchers have suggested frameworks and general descriptive, qualitative models, which are an important start toward a theoretical basis of behavior (e.g., Harper, 1977; Parker, Simpson, and Leck, 1989). The development of a theory of seed bank behavior, like that for seed dormancy and germination, would provide the basis of testable hypotheses to expand our knowledge of seed banks beyond the largely descriptive, discrete, and particular that now characterizes the science. The lack of theory is to some extent due to a lack of a critical mass of information to guide a more general view of seed bank dynamics.

## REFERENCES

Allard, R.W. (1965). Genetic systems associated with colonizing ability in predominantly self-pollinating species. In *Genetics of Colonizing Species* (Baker, H.G, and G.L. Stebbins, Eds.), pp. 50-76. Academic Press, New York.

Altieri, M.A., and W.H. Whitcomb. (1979). The potential use of weeds in the manipulation of beneficial insects. *HortScience* 14:12-18.

Ambrosio, L., J. Dorado, and J.P. Del Monte. (1997). Assessment of the sample size to estimate the weed seedbank in soil. *Weed Research* 37:129-137.

Baker, H.G. (1965). Characteristics and modes of origin of weeds. In *Genetics of Colonizing Species* (Baker, H.G, and G.L. Stebbins, Eds.), pp. 147-168. Academic Press, New York.

Ball, D.A., and S.D. Miller. (1989). A comparison of techniques for estimation of arable soil seedbanks and their relationship to weed flora. *Weed Research* 29:365-373.

Banks, P.A., P.W. Santelmann, and B.B. Tucker. (1976). Influence of long-term soil fertility treatments on weed species in winter wheat. *Agronomy Journal* 68:825-827.

Barrett, S.C.H., and B.J. Richardson. (1986). Genetic attributes of invading species. In *Ecology of Biological Invasions* (Groves, R.H., and J.J. Burdon, Eds.), pp. 21-33. Australian Academy of Science, Canberra.

Barrett, S.C.H., and J.S. Shore. (1989). Isozyme variation in colonizing plants. In *Isozymes in Plant Biology* (Soltis, D.E., and P.S. Soltis, Eds.), pp. 106-126. Dioscorides, Portland, OR.

Baskin, J.M., and C.C. Baskin. (1978). The seed bank in a population of an endemic plant species and its ecological significance. *Biological Conservation* 14:125-130.

Baskin, J.M., and C.C. Baskin. (1985). The annual dormancy cycle in buried weed seeds:a continuum. *Bioscience* 35(18):492-498.

Beauchamp, H., R. Lang, and M. May. (1975). Topsoil as a seed source for reseeding strip mine soils. *Research Journal of the Wyoming Agricultural Experiment Station* 90:8.

Brenchley, W.E. (1918). Buried weed seeds. *Journal of Agricultural Science, Cambridge* 9:1-31.

Brenchley, W.E., and K. Warington. (1930). The weed seed population of arable soil. I. Numerical estimation of viable seeds and observations on their natural dormancy. *Journal of Ecology* 18:235-272.

Brown, A.H.D., and D.R. Marshall. (1981). Evolutionary changes accompanying colonization in plants. In *Evolution Today* (Scudder, G.C.E., and J.L. Reveal, Eds.), pp. 351-363. Hunt Institute for Botanical Documentation, Carnegie-Mellon University, Pittsburgh, PA.

Buhler, D.D., R.G. Hartzler, and F. Forcella. (1997). Implications of weed seed and seedbank dynamics to weed management. *Weed Science* 45:329-336.

Buhler, D.D., R.G. Hartzler, F. Forcella, and J.L. Gunsolus. (1997a). Relative emergence sequence for weeds of corn and soybeans. *Iowa State University Extension Bulletin* SA-11.

Buhler, D.D., R.P. King, S.M. Swinton, J.L. Gunsolus, and F. Forcella. (1997b). Field evaluation of a bioeconomic model for weed management in soybean (*Glycine max*). *Weed Science* 45:158-165.

Buhler, D.D., and T.C. Mester. (1991). Effect of tillage systems on the emergence depth of giant foxtail (*Setaria faberi*) and green foxtail (*Setaria viridis*). *Weed Science* 39:200-203.

Burnside, O.C. (1978). Mechanical, cultural and chemical control of weeds in a sorghum-soybean (*Sorghum bicolor*)-(*Glycine max*) rotation. *Weed Science* 26:362-369.

Burnside, O.C., R.G. Wilson, S. Weisberg, and K.G. Hubbard. (1996). Seed longevity of 41 species buried 17 years in eastern and western Nebraska. *Weed Science* 44:74-86.

Burnside, O.C., G.A. Wicks, and C.R. Fenster. (1977). Longevity of shattercane seed in soil across Nebraska. *Weed Research* 17:139-143.

Cardina, J., and D.H. Sparrow. (1996). A comparison of methods to predict weed seedling populations from the soil seedbank. *Weed Science* 44:46-51.

Cardina, J., G.A. Johnson, and D.H. Sparrow. (1997). The nature and consequences of weed spatial distribution. *Weed Science* 45:364-373.

Cardina, J., D.H. Sparrow, and E.L. McCoy. (1995). Analysis of spatial distribution of common lambsquarters in no-till soybean. *Weed Science* 43:258-268.

Cavers, P.B. (1983). Seed demography. *Canadian Journal of Botany* 61:3578-3590.

Cavers, P.B., and D.L. Benoit. (1989). Seed banks in arable land. In *Ecology of Soil Seed Banks* (Eds. M.A. Leck, V.T. Parker, and R.L. Simpson). pp. 309-328. Academic Press Inc., New York.

Chancellor, R.J. (1965). Weed seeds in the soil. *Report of the Weed Research Organization for 1960-64*:15-19, UK.

Chancellor, R.J. (1981). The manipulation of weed behaviour for control purposes. *Philosophical Transactions of the Research Society London B.* 295:103-110.

Crist, T.O., and C.F. Friese. (1993). The impact of fungi on soil seeds: implications for plants and granivores in a semiarid shrub-steppe. *Ecology* 74(8):2231-2239.

Dekker, J. (1986). Iowa cover crop species screening project. *North Central Weed Control Conference Proceedings* 41:87.

Dekker, J. (1993). Pleiotropy in triazine resistant *Brassica napus*: Leaf and environmental influences on photosynthetic regulation. *Zeitschrift Naturforschung* 48c:283-287.

Dekker, J. (1997). Weed diversity and weed management. *Weed Science* 45:357-363.

Dekker, J., B.I. Dekker, H. Hilhorst and C. Karssen. (1996). Weedy adaptation in *Setaria* spp.: IV. Changes in the germinative capacity of *S. faberii* embryos with development from anthesis to after abscission. *American Journal of Botany* 83(8):979-991.

Dekker, J.H. and W.F. Meggitt. (1986). Field emergence of velvetleaf (*Abutilon theophrasti*) in relation to time and burial depth. *Iowa Journal of Research* 61:65-80.

Dyer, W.E. (1995). Exploiting weed seed dormancy and germination requirements through agronomic practices. *Weed Science* 43:498-503.

Egley, G.H., and R.D. Williams. (1990). Decline of weed seeds and seedling emergence over five years as affected by soil disturbances. *Weed Science* 38:504-510.

Ehrendorfer, F. (1965). Dispersal mechanisms, genetic systems, and colonizing abilities in some flowering plant families. In *Genetics of Colonizing Species* (Baker, H.G, and G.L. Stebbins, Eds.), pp. 331-351. Academic Press, New York.

Fay, P.K., and W.A. Olson. (1978). Technique for separating weed seed from soil. *Weed Science* 26:530-533.

Feast, P.M., and H.A. Roberts. (1973). Note on the estimation of viable weed seeds in soil samples. *Weed Research* 13:110-113.

Fisher, R.A. (1929) (revised 1958). *The Genetical Theory of Natural Selection.* Revised edition. Dover Press, New York.

Forbes, N. (1963). The survival of wild oat seeds under a long ley. *Experimental Husbandry* No. 9:10-13.

Forcella, F. (1992). Prediction of weed seedling densities from buried seed reserves. *Weed Research* 32:29-38.

Forcella, F., D.D. Buhler, and M.E. McGiffen. (1994). Pest management and crop residues. In *Advances in Soil Science: Crop Residue Mangement* (Hatfield, J.L., and B.A. Stewart, Eds.), pp. 173-189. Lewis Publishers, Boca Raton.

Forcella, F., B.R. Durgan, and D.D. Buhler. (1996). Management of weed seedbanks. In *Proceedings of the Second International Weed Control Congress* (Kudsk, P. Ed.), pp. 21-26. Copenhagen.

Forcella, F., and A.M. Gill. (1986). Manipulation of buried seed reserves by timing soil tillage in Mediterranean-type pastures. *Australian Journal of Experimental Agriculture* 26:71-78.

Forcella, F., R.G. Wilson, J. Dekker, R.J. Kremer, J. Cardina, R.L. Anderson, D. Alm, K.A. Renner, R.G. Harvey, S. Clay, and D.D. Buhler. (1997). Weed seed bank emergence across the corn belt. *Weed Science* 45:67-76.

Forcella, F., R.G. Wilson, K.A. Renner, J. Dekker, R.G. Harvey, D.A. Alm, D.D. Buhler and J.A. Cardina. (1992). Weed seedbanks of the U.S. corn belt: magnitude, variation, emergence, and application. *Weed Science* 40:636-644.

Gressel, J. (1978). Factors affecting the selection of herbicide resistant biotypes of weeds. *Outlook on Agriculture* 9:283-287.

Grime, J.P. (1979). *Plant Strategies and Vegetation Processes.* Wiley, Chichester, UK.

Grime, J.P. (1981). The role of seed dormancy in vegetation dynamics. *Annals of Applied Biology* 98:555-558.

Grime, J.P., and D.W. Jeffrey. (1965). Seedling establishment in vertical gradients of sunlight. *Journal of Ecology* 53:621-642.

Gross, K.L. (1990). A comparison of methods for estimating seed numbers in the soil. *Journal of Ecology* 78:1079-1093.

Grundy, A.C., A. Mead, and W. Bond. (1996). Modelling the effect of weed-seed distribution in the soil profile on seedling emergence. *Weed Research* 36:375-384.

Harper, J.L. (1957). The ecological significance of dormancy and its importance in weed control. *Proceedings of the 4th International Conference on Plant Protection,* Hamburg:415-420.

Harper, J.L. (1977). *Population Biology of Plants*, pp. 892. Academic Press Inc., New York.

Harper, J.L., P.H. Lovell, and K.G. Moore. (1970). The shapes and sizes of seeds. *Annual Review of Ecological Systems* 1:327-356.

Harper, J.L., and M. Obeid. (1967). Influence of seed size and depth of sowing on the establishment and growth of varieties of fiber and oil seed flax. *Crop Science* 7:527-532.

Hughes, R.G. 1974. Cereals as weeds. *Proceedings of the 12th British Weed Control Conference* 3:1023-1029.

Johnson, E.A. (1975). Buried seed populations in the subarctic forest east of Great Slave Lake, Northwest Territories. *Canadian Journal of Botany* 53:2933-2941.

Johnson, G.A., D.A. Mortensen, and C.A. Gotway. (1996). Spatial and temporal analysis of weed seedling populations using geostatistics. *Weed Science* 44:704-710.

Johnson, M.S., and A.D. Bradshaw. (1979). Ecological principles for the restoration of disturbed and degraded land. *Applied Biology* 4:141-200.

Jordan, N., D.A. Mortensen, D.M. Prenzlow, and K.C. Cox. (1995). Simulation analysis of crop rotation effects on seedbanks. *American Journal Botany* 82(3):390-398.

Kalisz, S. (1991). Experimental determination of seed bank age structure in the winter annual *Collinsia verna*. *Ecology* 72(2):575-585.

Kalisz, S., and M.A. McPeek. (1992). Demography of an age-structured annual: resampled projection matrices, elasticity analyses, and seed bank effects. *Ecology* 73(3):1082-1093.

Karssen, C.M. (1980). Environmental conditions and endogenous mechanisms involved in secondary dormancy of seeds. *Israel Journal of Botany* 29:45-64.

Kellman, M. (1978). Microdistribution of viable weed seed in two tropical soils. *Journal of Biogeography* 5:291-300.

Kivilaan, A.A., and R.S. Bandurski. (1981). The one hundred year period for Dr. W.J. Beal's seed viability experiment. *American Journal of Botany* 68:1290-1292.

Kropac, Z. (1966). Estimation of weed seeds in arable soil. *Pedobiologia* 6:105-128.

Le Baron, H.M. and J. Gressel. (1982). *Herbicide Resistance in Plants*, p. 401. Wiley-Interscience, New York.

Leck, M.A., V.T. Parker, and R.L. Simpson. (1989). *Ecology of Soil Seed Banks*, p. 462. Academic Press Inc., New York.

Le Page-Degivry, M-.T., P. Barthe, and G. Garello. (1990). Involvement of endogenous abscissic acid in onset and release of *Helianthus annuus* embryo dormancy. *Plant Physiology* 92: 1164-1168.

Levin, D.A., and J.B. Wilson. (1978). The genetic implications of ecological adaptations in plants. In *Structure and Functioning of Plant Populations* (Freysen, A.H.J., and J.W. Woldendorp, Eds.), pp. 75-100. North Holland, Amsterdam.

Lueschen, W.E., and R.N. Andersen. (1978). Effect of tillage and cropping systems on velvetleaf longevity. *North Central Weed Control Conference Proceedings* 33:110.

Lybecker, D.W., E.E. Schweizer, and P. Westra. (1994). *WEEDCAM Manual*. Fort Collins, CO: Colorado State University. p. 49.

Major, J., and W.T. Pyott. (1966). Buried viable seeds in two California bunchgrass sites and their bearing on the definition of a flora. *Vegetatio* 13:253-282.

McCormick, B., and J. Dekker. (1986). Cover crop interference in no-till corn and soybean production. *North Central Weed Control Conference Proceedings*. 41:85.

Miles, J. (1979). *Vegetation Dynamics*. Chapman and Hall, London, UK.

Montegut, J. (1975). Ecologie de la germination des mauvaises herbes. In *La Germination des Semences,* (Chaussat, R., and Y. le Deunff, Eds.). pp. 191-217, Gauthier-Villars, Paris.

Naylor, R.E.L. (1970). The prediction of blackgrass infestations. *Weed Research* 10:296-299.

Naylor, R.E.L. (1972). Aspects of the population dynamics of the weed *Alopecurus myosuroides* Huds. in winter cereal crops. *Journal of Applied Ecology* 9:127-139.

Oriade, C.A., R.P. King, F. Forcella, and J.L. Gunsolus. (1996). A bioeconomic analysis of site-specific management for weed control. *Review of Agricultural Economics* 18:523-535.

Pareja, M.R., D.W. Staniforth, and G.P. Pareja. (1985). Distribution of weed seed among soil structural units. *Weed Science* 33:182-189.

Parker, V.T, R.L. Simpson, and M.A. Leck. (1989). Pattern and process in the dynamics of seed banks. In *Ecology of Soil Seed Banks* (Eds. M.A. Leck, V.T. Parker, and R.L. Simpson). pp. 367-384. Academic Press Inc., New York.

Peter, A. (1893). Culturversuche mit "ruhenden" Samen. *Nachrichten von der Koeniglichen Gessellschaft der Wissenschaften zu Goettingen* 17:673-691.

Pigliucci, M. (1997). Ontogenetic phenotypic plasticity during the reproductive phase in *Arabidopsis thaliana* (Brassicaceae). *American Journal of Botany* 84(7): 887-895.

Putensen, H. (1882). Untersuchungen uber die im Ackerboden enthaltenen Unkrautsamereien. *Hannoverisches land-univers forstwirtsch Vereinbl* 21:514-524.

Rabotnov, T.A. (1978). On coenopopulations of plants reproducing by seeds. In

*Structure and Functioning of Plant Populations* (Freysen, A.H.J. and J.W. Woldendorp, Eds.). pp. 1-26. North Holland, Amsterdam.

Roberts, H.A. (1962). Studies on the weeds of vegetable crops. II. Effect of six years of cropping on the weed seeds in the soil. *Journal of Ecology* 50:803-813.

Roberts, H.A. (1970). Viable weed seeds in cultivated soils. *Report of the National Vegetation Research Station for 1969*:25-38.

Roberts, E.H. (1972). Dormancy: a factor affecting seed survival in the soil. In *Viability of Seeds* (Roberts, E.H., Ed.) pp. 321-359. Chapman and Hall, London, UK.

Roberts, H.A. (1981). Seed banks in soil. *Advances in Applied Biology* 6:1-55.

Roberts, H.A., and P.A. Dawkins. (1967). Effect of cultivation on the numbers of viable weed seeds in soil. *Weed Research* 7:290-301.

Roberts, H.A., and P.M. Feast. (1972). Fate of seeds of some annual weeds in different depths of cultivated and undisturbed soil. *Weed Research* 12:316-324.

Roberts, H.A., and P.M. Feast. (1973a). Emergence and longevity of seeds of annual weeds in cultivated and undisturbed soil. *Journal of Applied Ecology* 10:133-143.

Roberts, H.A. and P.M. Feast. (1973b). Changes in the numbers of viable weed seeds in soil under different regimes. *Weed Research* 13:298-303.

Roberts, H.A., and P.M. Lockett. (1978). Seed dormancy and periodicity of seedling emergence in *Veronica hederifolia* L. *Weed Research* 18:41-48.

Roberts, H.A., and M.E. Ricketts. (1979). Quantitative relationships between the weed flora after cultivation and the seed population in the soil. *Weed Research* 19:269-275.

Rothrock, P.E., E.R. Squires, and S. Sheeley. (1993). Heterogeneity and size of a persistent seedbank of *Ambrosia artemisiifolia* L. and *Setaria faberi* Herrm. *Bulletin. Torrey Botanical Club* 120(4):417-422.

Sagar, G.R., and A.M. Mortimer. (1976). An approach to the study of the population dynamics of plants with special reference to weeds. *Applied Biology* 1:1-47.

Salisbury, E.J. (1942). *The Reproductive Capacity of Plants.* Bell, London.

Schlichting, C. (1986). The evolution of phenotypic plasticity in plants. *Annual Review of Ecological Systems* 17:667-693.

Schweizer, E.E., and R.L. Zimdahl. (1979). Changes in the number of weed seeds in irrigated soil under two management systems. *Proceedings of the Western Society of Weed Science* 32:74.

Silvertown, J.W. (1984). Phenotypic variety in seed germination behavior: the ontogeny and evolution of somatic polymorphism in seeds. *American Naturalist* 124:1-16.

Strickler, G.S., and P.J. Edgerton. (1976). Emergent seedlings from coniferous litter and soil in eastern Oregon. *Ecology* 57:801-807.

Swinton, S.M. and R.P. King. (1994). A bioeconomic model for weed management in corn and soybeans. *Agricultural Systems* 44:313-335.

Taylorson, R.B. (1982). Anesthetic effects on secondary dormancy and phytochrome responses in Setaria faberi seeds. *Plant Physiology* 70:882-886.

Teo-Sherrell, C.P.A., D.A. Mortensen, and M.E. Keaton. (1996). Fates of weed seeds in soil: a seeded core method of study. *Journal of Applied Ecology* 33:1107-1113.

Thompson, K. (1978). The occurrence of buried viable seeds in relation to environmental gradients. *Journal of Biogeography* 5:425-430.

Thompson, K., and J.P. Grime. (1979). Seasonal variation in the seed banks of herbaceous species in ten contrasting habitats. *Journal of Ecology* 67:893-921.

Thompson, K., J.P. Grime, and G. Mason. (1977). Seed germination in response to diurnal fluctuations of temperature. *Nature, London* 267:147-149.

Thorsen, J.A., and G. Crabtree. (1977). Washing equipment for separating weed seed from soil. *Weed Science* 25:41-42.

Thurston, J.M. (1966). Survival of seeds of wild oats (*Avena fatua* L., and *Avena ludoviciana* Dur.) and charlock (*Sinapis arvensis* L.) in soil under leys. *Weed Research* 6:67-80.

Toole, E.H., and E. Brown. (1946). Final results of the Duvel buried seed experiment. *Journal of Agricultural Research* 72:201-210.

Trewavas, A.J. (1987). Timing and memory processes in seed embryo dormancy–a conceptual paradigm for plant development questions. *BioEssays* 6:87-92.

Twamley, B.E. (1967). Seed size and seedling vigour in birdsfoot trefoil. *Canadian Journal of Plant Science* 47:603-609.

Van der Pijl. (1969). *Principles of Dispersal in Higher Plants*. Springer-Verlag, Berlin.

Van der Valk, A.G., and C.B. Davis. (1976). The seed banks of prairie glacial marshes. *Canadian Journal of Botany* 54:1832-1838.

Van der Valk, A.G., and C.B. Davis. (1978). The role of seed banks in the vegetation dynamics of prairie glacial marshes. *Ecology* 59:322-335.

Van der Valk, A.G., and C.B. Davis. (1979). A reconstruction of the recent vegetational history of a prairie marsh, Eagle Lake, Iowa, from its seed bank. *Aquatic Botany* 6:29-51.

Van der Vegte, F.W. (1978). Population differentiation and germination ecology in *Stellaria media* (L.) Vill. *Oecologia (Berl.)* 37:231-245.

Wang, R.L. and J. Dekker. (1995). Weedy adaptation in *Setaria* spp.: III. Variation in herbicide resistance in *Setaria* spp. *Pesticide Biochemistry and Physiology* 51:99-116.

Wang, R.L., J. Wendel, and J. Dekker. (1995a). Weedy adaptation in *Setaria* spp.: I. Isozyme analysis of the genetic diversity and population genetic structure in *S. viridis*. *American Journal of Botany* 82(3):308-317.

Wang, R.L., J. Wendel, and J. Dekker. (1995b). Weedy adaptation in *Setaria* spp.: II. Genetic diversity and population genetic structure in *S. glauca, S. geniculata* and *S. faberii*. *American Journal of Botany* 82(8):1031-1039.

Warington, K. (1958). Changes in weed flora on Broadbalk permanent wheat field during the period 1930-55. *Journal of Ecology* 46:101-113.

Warr, S.J., K. Thompson, and M. Kent. (1993). Seed banks as a neglected area of biogeographical research: a review of literature and sampling techniques. *Progress in Physical Geography* 17(3):329-347.

Wesson, G., and P.F. Waring. (1967). Light requirements of buried seeds. *Nature, London* 213:600-601.

Wilson, B.J., and G.W. Cussans. (1975). A study of the population dynamics of *Avena fatua* L. as influenced by straw burning, seed shedding and cultivations. *Weed Research* 15:249-258.

Yenish, J.P., J.D. Doll, and D.D. Buhler. (1992). Effects of tillage on vertical distribution and viability of weed seed in soil. *Weed Science* 40:429-433.

Zorner, P.S., R.L. Zimdahl, and E.E. Schweizer. (1984). Sources of viable seed loss in buried dormant and non-dormant populations of wild oat (*Avena fatua* L.) seed in Colorado. *Weed Research* 24:143-150.

RECEIVED: 08/28/97
ACCEPTED: 05/20/98

# A Risk Management Perspective on Integrated Weed Management

J. L. Gunsolus
D. D. Buhler

**SUMMARY.** The variability inherent in agriculture influences many crop production decisions made by farmers, including weed management. This paper addresses how farmers perceive the variability, or risk, associated with integrated weed management systems in terms of yield, economic returns, and time and labor management. This paper addresses how key biological time constraints such as periodicity of weed emergence, rate of crop growth and development, and critical periods of weed control can influence the outcome of integrated weed management systems. A key component to developing successful integrated weed management systems lies in the ability of the crop producer to align individual time and labor management issues with existing biological time constraints. *[Article copies available for a fee from The Haworth Document Delivery Service: 1-800-342-9678. E-mail address: getinfo@haworthpressinc.com]*

**KEYWORDS.** Economic analysis, field working days, time management

---

J. L. Gunsolus, Department of Agronomy and Plant Genetics, University of Minnesota, St. Paul, MN 55108. D. D. Buhler, Research Agronomist, National Soil Tilth Laboratory, U.S. Department of Agriculture, Agricultural Research Service, 2150 Pammel Drive, Ames, IA 50011.

Address correspondence to: J. L. Gunsolus at the above address (E-mail: gunso001 @maroon.tc.umn.edu).

[Haworth co-indexing entry note]: "A Risk Management Perspective on Integrated Weed Management." Gunsolus, J. L., and D. D. Buhler. Co-published simultaneously in *Journal of Crop Production* (Food Products Press, an imprint of The Haworth Press, Inc.) Vol. 2, No. 1 (#3), 1999, pp. 167-187; and: *Expanding the Context of Weed Management* (ed: Douglas D. Buhler) Food Products Press, an imprint of The Haworth Press, Inc., 1999, pp. 167-187. Single or multiple copies of this article are available for a fee from The Haworth Document Delivery Service [1-800-342-9678, 9:00 a.m. - 5:00 p.m. (EST). E-mail address: getinfo@haworthpressinc.com].

© 1999 by The Haworth Press, Inc. All rights reserved.

## INTEGRATED WEED MANAGEMENT–AN OVERVIEW

Integrated weed management (IWM) has been defined as the integration of effective, environmentally safe, and sociologically acceptable control tactics that reduce weed interference below the economic injury level (Thill et al., 1991). Walker and Buchanan (1982) stress that IWM must have a broader focus than weed control alone and must be integrated with all other crop production practices that influence the ecosystem. Current issues such as the development of herbicide-resistant weeds confirm these authors' view that we need to broaden our perspective of IWM to include a greater understanding of how weed management tactics interact with the biological, ecological, and socioeconomic processes that drive the cropping system (Roush, Radosevich, and Maxwell, 1990). Powles et al. (1997) consider IWM as essential to preserving the diversity of efficacious herbicide modes of action. The challenge is to convince herbicide .manufacturers and users in a free-market society to consciously restrain herbicide use patterns (i.e., employ IWM) to minimize or delay the development of herbicide-resistant weeds (Powles et al., 1997).

When growing row crops such as corn (*Zea mays* L.) and soybean [*Glycine max* (L.) Merrill], many farmers combine several weed control tactics, such as tillage, planting time, crop rotation and row spacing, mechanical weed control, and rotation of herbicides with different modes of action. However, current trends in corn and soybean management practices indicate a movement away from inter-row cultivation, toward less rotation of herbicide modes of action, and toward a greater reliance on single-pass herbicide application programs (personal communications with Extension Weed Scientists in the Midwestern states of the USA). As weed scientists and other applied agronomists develop alternative weed management tactics and decision aids, they must identify and take under consideration the reasons why farmers are becoming increasingly reluctant to use various existing weed management tactics.

## RISK ASSESSMENT AND IWM

### Perception of Risk

Perhaps the key limiting factor to IWM adoption is the farmer's perception of risk. Olson and Eidman (1992), in developing a theoretical model of a corn/soybean and a corn/soybean/alfalfa rotation to describe the connection between a farmer's use of herbicides and government policy, found that neither a bonus of higher government payments nor a tax on herbicides could be expected to cause many farmers to switch from chemical to mechanical weed control. Although the study was limited by the number of farmer situations and available weed control tactics, the authors were able to draw two major conclusions:

1. Variability in returns can influence farmers to choose to use herbicides even though the expected return is greater for a non-herbicide cropping system.
2. Variability in yields without herbicides is great enough that farmers seek to avoid alternatives to herbicides even when the payments for not using herbicides or the additional costs of using herbicides (i.e., taxes) are obvious.

Based on this research, one may consider the main value of herbicides to a farmer to be their ability to reduce yield variability and, thus, income variability. Olson and Eidman (1992) suggest that future research on emerging weed management options should address reduction in the variability of returns (i.e., risk).

### Impact of Integrated Weed Management Strategies on Soybean Grain Yield, Economic Returns, Risk, and Time Constraints

A better understanding of a farmer's perspective of risk can be gained by analyzing the data collected in two IWM studies conducted in soybean (Buhler, Gunsolus, and Ralston 1992 and 1993). The following conventional IWM practices were used: herbicides at label and reduced rates applied by banding and broadcasting, rotary hoeing, and inter-row cultivation. The primary goal of these studies was to reduce herbicide inputs without sacrificing profitability.

#### Impact on Grain Yield

IWM studies conducted on fields infested with common lambsquarters (*Chenopodium album* L.), giant foxtail (*Setaria faberi* Herrm.), redroot pigweed (*Amaranthus retroflexus* L.), and Powell amaranth (*Amaranthus powellii* S. Wats.) focused on reducing the input of alachlor [2-chloro-N-(2,6-diethylphenyl)-N-(methoxymethyl)acetamide] and metribuzin [4-amino-6-(1,1-dimethylethyl)-3-(methylthio)-1,2,4-triazin-5(4H)-one] (Buhler, Gunsolus, and Ralston, 1992). The following IWM tactics were evaluated and all treatments were applied in factorial combination with 0, 1, and 2 inter-row cultivation passes:

a. Full label rate of alachlor (3.3 kg/ha) and metribuzin (0.6 kg/ha) broadcast over the soil surface (full-rate broadcast).
b. One-half of the labeled rate of alachlor and metribuzin broadcast over the soil surface (one-half rate broadcast).

   c. Full labeled rate of alachlor and metribuzin applied in a 0.38-m band over the soybean row (full-rate band).

   d. One-half of the labeled rate of alachlor and metribuzin applied in a 0.38-m band over the soybean row (one-half rate band).

   e. No herbicide application.

   f. One field pass with a rotary hoe.

   g. Two field passes with a rotary hoe.

Integrated weed management treatments that produced soybean grain yields equivalent to the weed-free control were consistent over the two years of the study and are shown in Table 1. One-half of the full labeled rates of herbicides combined with one or two inter-row cultivation passes resulted in a soybean yield equivalent to that obtained for the weed-free control. A 75% reduction from the full labeled rates of herbicides (i.e., the one-half rates of herbicides applied in a 0.38-m band over the crop row) combined with two inter-row cultivation passes also resulted in a soybean yield equivalent to that obtained for the weed-free control. Two field passes with a rotary hoe reduced weed densities and increased the effectiveness of subsequent cultivation passes; however, in 1 of 3 years high weed densities resulted in reduced

TABLE 1. Integrated weed management treatments for soil-applied grass and broadleaf herbicides and inter-row cultivation that resulted in 1989 and 1990 soybean grain yields equivalent to the weed-free treatment at Rosemount, MN (Adapted from Buhler, Gunsolus, and Ralston 1992).

| Herbicide + Cultivation | Grain Yield | |
| --- | --- | --- |
| Treatment | 1989 | 1990 |
| | - - - - - $Mg\ ha^{-1}$ - - - - - - - | |
| Full-rate broadcast + | | |
|   No cultivation | 1.97 | 2.60 |
|   One cultivation | 2.05 | 2.41 |
|   Two cultivations | 1.99 | 2.50 |
| One-half rate broadcast + | | |
|   One cultivation | 1.95 | 2.56 |
|   Two cultivations | 2.05 | 2.53 |
| Full-rate band + | | |
|   One cultivation | 2.07 | 2.68 |
|   Two cultivations | 2.17 | 2.62 |
| One-half rate band + | | |
|   Two cultivations | 1.94 | 2.43 |
| Weed-free control | 2.11* | 2.88* |

*Means equivalency was determined via Dunnett's test at P = 0.05.

soybean yields under the mechanical weed control system as compared to those for the herbicide-containing treatments (data not shown).

Buhler, Gunsolus, and Ralston (1993) also conducted IWM studies on fields infested with common cocklebur (*Xanthium strumarium* L.). These studies focused on reducing the input of bentazon [3-(1-methylethyl)-(1H)-2,1,3-benzothiadiazin-4(3H)-one-2,2-dioxide]. The following IWM tactics were evaluated, and with the exception of the two sequential postemergence treatments, all tactics were applied in factorial combination with 0, 1, and 2 inter-row cultivation passes:

   a. Full labeled rate of bentazon broadcast over 1- to 6-cm tall common cocklebur (full-rate broadcast).
   b. One-half of the labeled rate of bentazon broadcast over 1- to 6-cm tall common cocklebur (one-half rate broadcast).
   c. Full labeled rate of bentazon applied in a 0.38-m band over the soybean row (full-rate band).
   d. One-half of the labeled rate of bentazon applied in a 0.38-m band over the soybean row (one-half rate band).
   e. No herbicide application.
   f. Full labeled rate of bentazon broadcast over 1- to 6-cm tall common cocklebur followed by a second application at the full labeled rate to 5-8-cm tall common cocklebur that had escaped the first application (full rate broadcast–two pass).
   g. One-half of the labeled rate of bentazon broadcast over 1- to 6-cm tall common cocklebur followed by a second application at the one-half labeled rate to 5- to 8-cm tall common cocklebur that had escaped the first application (one-half rate broadcast–two pass).

Integrated weed management treatments that produced soybean grain yields equivalent to the weed-free control were not the same over the three years of the common cocklebur control study (Table 2). In 1989, below-normal precipitation prior to the first bentazon application caused a high percentage of the common cocklebur plants to emerge after bentazon treatment. In this situation, sequential bentazon treatments improved common cocklebur control and resulted in greater soybean yields than did combinations of bentazon and inter-row cultivation. In 1990 and 1991, common cocklebur emergence patterns were more uniform. In these years, the full labeled rate and the one-half labeled rate of bentazon applied broadcast with or without inter-row cultivation passes; the full labeled rate applied in a 0.38-m band and followed by one or two inter-row cultivation passes; and the one-half labeled rate applied in a 0.38-m band followed by two cultivation passes resulted in soybean grain yields that were equivalent to the weed-free control.

In these two studies, the addition of inter-row cultivation allowed the

TABLE 2. Integrated weed management treatments for postemergence common cocklebur herbicides and inter-row cultivation that resulted in 1989 and 1990-1991 soybean grain yields equivalent to the weed-free treatment at Rosemount, MN (Adapted from Buhler, Gunsolus, and Ralston 1993).

| Herbicide + Cultivation | Grain Yield | |
|---|---|---|
| Treatment | 1989 | 1990-91 |
| | - - - - $Mg\,ha^{-1}$ - - - - - - | |
| Full-rate broadcast–two pass | 2.50* | 2.46* |
| One-half rate broadcast–two pass | 2.28 | 2.50 |
| Full-rate broadcast + | | |
| No cultivation | ---** | 2.40 |
| One cultivation | --- | 2.43 |
| Two cultivations | --- | 2.40 |
| One-half rate broadcast + | | |
| No cultivation | --- | 2.41 |
| One cultivation | --- | 2.40 |
| Two cultivations | --- | 2.51 |
| Full-rate band + | | |
| One cultivation | --- | 2.27 |
| Two cultivations | --- | 2.39 |
| One-half rate band + | | |
| Two cultivations | --- | 2.23 |
| Weed-free control | 2.61 | 2.48 |

*Means equivalency was determined via Dunett's test at P = 0.05.
**Indicates yield less than weed-free treatment.

herbicide rate, the herbicide area of application, or both to be reduced without reducing soybean grain yield. The only exception to this finding was in the control of common cocklebur by the one-half labeled rate of bentazon applied broadcast in 1990 and 1991, in which no inter-row cultivation was necessary. These studies also demonstrate that high weed densities and variable weed emergence patterns can result in inconsistent weed control from year to year. Therefore, scouting fields and applying IWM tactics over time will help reduce the impact of biological and environmental variability on crop yield.

## Impact on Economic Returns and Risk

Lybecker, Schweizer, and King (1988) evaluated and ranked IWM systems based on a mean-variance economic analysis model that compared the mean adjusted gross returns to the standard errors for each IWM system. The adjusted gross return was defined as gross revenue less unique variable costs. Gross revenue was defined as crop yield multiplied by crop price, and unique variable costs were defined as those costs not common to all weed management systems. Young, Kwon, and Young (1994) used a similar approach in the analysis of twelve farming systems in the Palouse region of southeastern Washington. This model was applied to the data collected in the studies by Buhler, Gunsolus, and Ralston (1992 and 1993). The soybean price was set at $0.21/kg. The variable weed management costs (labor, depreciation, and fuel) in the studies included costs of herbicide and herbicide application, cultivation, and rotary hoeing (Table 3). Herbicide costs were obtained from a report by Durgan, Gunsolus and Becker (1994). Herbicide application and mechanical weed control costs were obtained from Lazarus et al. (1994).

The standard error of the mean adjusted gross return is a measurement of variability and represents risk in the mean-variance model. The higher the standard error for a particular IWM system, the higher the assumed risk. Lybecker, Schwiezer, and King (1988) point out that changes in crop prices affected both the adjusted gross return (profit) and its' standard error (risk) for the various IWM systems. An increase in crop price increases both the adjusted gross return and its' standard error. Similarly, a decrease in crop price decreases both the adjusted gross return and standard error. Therefore,

TABLE 3. Variable weed management costs for the soil-applied and postemergence herbicide studies conducted at Rosemount, MN (Herbicide costs from Durgan, Becker, and Gunsolus, 1994, and herbicide application and mechanical weed control costs from Lazarus et al., 1994).

| Treatment | Cost |
|---|---|
| | - - $ ha$^{-1}$ - - - |
| Cultivation | 11.12 |
| Rotary hoe | 7.04 |
| Herbicide application | 9.14 |
| Soil-applied herbicides | |
|     Full-rate broadcast | 90.72 |
|     One-half rate broadcast | 45.36 |
|     Full-rate band | 45.36 |
|     One-half rate band | 22.68 |
| Postemergence herbicides | |
|     Full-rate broadcast–two pass | 48.48 |
|     One-half rate broadcast–two pass | 27.28 |
|     Full-rate broadcast | 24.24 |
|     One-half rate broadcast | 13.64 |
|     Full-rate band | 12.12 |
|     One-half rate band | 6.82 |

the order of profitability and the risk associated with IWM systems depend upon the ratio of crop price to unique variable costs.

When comparing several IWM systems, an IWM system is termed risk efficient if there is no other system that has both a higher adjusted gross return (profit) and a lower standard error (risk) (Boehlje and Eidman, 1984). It is possible to have more than one IWM system that is risk efficient provided that the adjusted gross income (profit) and standard deviation (risk) of a particular system are both either higher or lower than any other system. This set of IWM systems is referred to as an efficiency frontier (Boehlje and Eidman, 1984). An efficiency frontier indicates the maximum expected income for any given level of risk (standard deviation), or alternatively, it shows the minimum level of risk for any level of expected income. Boehlje and Eidman (1984) note that risk neutral or risk averse decision-makers will prefer a system that is on the efficiency frontier to one that is below the frontier. Further, risk neutral decision-makers will consider only the expected income without regard to the level of risk. Olson and Eidman (1992) indicate that most farmers are risk averse.

The IWM systems in the soil-applied herbicide studies that were considered to be risk efficient are listed in Table 4. Standard deviations of adjusted gross returns were calculated over replications (four) and years (two). The main points to note in the soil-applied herbicide study are as follows:

a. Full herbicide rate broadcast treatments were not risk efficient. Although the yields of all of these treatments were statistically equivalent (p = 0.05) to the weed-free treatment (Table 1), the higher cost of these treatments reduced their adjusted gross returns.

b. The only treatments with yields that were statistically equivalent (p = 0.05) to the weed-free treatment (Table 1) and considered risk efficient were the labeled rates of the herbicides applied in a band over the soybean row followed by one and two inter-row cultivations. Note that sequential inter-row cultivation of either one or two passes was necessary to be risk efficient. This should be expected since banding of herbicides over the row implies the need for cultivation.

c. Treatments with yields statistically lower than the weed-free yield considered to be risk efficient included: two passes with a rotary hoe followed by two inter-row cultivation passes; one-half of the labeled herbicide rates broadcast without inter-row cultivation; and one-half of the labeled herbicide rates applied in a band over the soybean row with one inter-row cultivation. The cost cutting measures of reducing herbicide rates or banding combined with a reduction in inter-row cultivation costs contributed to the latter two treatments' high ranking of risk efficiency. The two pass rotary hoe followed by two inter-row cultivation passes also substantially reduced input costs.

d. It is difficult to make any significant conclusions about the relationship of the five IWM strategies to their corresponding standard deviations (risk). This is most likely due to the need for a greater sample size before this approach will adequately represent the true mean and standard deviations of the various IWM strategies (Collander, 1989).

The efficiency frontier for the integrated weed management treatments from the postemergence herbicide study is illustrated in Figure 1. Adjusted gross return standard deviations were plotted against adjusted gross returns. Standard deviations were calculated over replications (four) and years (three). The main points to note in the postemergence herbicide study are as follows:

a. Only the full and one-half rate-two pass sequential broadcast applications of bentazon were risk efficient treatments (Figure 1). The sequential broadcast applications prevailed because the nonuniform common cocklebur emergence patterns in 1989 resulted in only the sequential broadcast treatments producing grain yields equivalent to the weed-free treatment in that year (Table 2). Applying the postemergence weed control tactic sequentially over time helped reduce the impact that biological variation (weed emergence) had on yield.
b. Excluding the sequential applications, all broadcast bentazon applications (full or one-half of label rate) and number of cultivations were similar in economic efficiency (Figure 1).
c. Banded bentazon applications were as risk efficient as broadcast bentazon applications if they were followed by one or two inter-row cultivation passes.

A biological aspect of risk that must be considered when evaluating IWM systems is the number of weeds that survive a particular weed management strategy and their potential to produce seeds that return to the seed bank. Most farmers have a low tolerance for weed survival in their fields because of the potential impact of these plants' seeds in future years. As a result many weed economic thresholds are lowered to account for seed longevity and future impacts on the crop (Bauer and Mortensen, 1992; Cardina, Regnier, and Sparrow, 1995).

The number of weeds that survived the particular IWM treatments that produced grain yields equivalent to the weed-free treatment or that were considered risk efficient are listed in Tables 5 and 6. In the soil-applied herbicide study (Table 5), the IWM strategies that left the most weeds to produce seed were those that used one-half of the labeled rate of herbicide. The addition of inter-row cultivations to the one-half labeled rate treatments reduced the number of remaining weeds, but not always to the level remain-

TABLE 4. Risk efficient integrated weed management treatments from the soil-applied herbicide study conducted at Rosemount, MN from 1989-1990.

| Herbicide + Cultivation Treatment | Adjusted Gross Return | Standard Deviation of Adjusted Gross Return |
|---|---|---|
| | - - - - - - - $ ha$^{-1}$ - - - - - - - | |
| Full-rate band + | | |
| One cultivation | 438.19 | 81.26 |
| Two cultivations | 431.04 | 73.92 |
| Rotary hoe–two pass + | | |
| Two cultivations | 420.24 | 64.86 |
| One-half rate broadcast + | | |
| No cultivation | 395.85 | 51.55 |
| One-half rate band + | | |
| One cultivation | 395.57 | 31.03 |

ing when the full labeled rate of herbicide was used. Two passes with a rotary hoe followed by two inter-row cultivations left fewer remaining weeds than many of the one-half labeled rate herbicide treatments but more weeds than the full labeled rate herbicide treatments. In the postemergence herbicide study (Table 6), the sequential postemergence bentazon treatment that used the full labeled rate consistently resulted in fewer common cocklebur plants at the end of the growing season than did the one-half labeled rate sequential treatment.

*Impact on Timeliness*

In a review of the last 40 years of scientific research on cultural and mechanical weed control in corn and soybean, Gunsolus (1990) noted that crop producers make weed control decisions based on many factors, such as effectiveness of control, costs, risks associated with treatment failure, and time and labor constraints. Estimates of the time required to complete each of the IWM tactics used by Buhler, Gunsolus, and Ralston (1992 and 1993) on 40 ha of soybean were calculated as specified by Gunsolus (1990) and Mulder and Doll (1993). These estimates are given in Table 7. Time estimates to complete 40 ha for the IWM treatments with yields equivalent to the weed-free treatment or considered risk efficient in the soil-applied and postemergence herbicide studies are given in Tables 5 and 6. In general, the IWM treatments that used the greatest amounts of herbicides required the least amount of time in the field. Mechanical weed control or band herbicide application increased the amount of time spent in the field (Table 5). The IWM treatments in the postemergence herbicide study that were efficient

FIGURE 1. Efficiency frontier for the integrated weed management treatments from the postemergence herbicide common cocklebur study (Adapted from Buhler, Gunsolus, and Ralston 1993) conducted at Rosemount, MN in 1989-1991.

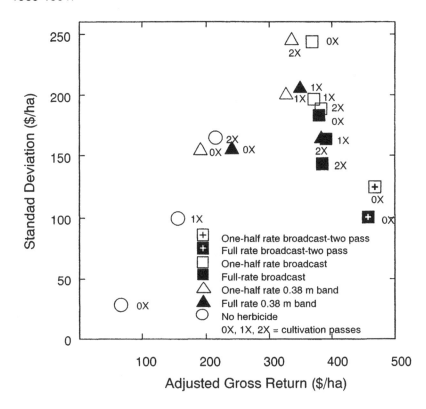

across all of the criteria (efficacy, cost, risk, and time management) were the sequential bentazon treatments at the full-labeled rate and at the one-half labeled rates (Table 6). In the soil-applied herbicide study only the full-labeled rate of herbicides banded over the crop row followed by one or two inter-row cultivation passes were efficient across all of these parameters (Table 5). Therefore, for soil-applied and postemergence herbicides, a 50% reduction in herbicide inputs was feasible from an efficacy, cost, risk, and time management perspective. However, this successful reduction was dependent upon weed species, density, and emergence patterns as well as on the effects of early-season environmental conditions on weed and crop growth and development and field working days. Mulder and Doll (1993) and Van-

TABLE 5. Weed densities 60 days after planting (DAP) and estimates of field work time from the soil-applied herbicides and inter-row cultivation treatments for which the grain yields were equivalent to the weed-free control or that were considered risk efficient, Rosemount, MN, 1989-1990.

| Herbicide + Cultivation Treatment | Weed Density[1] 60 DAP | Hours to Cover 40 ha |
|---|---|---|
| | no. m$^{-2}$ | |
| Soil-applied herbicides | | |
| Full-rate broadcast + | | |
| No cultivation[2] | 4 | 2.5 |
| One cultivation[2] | 12 | 16.5 |
| Two cultivations[2] | 7 | 26.5 |
| One-half rate broadcast + | | |
| No cultivation[3] | 74 | 2.5 |
| One cultivation[2] | 44 | 16.5 |
| Two cultivations[2] | 22 | 26.5 |
| Full rate band + | | |
| One cultivation[2,3] | 8 | 24 |
| Two cultivations[2,3] | 10 | 34 |
| One-half rate band + | | |
| One cultivation[3] | 49 | 24 |
| Two cultivations[2] | 14 | 34 |
| Rotary hoe–two pass + | | |
| Two cultivations[3] | 26 | 38 |

[1]Unweeded control averaged 160 weeds m$^{-2}$.
[2]Treatment grain yield equivalent to weed-free yield (Table 1).
[3]Treatment considered to be risk efficient (Table 4).

gessel et al. (1995) have determined that similar reductions in herbicide inputs in corn are possible if properly integrated with mechanical weed control practices.

## IMPACT OF FIELD WORKING DAYS ON IWM

The time available to complete field operations (field working days) has a significant impact on optimal machine size and field acreage planted, profitability, and risk (Apland, 1993). The number of available field working days can play a major role in determining which IWM strategy or strategies are optimal for a particular farming situation. The effect of field working days on the economics of cropping decisions is complicated by the significant variation that occurs in the number of field working days from season to season

TABLE 6. Weed densities remaining 60 days after planting (DAP) and estimates of field work time from the postemergence herbicide and inter-row cultivation treatments for which the grain yields were equivalent to the weed-free control or that were considered risk efficient, Rosemount, MN, 1989-1991.

| Herbicide Treatment | Weed Density[1] 60 DAP 1989 1990-91 | | Hours to Cover 40 ha |
|---|---|---|---|
| | - - no. $m^2$ - - | | |
| Full-rate broadcast–two pass[2,3] | 1 | 2 | 5 |
| One-half rate broadcast–two pass[2,3] | 25 | 3 | 5 |

[1]Unweeded control averaged 73 and 51 common cocklebur $m^{-2}$ in 1989 and 1990-91, respectively.
[2]Treatment grain yield equivalent to weed-free yield (Table 2).
[3]Treatment considered to be risk efficient (Figure 1).

TABLE 7. Estimates of field work time for various weed control treatments used in the integrated weed management soil-applied and postemergence herbicide studies at Rosemount, MN (Gunsolus, 1990 and Mulder and Doll, 1993).

| Control Treatments | Width | Speed | Hours to Cover 40 ha |
|---|---|---|---|
| | m | $m\ s^{-1}$ | |
| Broadcast herbicide | 18 | 2.7 | 2.5 |
| Banded herbicide | 4.5 | 2.7 | 10 |
| Inter-row cultivation | | | |
| First pass | 4.5 | 1.8 | 14 |
| Second pass | 4.5 | 2.7 | 10 |
| Rotary hoe | 4.5 | 3.6 | 7 |

and from year to year (Apland, 1993 and Seeley, 1995). If, for example, a farmer plans to substitute mechanical weed control practices for herbicides, the possible need to buy larger equipment or to put in longer hours (if this is possible) must be considered. From a farmer's perspective, the objective is to get the right-sized equipment to get the job done in the time the weather allows.

Apland (1993) addressed the impact of field working days on profit (net revenue) and risk (variance of the net revenue) using a representative Midwestern U.S. corn and soybean farm. Apland developed two linear programming models that treat field working days as fixed and random variables. Similar production activities and constraints existed in both models. The major difference between the two models is that a single field working day value is used to define resource availability in each intra-year time period (e.g., field preparation, planting, harvesting, etc.) in the deterministic model, but in the stochastic model intra-year time period decisions are made with only probabilistic knowledge of future field working days (i.e., more reflective of actual farming situations). Expected net revenue and standard deviation of net revenue for the deterministic and stochastic models were quite similar at farm sizes of 243, 283 and 324 ha. As acreage increased above 324 ha, expected net revenue increased relatively slowly in the stochastic model while its standard deviation increased quickly. In other words, as farm size increased, field working days had a greater impact on profit (a slower rate of increase in net revenue) and risk (greater variability in net revenue from year to year). The differences in net revenues and their standard deviations between these two models were attributed to differences in the timing of production activities. According to this research, an IWM decision-making process that did not account for field working days could easily overstate the contribution of additional crop land to expected net revenue and understate the amount of risk (variability) that would be assumed. The key relationship is that of timeliness in crop production operation (e.g., weed control) to field working days. Further research is warranted to determine practical means of accounting for field working days when making IWM decisions.

Seeley (1995) reported on a diary of field working days that had been kept by the staff at the University of Minnesota Southwest Experiment Station at Lamberton, MN from 1974 to 1992. Field working days encompassed days when the staff was actually working in the field for the entire day or a portion of the day as well as days when conditions were favorable for field work yet the staff had no particular reason to be working in the field. Figure 2 represents the actual and smoothed (3-day and 10-day) field working days as a probability distribution for the months of April, May, and June averaged over the period 1974 to 1992. Seeley (1995) noted the following observations:

1. Despite the variability evident in the daily probability values, the smoothed lines show a clear temporal trend of increasing field working day probability throughout the month of April and into the first week of May.
2. Negative deviations from this trend (a reduced probability) are noted around April 13 and 28, and May 14.
3. Peak field working day probabilities occurred around the dates of May 6 and May 11. These dates are very close and on either side of the historical average planting date of May 9 for corn in southwest Minnesota.

FIGURE 2. Actual and smoothed field working day probability–April to June–Southwest Experiment Station in Lamberton, MN.

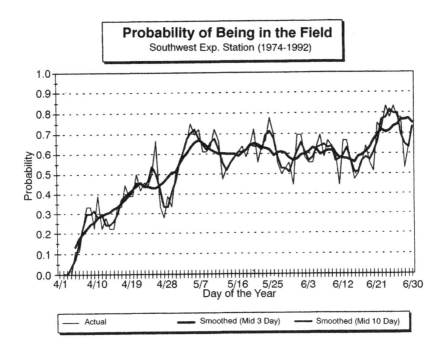

Seeley (1995) noted that previous research suggests that field working days appear to be closely associated with soil moisture conditions. There was an association between soil moisture and field working days recorded in the diary but the relationship was not as strong as expected. Seeley (1995) notes that several factors could account for this. Some field operations may have been performed regardless of the weather in order to meet certain criteria associated with experimental design. Errors in soil moisture measurements in conjunction with the spatial variability in soil moisture could also have produced unrepresentative values with respect to the experiment station at large. If the relationship between temporal patterns of daily soil moisture and field working days can be refined it may be possible to develop a greater historical data base of field working days by using existing soil moisture data bases or indirectly through rainfall data bases. This information would enhance the decision-making process in regard to the timeliness of an IWM practice in a cropping system.

## IMPACT OF TIMELINESS ON IWM

Timeliness of weed control is best expressed by the concept of the critical period of weed interference. Zimdahl (1988) identifies two key components of this critical period. The first is the period of time that weeds can remain in the crop before they interfere with crop yield. The second is the period of time over which weed control efforts must be maintained before the crop can effectively compete with weeds and prevent crop yield loss. Swanton and Weise (1991) discuss the ways in which this information can be used in an IWM program to reduce the need for long-term residual weed control by herbicides, to optimize the dose of postemergence herbicides and the timeliness of their application, and to influence the timing of cover crop seeding and mechanical cultivations. The focus of the critical period is on crop yield losses, not on the mechanisms or timing of weed-crop interference. Weaver (1984) and Zimdahl (1988) note that the critical period varies with the crop, its cultivar, the geographic location where the work was done, the season, the weed, and the density of the crop.

The variability inherent in the concept of the critical period may appear to negate its usefulness in IWM programs. In the context of this paper, variability does indicate the potential increase in risk of an IWM program. Berti et al. (1996) addressed this issue of variability by developing a model that accounts for varying times of weed emergence and permits the timing of weed control to minimize economic loss due to weeds that emerge both before and after weed control treatments have been carried out. They attempted to integrate aspects of weed-crop interference, weed biology, and economics to better answer the questions of when and how often to control weeds and to maximize net margins. The model is based on the concept of time density equivalent. The time density equivalent is defined as the density of weed plants that germinate with the crop and compete until harvest that causes the same yield loss caused by other groups of weeds with a given density, time of emergence, and time of removal. The concept is best illustrated in the context of application of a non-residual, broad-spectrum postemergence herbicide [e.g., glyphosate (N-(phosphonomethyl)glycine)]. Three groups of weeds must be considered in relation to a single postemergence application: those that emerge before the treatment and controlled by it, emerged plants that escaped the treatment, and those emerged after the treatment. The relative importance of these groups of weeds to crop yield depends upon the time of weed emergence and the time of postemergence herbicide application. The time density equivalent approach accounts for the impact of time of weed emergence and time of weed removal on crop yield.

Berti et al. (1996) propose that their model currently be used to identify areas in which more research is needed rather than as a decision-making tool; however, their modeling efforts have produced some results that are pertinent to IWM decision making. They found that with velvetleaf (*Abutilon theophrasti* Medicus) interference in corn and *Amaranthus cruentus* L. interference in soybean, the best strategy (based on maximum net margin) for both crops changed from two postemergence herbicide applications to one preemergence application followed by two postemergence herbicide applications as total weed density increased. The net margins obtained using the model were time dependent. Selection of the best control strategy based only on net margin could be misleading, as some strategies may be associated with a lower maximum net margin but with a flatter net margin curve, which allows more time for weed control. For example, in both corn and soybean, single postemergence herbicide applications were associated with a larger maximum net margin than were preemergence followed by postemergence treatments; however, with the latter sequential strategy, the period of time over which the postemergence herbicide could be applied was longer. This modeling approach has a significant impact on IWM in relation to field size and the variability of field working days. Berti et al. (1996) highlight the need to understand and predict weed emergence patterns if we hope to improve our decision-making capabilities and diversify our IWM strategies.

## DECISION AIDS AND IWM

The purpose of a weed management decision aid is to integrate a wide array of information to make better informed and profitable weed management decisions. Several groups have developed personal computer-based weed management decision aids to facilitate this process (Buhler et al., 1996; Forcella et al., 1996; Oriade et al., 1996). Buhler et al. (1996) grouped the current computer-based decision aids into the following three program types:

1. Herbicide efficacy models designed to aid in herbicide product selection.
2. Economic threshold models designed to aid in postemergence weed control decisions by quantifying emerged weed seedlings and determining economic loss due to existing weeds based on economic thresholds.
3. Bioeconomic models that integrate weed seed bank populations, predicted weed seedling emergence, and economic thresholds to predict weed seed bank dynamics over time and to utilize a range of herbicide application times and mechanical weed control operations.

Swinton and King (1994a) addressed the value of weed seed and emerged weed seedling populations in bioeconomic models from a risk management perspective. These authors found weed density to be of greatest value when weed populations were high with a greater importance on the decision of how to control, rather than whether to control. When weed populations are high, less than optimal weed control can have more serious repercussions than when weed populations are low (Swinton and King, 1994a). Swinton and King explain that when weed populations are low, an information-based strategy occasionally calls for no control, which can cause a sharp drop in net returns, an increase in yield variance, or both (i.e., risk). When weed populations are high, an information-based strategy is likely to recommend herbicides, thereby reducing the variance in net return. Swinton and King (1994a) and Forcella et al. (1996) reported economic advantages to the use of information-based weed management models. However, Buhler et al. (1996) did not find an information-based model to greatly alter economic returns. Buhler et al. (1996) concluded that the lack of economic benefit was due to high weed populations that negated any opportunity to reduce the frequency of herbicide use.

Swinton and King (1994b) have incorporated a bioeconomic model into a whole-farm context, where time, land, labor, and machinery resources can be limiting. This whole-farm model can incorporate the constraints of field working days as it relates to field preparation, soil-applied and postemergence weed control, and planting date. Individual cropping areas, equipment capacity, and labor needs can be accounted for. Biological parameters that the model simulates includes: yield penalties for delayed planting, crop and weed growth rates, and size thresholds, beyond which specific weed control treatments are ineffective. Recommendations from the whole-farm model have appeared plausible but as Buhler et al. (1996) indicate, further research under a wide range of production, weed population, and economic conditions is needed to define bioeconomic and whole-farm models utility to producers and crop consultants.

All of the models discussed have been successful to some degree; however, crop consultants and producers are reluctant to use them to any great extent because of the time required to gather the appropriate information and the time-sensitive nature of many weed control recommendations. In other words, the potential users of the bioeconomic models perceive that the costs (in time) required to gather the model inputs exceed the benefits derived from their use. Swinton and King (1994a) indicate in their assessment of a representative Minnesota corn and soybean farm that weed seedling counts are highly valuable to bioeconomic models but that weed seed counts are of relatively low value. Based on this analysis, it is unlikely that the cost of weed seed bank analysis would be cost-effective to crop consultants and producers.

Several researchers have explored the potential of decision-aids and alternative cropping systems to address the environmental impacts and costs associated with various weed control tactics (Teague, Mapp, and Bernardo, 1995; Forcella et al., 1996; Oriade et al., 1996). To date, the factor that limits this research is a lack of consensus as to what is the most appropriate method for computing the environmental impact of a pesticide. Teague, Mapp, and Bernardo (1995) evaluated the environmental risk and economic return associated with alternative production systems in western Oklahoma using three alternative environmental risk indices. In general, the environmental risk-cost frontiers suggested that environmental risk could be reduced without greatly reducing net returns. However, this relationship between risk and net return was dependent upon the environmental index applied. Therefore, a risk index must be developed before a thorough economic analysis of IWM strategies can begin.

## CONCLUSION

Two following questions formed the basis for this discussion on risk management as it relates to IWM: (1) In which direction(s) should we broaden our perspective of IWM? (2) How can farmers be convinced that it is in their best interests to diversify their weed management tactics? It is, perhaps, at the interface between the biology and economics of crop production where real progress toward the implementation of IWM practices can begin.

An analysis of the impact of different IWM strategies in soybean from a grain yield, economic return, risk, and time constraint perspective demonstrated that a 50% reduction in herbicide inputs was feasible. The discussion of field working days suggests that successful implementation of IWM strategies depends as much on the size of the farming operation and on time and labor constraints as it does on the efficacy of weed management tactics.

Further research is needed to evaluate weed management tactics and decision aids within the context of field working days. An important unanswered question is, given the time constraints of particular weed management tactics and the short time frame under which many decision aids would be of value to a farmer, what is the probability that the tactics and decision aids could be completed in a timely manner? Acceptance of new weed management tactics or decision-aids by the farmer hinges to a great extent on this timeliness factor.

The authors believe that approaching IWM from a risk management perspective is a logical and productive strategy that can broaden our understanding of IWM and provide information that farmers can use as they evaluate the advantages and disadvantages of diversifying their weed management tactics. Weed scientists must address risk management as we attempt to diversify our weed management tactics to a point at which truly sustainable IWM programs can be developed.

## REFERENCES

Apland, J. (1993). The use of field days in economic models of crop farms. *Journal of Production Agriculture* 6:437-444.

Bauer, T.A., and D.A. Mortensen. (1992). A comparison of economic and economic optimum thresholds for two annual weeds in soybeans. *Weed Technology* 6:228-235.

Berti, A., C. Dunan, M. Sattin, G. Zanin, and P. Westra. (1996). A new approach to determine when to control weeds. *Weed Science* 44:496-503.

Boehlje, M.D., and V.R. Eidman. (1984). Consideration of risk and uncertainty. In *Farm Management,* New York, NY: John Wiley & Sons, pp. 438-494.

Buhler, D.D., J.L. Gunsolus, and D.F. Ralston. (1992). Integrated weed management techniques to reduce herbicide inputs in soybean. *Agronomy Journal* 84:973-978.

Buhler, D.D., J.L. Gunsolus, and D.F. Ralston. (1993). Common cocklebur (*Xanthium strumarium*) control in soybean (*Glycine max*) with reduced bentazon rates and cultivation. *Weed Science* 41:447-453.

Buhler, D.D., R.P. King, S.W. Swinton, J.L. Gunsolus, and F. Forcella. (1996). Field evaluation of a bioeconomic model for weed management in corn (*Zea mays*). *Weed Science* 44:915-923.

Cardina, J., E. Regnier, and D. Sparrow. (1995). Velvetleaf (*Abutilon theophrasti*) competition and economic thresholds in conventional and no-tillage corn (*Zea mays*). *Weed Science* 43:81-87.

Collander, R.N. (1989). Estimation of risk in farm planning under uncertainty. *American Journal of Agricultural Economics* 71:996-1002.

Durgan, B.R., J.L. Gunsolus, and R.L. Becker. (1994). *Cultural and Chemical Weed Control in Field Crops.* BU-3157. University of Minnesota, St. Paul: Minnesota Extension Service.

Forcella, F., R.P. King, S.M. Swinton, D.D. Buhler, and J.L. Gunsolus. (1996). Multi-year validation of a decision aid for integrated weed management in row crops. *Weed Science* 44:650-661.

Gunsolus, J.L. (1990). Mechanical and cultural weed control in corn and soybeans. *American Journal of Alternative Agriculture* 5:114-119.

Lazarus, B., A. Brudelie, J. Christensen, E. Fuller, V. Richardson, D. Talley, E. Weness, and L. Westman. (1994). *Minnesota Farm Custom Rate Survey for 1994.* FO-3700. University of Minnesota, St. Paul: Minnesota Extension Service.

Lybecker, D.W., E.E. Schweizer, and R.P. King. (1988). Economic analysis of four weed management systems. *Weed Science* 36:846-849.

Mulder, T.A., and J.D. Doll. (1993). Integrating reduced herbicide use with mechanical weeding in corn (*Zea mays*). *Weed Technology* 7:382-389.

Olson, K.D., and V.R. Eidman. (1992). A farmer's choice of weed control method and the impacts of policy and risk. *Review of Agricultural Economics* 14:125-137.

Oriade, C.A., R.P. King, F. Forcella, and J.L. Gunsolus. (1996). A bioeconomic analysis of site-specific management for weed control. *Review of Agricultural Economics* 18:523-535.

Powles, S.B., C. Preston, I.B. Bryan, and A.R. Jutsum. (1997). Herbicide resistance: impact and management, ed. D.L. Sparks. In *Advances in Agronomy.* San Diego, CA: Academic Press, Vol. 58:57-93.

Roush, M.L., S.R. Radosevich, and B.D. Maxwell. (1990). Future outlook for herbicide-resistance research. *Weed Technology* 4:208-214.

Seeley, M. (1995). Some applications of temporal climate probabilities to site-specific management of agricultural systems. In *Site-Specific Management for Agricultural Systems,* eds. P.C. Robert, R.H. Rust, and W.E. Larson. Madison, WI: ASA-CSSA-SSSA, pp. 513-530.

Swanton, C.J., and S.F. Weise. (1991). Integrated weed management: the rationale and approach. *Weed Technology* 5:657-663.

Swinton, S.M., and R.P. King. (1994a). The value of pest information in a dynamic setting: the case of weed control. *American Journal of Agricultural Economics* 76:36-46.

Swinton, S.M., and R.P. King. (1994b). A bioeconomic model for weed management in corn and soybean. *Agricultural Systems* 44:313-335.

Teague, M.L., H.P. Mapp, and D.J. Bernardo. (1995). Risk indices for economic and water quality tradeoffs: an application to great plains agriculture. *Journal of Production Agriculture* 8:405-415.

Thill, D.C., J.M. Lish, R.H. Callihan, and E.D. Bechinski. (1991). Integrated weed management–A component of integrated pest management: a critical review. *Weed Technology* 5:648-656.

Vangessel, M.J., E.E. Schweizer, D.W. Lybecker, and P. Westra. (1995). Compatibility and efficiency of in-row cultivation for weed management in corn (*Zea mays*). *Weed Technology* 9:754-760.

Walker, R.H. and G.A. Buchanan. (1982). Crop manipulation in integrated weed management systems. *Weed Science (Suppl.)* 30:17-24.

Weaver, S.E. (1984). Critical period of weed competition in three vegetable crops in relation to management practices. *Weed Research* 24:317-325.

Young, D.L., T.J. Kwon, and F.L. Young. (1994). Profit and risk for integrated conservation farming systems in the Palouse. *Journal of Soil and Water Conservation* 49:601-606.

Zimdahl, R.L. (1988). The concept and application of the critical weed-free period. In *Weed Management in Agroecosystems: Ecological Approaches*, eds. M.A. Altieri and M. Liebman. Boca Raton, FL: CRC Press, pp. 145-155.

RECEIVED: 09/22/97
ACCEPTED: 04/30/98

# Maximizing Efficacy and Economics of Mechanical Weed Control in Row Crops Through Forecasts of Weed Emergence

## Caleb Oriade
## Frank Forcella

**SUMMARY.** In row crops of the North American Corn Belt two important forms of postplant mechanical weed control are rotary hoeing and inter-row cultivation. Unfortunately, the efficacies of these two control technologies are variable, which leads to high levels of economic risk. We hypothesized that efficacies and profitability of rotary hoeing and inter-row cultivation would increase, and risk would decrease, if the timing of control was based more on weed emergence times, than on rule-of-thumb calendar dates. Field research was conducted in soybean (*Glycine max* [L.] Merr.) for two years in Minnesota wherein four dates of rotary hoeing and three dates of inter-row cultivation, alone or supplemented by grass or broadleaf herbicides, were examined for weed control, crop yield, and net returns. Results indicate that timing influences the efficacy of mechanical control operations, but blanket optimal calendar windows that are generally applicable cannot be es-

Caleb Oriade, Department of Agricultural Economics, University of Arkansas, Fayetteville, AR 72701. Frank Forcella, U.S. Department of Agriculture, Agricultural Research Service, North Central Soil Conservation Research Laboratory, Morris, MN 56267.

Address correspondence to: Frank Forcella, U.S. Department of Agriculture, Agricultural Research Service, North Central Soil Conservation Research Laboratory, Morris, MN 56267 (E-mail: fforcella@mail.mrsars.usda.gov).

[Haworth co-indexing entry note]: "Maximizing Efficacy and Economics of Mechanical Weed Control in Row Crops Through Forecasts of Weed Emergence." Oriade, Caleb, and Frank Forcella. Co-published simultaneously in *Journal of Crop Production* (Food Products Press, an imprint of The Haworth Press, Inc.) Vol. 2, No. 1 (#3), 1999, pp. 189-205; and: *Expanding the Context of Weed Management* (ed: Douglas D. Buhler) Food Products Press, an imprint of The Haworth Press, Inc., 1999, pp. 189-205. Single or multiple copies of this article are available for a fee from The Haworth Document Delivery Service [1-800-342-9678, 9:00 a.m. - 5:00 p.m. (EST). E-mail address: getinfo@haworthpressinc.com].

© 1999 by The Haworth Press, Inc. All rights reserved.

tablished, as such decisions may be location-specific and/or time-dependent. In contrast, efficacies appear more consistent if emergence percentages are used to decide the time of mechanical operations, e.g., rotary hoe at 30% and cultivate at 60% green foxtail *(Setaria viridis* [L.] Beauv.) emergence. The results also suggest that while it is possible for exclusive mechanical weed control to be optimal in some instances, consistently profitable weed control strategies will inevitably involve some herbicide usage. *[Article copies available for a fee from The Haworth Document Delivery Service: 1-800-342-9678. E-mail address: getinfo@ haworthpressinc.com]*

**KEYWORDS.** Economic returns, interrow cultivation, rotary hoeing

## INTRODUCTION

The resurgence of interest in alternative nonchemical or low input weed control strategies is stimulated by cost considerations and concerns about medical and environmental implications of pesticide use. Herbicides account for over two-thirds of pesticide use in U.S. agriculture (Gianessi and Puffer, 1991). Because the bulk of these herbicides is employed to control weeds in corn *(Zea mays* L.) and soybean fields, appropriate efforts toward reducing herbicide use should include the reevaluation of conventional weed control practices in these crops.

Mechanical weeding, which includes tillage and cultivation practices, is one of the few suitable weed management strategies that have the potential to curtail herbicide use while effectively controlling weeds in soybean (Gunsolus, 1990). Its efficacy in controlling weeds, with or without herbicides, has been demonstrated (Bowman, 1997; Buhler and Gunsolus, 1996; Buhler, Gunsolus, and Ralston, 1992; Fernholz, 1990; Lovely, Weber, and Staniforth, 1958; Mt. Pleasant, Burt, and Frisch, 1994; Mulder and Doll, 1993). However, one feature that tends to influence the effectiveness of mechanical weeding, when used either solely or partly to control weeds, is the timeliness in the performance of such operations. Lovely, Weber, and Staniforth (1958) state that timely rotary hoeing could eliminate over two-thirds of potential weed populations. For these authors timely hoeing refers to the sequence of rotary hoeing operations commencing after weed seed germination but before seedling emergence. In contrast, poorly timed and less effective rotary hoeing includes operations performed after emergence when weeds are between the 1- and 3-leaf stages of growth.

The standard guide for attaining high efficacy is to rotary hoe twice, initially at seven days after planting (DAP) and again at 14 DAP; or at 10

DAP if only one pass of the rotary hoe is possible (Lovely, Weber, and Staniforth, 1958; Gunsolus, 1990). Another standard guide is to rotary hoe when radicals of germinated seeds are visible in the top few centimeters of soil but before shoots emerge. These time frames or visual windows implicitly assume that weeds germinate simultaneously within and among species at some time, usually between 7 and 14 DAP, and emerge in the same manner after 14 DAP. Despite weed scientists' recognition that such assumptions often fail, these rules-of-thumb have allowed producers to easily schedule relatively effective rotary hoeing operations.

Emergence of summer annual weeds does not occur *en masse*. Instead, it occurs over a considerable time period in spring and early summer (Stoller and Wax, 1973). Cumulative seedling emergence, when plotted against time, often approximates a sigmoidal curve with aberrations. The aberrations are caused primarily by rate-limiting fluctuations of soil temperature and water potential (e.g., Forcella, 1993). Theoretically, a point or points should be reached along this sigmoidal curve at which a weed is most sensitive to varying levels of soil disturbance caused by implements such as rotary hoes, harrows, and inter-row cultivators. A hypothetical relationship can be envisioned of how the timing of rotary hoeing, for example, and weed emergence timing interact to affect weed control (Figure 1a). Here, relative weed density remains high and weed control is low if rotary hoeing occurs before about 20% or after about 40% emergence. Prior to 20% emergence, rotary hoeing might serve primarily to move ungerminated weed seeds from one soil location to another. After 40% emergence, emerged weed seedlings might be too large and well-established to be uprooted by a rotary hoe. However, rotary hoeing between 20 to 40% emergence results in low relative weed densities and weed control is maximized (Figure 1b). The relationship emergence percentage at rotary hoeing and eventual relative weed density is depicted as a negative parabola in Figure 1b, but the exact nature of this relationship is not yet known.

Recently, development of emergence models for 15 weed species (Reese and Forcella, 1997), important throughout the Corn Belt of North America, provide the means to predict the timing and magnitude of weed emergence. These models use estimates of daily soil temperature and water potential to generate emergence predictions. Soil temperature and water potential estimates at 5 cm depth are derived from user-supplied measurements of daily minimum and maximum air temperature and rainfall, and simple tillage, residue type, and soil type information.

Emergence models should be able to contribute the required forecasting capabilities for matching weed development with optimal timing of mechanical weed control, as has been demonstrated for seedling growth models and application timing of postemergence herbicides (Forcella and Banken, 1996). What needs to be demonstrated is that optimal development-time windows

FIGURE 1. (A) Hypothetical sequence of weed seedling emergence (heavy line) with aberrations caused by weather fluctuations, and relative weed density (light line) resulting from rotary hoeing at the dates depicted on the X-axis. (B) Possible relationship between late-season relative weed density when rotary hoeing and rotary hoeing implemented at various emergence percentages of the weed. The relationship depicted is parabolic, but could take any of several similar forms.

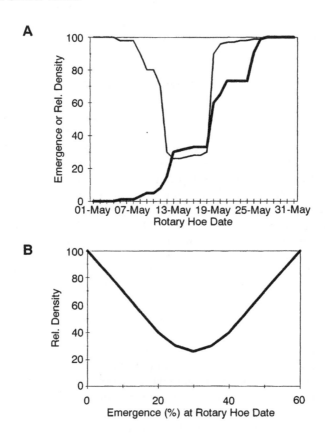

exist and can be reasonably predicted for mechanical weed control. Thus, the objectives of this study were to evaluate the implications of different post-plant tillage schedules (calendar dates and emergence times) on the efficacy and economics of mechanical weeding as a partial or total weed control tool in soybean.

## MATERIALS AND METHODS

### Agronomy

Data for the study were obtained from a mechanical weed control experiment conducted at the West Central Experiment Station, University of Minnesota, Morris, MN, in 1995 and 1996. The soil was an Aastad clay loam (Pachic Udic Haploboroll, fine-loamy, mixed, mesic). Seedbed preparation was accomplished through a combination of fall moldboard plowing followed by one spring cultivation prior to planting. Soybean was sown on May 24, 1995 and May 22, 1996 at a rate of about 300,000 seeds ha$^{-1}$ in rows spaced 76 cm apart. Corn was the previous crop in both years.

Experimental design was a randomized complete block design with a split-plot factorial arrangement of treatments in four replications. Main plot sizes of 9 by 30 m were established for four schedules of rotary hoeing, i.e., target times of 5, 10, 15 and 20 days after planting (DAP) in both years (and 12 DAP in 1995). Actual rotary hoeing dates varied somewhat because of weather conditions (Table 1). No rotary hoeing served as a weedy check in each year. The 4.6 m wide rotary hoe (John Deere 400[1]) consisted of two gangs of hoes offset from one another by 18 cm (front/back) and 9 cm laterally. Each hoe was equipped with 16 spoon-shaped tines whose 1.6 cm wide tips were spaced 10 cm from one another around the circumference of each hoe. The tractor-mounted implement was operated at about 25 km hr$^{-1}$.

Subplots were comprised of three dates of inter-row cultivation; target dates were 21, 28, and 35 DAP. A second cultivation occurred in all plots at about 45 DAP. The tractor-mounted cultivator (Buffalo Till Accuflex[1]) was 3.1 m wide and consisted of three series of devices. The first series contained two 33 cm diameter angled disks per inter-row, the second held a single 41 cm diameter coulter per inter-row, and the third accommodated a single 38 cm wide duckfoot sweep per inter-row. Use of this implement at about 5 km hr$^{-1}$ resulted in a 20 cm wide shielded band of relatively undisturbed soil centered on the crop row interspersed by 56 cm of soil highly fractured to a depth of about 10 cm.

Weed management subsubplots comprising three herbicide treatments were established within each subplot. Postemergence herbicide treatments in the subsubplots were sethoxydim at 0.17 kg ha$^{-1}$ to control grass weeds and allow broadleaf weeds to persist, bentazon at 1.12 kg ha$^{-1}$ to control broadleaf weeds and allow grasses to persist, and no herbicide to allow all weeds to persist. Recalcitrant perennial weeds and common cocklebur (*Xanthium strumarium* L.) were hand-hoed. Subsubplots with no herbicide treatment pro-

---

1. Trade names are for clarity of information and do not represent endorsement by USDA.

TABLE 1. Schedule of rotary hoe (R.H.) and inter-row cultivation (Cult.) operations for soybean crops sown on 24 May 1995 and 22 May 1996 in Morris, Minnesota, in terms of target date for days after planting (DAP); actual days after planting, calendar date; cumulative growing degree days (GDD, base 10 C); cumulative rainfall since sowing, and rainfall within 24 hours after the mechanical weeding operation; and forecasted emergence percentages of green foxtail (Grft), common lambsquarters (Colq), Pennsylvania smartweed (Pesw), and redroot pigweed (Rrpw).

| Year | Target Date | Actual Date | Calendar Date | Thermal Time | Rainfall Cum. | Rainfall 24 h | Emergence Grft | Colq | Pesw | Rrpw |
|---|---|---|---|---|---|---|---|---|---|---|
| | – – DAP – – | | day/Month | GDD | – – (mm) – – | | – – – – (%) – – – – | | | |
| **1995** | | | | | | | | | | |
| R.H. | 5 | 6 | 30 May | 32 | 17.8 | 0 | 2.0 | 2.4 | 3.8 | 1.1 |
| | 10 | 9 | 2 June | 60 | 17.8 | 0 | 6.3 | 10.6 | 14.4 | 5.7 |
| | 12 | 12 | 5 June | 98 | 17.8 | 3.3 | 17.1 | 15.8 | 14.0 | 9.1 |
| | 15 | 15 | 8 June | 128 | 33.5 | 12.4 | 28.7 | 24.9 | 16.8 | 15.7 |
| | 20 | 19 | 12 June | 147 | 58.9 | 0 | 38.0 | 38.4 | 36.7 | 26.7 |
| Cult. | 21 | 19 | 12 June | 147 | 58.9 | 0 | 38.0 | 38.4 | 36.7 | 26.7 |
| | 28 | 26 | 19 June | 252 | 58.9 | 0 | 73.6 | 59.8 | 56.7 | 55.7 |
| | 35 | 33 | 26 June | 351 | 66.0 | 5.3 | 90.6 | 72.0 | 56.7 | 72.3 |
| **1996** | | | | | | | | | | |
| R.H. | 5 | 4 | 26 May | 21 | 0.1 | 0.1 | 1.2 | 2.2 | 1.0 | 5.0 |
| | 10 | 8 | 30 May | 46 | 0.1 | 0 | 3.5 | 4.0 | 1.9 | 5.0 |
| | 15 | 15 | 6 June | 91 | 40.6 | 0.1 | 12.1 | 9.1 | 14.9 | 4.8 |
| | 20 | 20 | 11 June | 146 | 41.7 | 0 | 12.1 | 9.1 | 14.9 | 4.8 |
| Cult. | 21 | 23 | 14 June | 184 | 41.7 | 0.3 | 12.1 | 9.1 | 14.9 | 4.8 |
| | 28 | 28 | 19 June | 235 | 42.2 | 0.1 | 12.1 | 9.1 | 14.9 | 4.8 |
| | 35 | 34 | 25 June | 281 | 60.5 | 0 | 19.4 | 15.1 | 26.8 | 8.6 |

vided data on the effectiveness of sole reliance on mechanical weed control. The choice of herbicides and other management practices was influenced by farmers' preferences in the northwestern Corn Belt (Durgan et al., 1996).

Weed densities were determined at approximately 70 DAP and after all rotary hoeing and cultivation operations were concluded. Weeds were counted in six 0.1 m² quadrats in the central two crop rows in each subsubplot. Soybean yields in the same two rows were determined by combine-harvesting in late September.

Analysis of variance (ANOVA) tests were conducted in order to detect the statistical significance of various treatments. Significant year and treatment interactions occurred for rotary hoeing and herbicide treatments at the 5% level. Therefore, the data were analyzed and presented separately for these treatments in each year. Only the four rotary hoe schedules that were consistent between years were examined for yield effects (i.e., 12 DAP in 1995 excluded). Also, the means of soybean yields were pooled across the cultivation subplots as the effects of timing of first inter-row cultivation were not significant. For the treatments whose effects were significant as indicated by the overall F-statistics, mean separation was done using the Fisher's protected least significant difference (LSD) test at p = 0.05.

Daily values for precipitation and minimum and maximum air temperatures were accessed for the West Central Experiment Station (Anon., 1995, 1996). These data, beginning on the day of crop planting each year and for 60 days thereafter, were inserted into the *WeedCast* software (Reese and Forcella, 1997) to predict emergence timing of green foxtail. Percent seedling emergence at each date of mechanical weed control operation in 1995 and 1996 were recorded. Weed species other than green foxtail emerged at such low densities that *WeedCast* predictions were not merited. Analogous weather data (Anon., 1989-91) for the University of Minnesota Rosemount Experiment Station, Minnesota, for 1989, 1990, and 1991 were used in *WeedCast* to determine giant foxtail (*Setaria faberi* Herrm.) emergence percentages at each date of mechanical weed control operation cited by Buhler, Gunsolus, and Ralston (1992) and Buhler and Gunsolus (1996). Similarly, weather data for Aurora, New York, from 1992-94, were obtained from C.L. Mohler and were used in *WeedCast* to determine yellow foxtail (*Setaria glauca* [L.] Beauv.) emergence percentages at each date of inter-row cultivation reported by Mohler, Frisch, and Mt. Pleasant (1997).

## *Economics*

The enterprise budgeting technique was used to assess the economic implications of alternative rotary hoeing schedules and herbicide treatments for soybean production. The budgets, which delineate the structure of costs and returns associated with these practices, were generated with the aid of the Mississippi State Budget Generator (MSBG) (Spurlock and Laughlin, 1992). MSBG is a computer-based budgeting program that can produce the cost and returns for specified crop or livestock enterprises. The program is driven by the associated input quantities and costs, as well as output levels and prices specified by the user.

Net returns for soybean production that denote the excess of gross revenue over the estimated production costs were calculated for different weed management options. A moderate soybean price of $240 $Mg^{-1}$ ($6.50 per bushel) was applied to soybean yields to obtain gross revenue per hectare. This price closely approximated the average soybean price that prevailed in the

Midwest region in both years. The uniform average price was used, rather than the average seasonal price in each year, so that differences in returns can be solely attributed to the effects of alternative production systems.

Information on most fixed and variable costs were obtained from annual crop enterprise budgets produced by Fuller, Lazarus, and Carrigan (1994) and Olson et al. (1994). Herbicide costs were estimated from values provided by Durgan et al. (1996). Variable costs are direct expenses that are dependent on a particular production system. These expenses were estimated from average published costs for seed, fertilizer, pesticides, custom hire, repairs, maintenance, fuel, and other operating expenses. The farm machinery economic cost estimates by Lazarus (1996), which are based on the American Society of Agricultural Engineers economic engineering concept, were used to determine other land preparation, planting, tillage, crop maintenance, and harvesting costs. The fixed costs included depreciation, insurance, crop insurance, real estate and property taxes, and interest on capital invested in farm machinery. Total costs included both the fixed and variable costs. Total costs did not include charges for land and management.

Economic evaluation based solely on yields and associated profits has implicitly assumed that the outcome of the decision-making process is known with certainty. However, if mechanical weeding is expected to gradually replace herbicides as a preferred strategy for controlling weeds, growers' attitudes toward yield variability and fluctuations in production and economic environments becomes important considerations in evaluating alternative strategies. As a result, sensitivity analysis was conducted to identify preferred production strategies under varying levels of soybean prices and weed control costs.

## RESULTS AND DISCUSSION

### Weed Densities and Emergence

In the absence of a grass herbicide, green foxtail (with some yellow foxtail) densities were 47 plants $m^{-2}$ in 1995 and 194 plants $m^{-2}$ in 1996. Where a broadleaf herbicide was lacking, the 1995 and 1996 densities of common lambsquarters (*Chenopodium album* L.), Pennsylvania smartweed (*Polygonum pensylvanicum* L.), and redroot pigweed (*Amaranthus retroflexus* L.) were 7 and 14, 2 and 1, and 14 and 5 plants $m^{-2}$, respectively. Because of the low densities of broadleaf weeds in the study, the results for grasses (foxtail) will be emphasized in the following discussion of rotary hoe and cultivation schedules.

The times of rotary hoeing and inter-row cultivation (Table 1), as determined by DAP, were not related clearly to the relative density of foxtails in either 1995 or 1996 (Figure 2). One reason for the lack of association between weed density and DAP is apparent in the 1996 data. Lack of rainfall and dry soil conditions inhibited foxtail emergence at 12% development between 15 and 28 DAP during 1996, despite cumulative growing degree days increasing from 91 to 235 (Table 1). With so few seeds having germinated and emerged by these dates, mechanical control operations during that time period had only minor effects on relative weed density in 1996.

In contrast to DAP, times of rotary hoeing and cultivation, as determined by *WeedCast*-generated green foxtail emergence percentages, were related to

FIGURE 2. (A, B) Comparison of relationships between relative foxtail density after rotary hoeing was implemented at specified green foxtail emergence times (A) and days after planting (B). (C, D) Comparison of relationships between relative foxtail density after inter-row cultivation was implemented at specified foxtail emergence times (C) and days after planting (D). Results for green foxtail in western Minnesota are represented solid squares (1995) and open squares (1996); whereas the closed circles (C) represent results for yellow foxtail in New York (Adapted from Mohler, Frisch, and Mt. Pleasant, 1997).

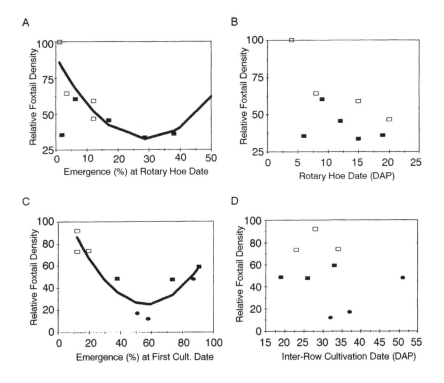

relative densities of foxtails (Figure 2). Relationships between relative foxtail densities and predicted emergence percentages at times of rotary hoeing and inter-row cultivation seemed parabolic, as hypothesized in Figure 1b, and conformed to simple quadratic relationships (Figure 2). The predicted emergence percentage at which rotary hoeing appeared to minimize relative weed densities (maximize weed control) was about 30% for rotary hoeing and 60% for first inter-row cultivation.

If these results were consistent across sites and years they would be encouraging for establishing a general biologically-based schedule for mechanical weed control operations in row crops plagued by foxtail weeds. In this regard, Figure 3 depicts relative density of giant foxtail as a function of *WeedCast*-generated giant foxtail emergence percentages at the time of the last rotary hoeing in row crops in eastern Minnesota (Buhler, Gunsolus, and Ralston, 1992; Buhler and Gunsolus, 1996). The results for giant foxtail

FIGURE 3. Relationship between relative giant foxtail density after rotary hoeing was implemented at specified giant foxtail emergence times at Rosemount, Minnesota. Data derived from density values cited in Buhler and Gunsolus (1996) for early (squares) and late (triangles) plantings in 1989, 1990, and 1991; and from Buhler, Gunsolus, and Ralston (1992) for a nearby site for 1989 and 1990 (circles). The outlying point at 2.6% emergence and 25 relative density is based on abnormally low weed densities, where the weedy check and rotary hoe treatments had only 8 and 2 giant foxtails m$^{-2}$, respectively.

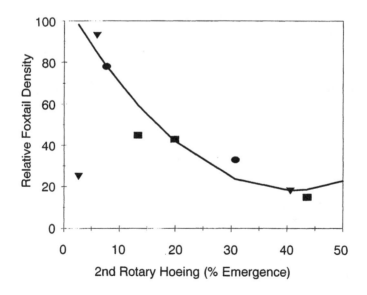

closely approximate those for green foxtail. Furthermore, the effects of inter-row cultivation timing on yellow foxtail densities in New York (Mohler, Frisch, and Mt. Pleasant, 1997) appear to follow the same relationship as that for green foxtail in western Minnesota (Figure 2c).

## Soybean Yields

High soybean yields (i.e., yields higher than one LSD unit above those of the "No Rotary Hoeing" treatment) in 1995 were associated with any rotary hoeing schedule when broadleaf weeds were controlled by bentazon (Table 2). Where no herbicide was used, or when sethoxydim was applied to control grasses, high yields tended to be associated with any but the latest rotary hoeing schedule. In contrast, during 1996, high soybean yields were coupled only with late rotary hoeing schedules regardless of herbicide treatments.

The differences in mean yields within years (Table 2) can be attributed to several factors: spectra of effectiveness of the two herbicides, and the composition and densities of the weed populations. However, the timing of emergence of the weed populations may be especially important. Emergence of weeds clearly progressed more rapidly during the first two weeks after planting in 1995 than in 1996 (Table 1). For example, by 15 DAP in 1995, the predicted emergence percentages were about 29, 25, 17, and 16% (green foxtail, common lambsquarters, Pennsylvania smartweed, and redroot pigweed, respectively), whereas at this same time in 1996 predicted emergence percentages ranged only from 4 to 16% (Table 1). Inhibition of germination and emergence in 1996 compared to 1995 probably occurred because of low soil temperatures and lack of rainfall during this period in 1996 (Table 1). Consequently, prior to 15 DAP too few weeds had germinated and emerged in 1996 for early rotary hoeing to be effective. In contrast, continuously moist and warm soils during the first 15 DAP in 1995 elicited immediate and rapid weed seedling emergence, and these seedlings were relatively susceptible to nearly any rotary hoeing operation between 5 and 20 DAP.

These results illustrate that fixed calendar schedules, such as DAP, for rotary hoeing may not be appropriate for maximizing soybean yields. Drought or cold soils may inhibit weed germination for considerable periods, rendering fixed schedules ineffective. Instead, soybean yields can be maximized, with or without herbicides, when the use of a rotary hoe is coupled with an understanding of the germination and emergence of the weeds that require control.

## Net Returns

When prices and costs were applied to grain yields in order to calculate net returns (Table 3), the results closely mimicked those for yields (Table 2). The

TABLE 2. Summary statistics of soybean yields associated with alternative weed control treatments. Within each herbicide treatment for each year, soybean yields significantly greater than the "No Rotary Hoeing" treatment are highlighted.

| Weed Control & Target Date | 1995 | | | | 1996 | | | |
|---|---|---|---|---|---|---|---|---|
| | Mean Yield | Std. Dev. | Max. Yield | Min. Yield | Mean Yield | Std. Dev. | Max. Yield | Min. Yield |
| | - - - - - - - - - - - - Mg ha$^{-1}$ - - - - - - - - - - - - | | | | | | | |
| A: Bentazon and: | | | | | | | | |
| No Rotary Hoeing | 1.99 | 0.79 | 2.91 | 0.84 | 2.09 | 0.22 | 2.44 | 1.79 |
| Rotary Hoeing 5 DAP | **2.59** | 0.41 | 3.28 | 2.03 | 2.11 | 0.33 | 2.48 | 1.34 |
| Rotary Hoeing 10 DAP | **2.26** | 0.66 | 3.15 | 1.20 | **2.29** | 0.21 | 2.64 | 1.94 |
| Rotary Hoeing 15 DAP | **2.61** | 0.52 | 3.40 | 1.66 | **2.51** | 0.34 | 2.93 | 1.78 |
| Rotary Hoeing 20 DAP | **2.42** | 0.26 | 2.86 | 2.06 | **2.47** | 0.34 | 3.16 | 1.92 |
| LSD (p = 0.05) | 0.22 | | | | 0.19 | | | |
| B. Sethoxydim and: | | | | | | | | |
| No Rotary Hoeing | 2.26 | 0.43 | 2.90 | 1.54 | 2.44 | 0.32 | 3.07 | 1.92 |
| Rotary Hoeing 5 DAP | **2.64** | 0.49 | 3.59 | 1.90 | 2.64 | 0.50 | 3.20 | 1.66 |
| Rotary Hoeing 10 DAP | 2.37 | 0.50 | 2.90 | 1.59 | 2.62 | 0.51 | 3.18 | 1.41 |
| Rotary Hoeing 15 DAP | **2.51** | 0.35 | 2.92 | 1.68 | **2.80** | 0.37 | 3.27 | 2.16 |
| Rotary Hoeing 20 DAP | 2.10 | 0.48 | 3.10 | 1.32 | **3.06** | 0.28 | 3.46 | 2.38 |
| LSD (p = 0.05) | 0.24 | | | | 0.32 | | | |
| C. No Herbicide with: | | | | | | | | |
| No Rotary Hoeing | 2.03 | 0.62 | 2.84 | 0.89 | 2.23 | 0.49 | 3.00 | 1.75 |
| Rotary Hoeing 5 DAP | **2.49** | 0.55 | 3.48 | 1.40 | 2.09 | 0.39 | 2.70 | 1.61 |
| Rotary Hoeing 10 DAP | **2.56** | 0.45 | 3.37 | 1.77 | 2.27 | 0.35 | 2.79 | 1.59 |
| Rotary Hoeing 15 DAP | 2.41 | 0.60 | 3.08 | 1.19 | **2.47** | 0.40 | 3.07 | 1.75 |
| Rotary Hoeing 20 DAP | 1.94 | 0.43 | 2.74 | 1.42 | **2.55** | 0.39 | 3.13 | 1.68 |
| LSD (p = 0.05) | 0.26 | | | | 0.23 | | | |
| Overall LSD (p = 0.05) | 0.25 | | | | 0.17 | | | |

DAP = days after planting; Std. Dev. = standard deviation; Max. = maximum, and Min. = minimum.

sole exception was that in 1996 only the very latest rotary hoeing operation provided net revenues significantly greater than those of the "No Rotary Hoeing" treatments in the absence of bentazon.

Mechanical weeding with rotary hoeing 10 DAP produced the highest net returns in terms of dollars per hectare in 1995. This was followed by the weed control strategy that produced highest mean yields (5 DAP with sethoxydim). This should not be surprising, as the yield difference of about 0.08 Mg ha$^{-1}$ could hardly pay for the additional herbicide material and application costs needed to achieve the level of weed control that would produce the higher yields. Therefore, evaluation of the potential of mechanical weeding based

TABLE 3. Cost and returns associated with alternative weed control treatments. Within each herbicide treatment for each year, revenues significantly greater than the "No Rotary Hoeing" treatment are highlighted.

| Weed Control & Target Date | 1995 | | | | 1996 | | | |
|---|---|---|---|---|---|---|---|---|
| | Gross Rev. | Var. Costs | Total Costs | Net. Rev. | Gross Rev. | Var. Costs | Total Costs | Net. Rev. |
| | | | | — ($ ha$^{-1}$) — | | | | |
| A: Bentazon and: | | | | | | | | |
| No Rotary Hoeing | 480 | 168 | 397 | 83 | 504 | 171 | 405 | 99 |
| Rotary Hoeing 5 DAP | **623** | 179 | 408 | **215** | 507 | 183 | 416 | 91 |
| Rotary Hoeing 10 DAP | **546** | 179 | 408 | **137** | **551** | 183 | 416 | 135 |
| Rotary Hoeing 15 DAP | **628** | 179 | 408 | **219** | **604** | 183 | 416 | **187** |
| Rotary Hoeing 20 DAP | **584** | 179 | 408 | **175** | **595** | 183 | 416 | **179** |
| LSD (p = 0.05) | 53 | | | | 46 | | | |
| B. Sethoxydim and: | | | | | | | | |
| No Rotary Hoeing | 544 | 153 | 382 | 162 | 588 | 156 | 390 | 198 |
| Rotary Hoeing 5 DAP | **635** | 164 | 393 | **242** | 636 | 168 | 401 | 234 |
| Rotary Hoeing 10 DAP | 572 | 164 | 393 | 179 | 632 | 168 | 401 | 231 |
| Rotary Hoeing 15 DAP | **605** | 164 | 393 | **212** | **675** | 168 | 401 | 273 |
| Rotary Hoeing 20 DAP | 506 | 164 | 393 | 113 | **736** | 168 | 401 | **335** |
| LSD (p = 0.05) | 58 | | | | 77 | | | |
| C. No Herbicide with: | | | | | | | | |
| No Rotary Hoeing | 489 | 125 | 354 | 134 | 538 | 128 | 362 | 176 |
| Rotary Hoeing 5 DAP | **601** | 137 | 366 | **235** | 504 | 139 | 373 | 131 |
| Rotary Hoeing 10 DAP | **616** | 137 | 366 | **251** | 548 | 139 | 373 | 175 |
| Rotary Hoeing 15 DAP | **581** | 137 | 366 | **215** | **594** | 139 | 373 | 221 |
| Rotary Hoeing 20 DAP | 469 | 137 | 366 | 103 | **614** | 139 | 373 | **241** |
| LSD (p = 0.05) | 62 | | | | 55 | | | |
| Overall LSD (p = 0.05) | 60 | | | | 41 | | | |

DAP = days after planting; Gross Rev. = Gross revenue which is the product of soybean yield and soybean price at $240 Mg$^{-1}$ ($6.50 per bushel); Var. Cost = Variable cost of production; Tot. cost = Total cost of production which is the sum of variable and fixed costs; Net Rev. = net revenue or net return which is the residual of gross revenue over total cost.

purely on crop yield considerations might sometimes be biased in favor of herbicide use as the higher yields of such strategy could mask the potential cost savings of mechanical weed control.

In 1996, the order of crop yields was preserved when translated to net returns, with the same weed control strategy of grass herbicide and rotary hoeing 20 DAP posting both the highest crop yields and net returns (Table 3). In this case, the value of its yield advantage of over 0.47 Mg ha$^{-1}$ when compared to the best yields obtained without herbicide use would easily surpass any herbicide cost. The consistency of high yields and high net returns associated with sethoxydim application in 1996 compared to 1995 probably reflects the fact that foxtail densities were 4-fold greater in 1996 than in 1995. Thus, even though the latest rotary hoeing (20 DAP) in 1996 was effective in reducing relative foxtail densities (Figure 2), absolute densities were still high enough to reduce soybean yield and net returns (Table 3). In both years, forgoing weed control completely was not advisable as the value of yield reductions in weedy check plots ("No Rotary Hoeing" treatments) was between 15 and 20%, which exceeded the potential costs of weed control.

### Sensitivity Analysis

A careful inspection of the standard deviations, and maximum and minimum yields in Table 2 indicate that these measures of central dispersion are as important as the mean values for yield in evaluating alternative weed control strategies. Most growers tend to prefer strategies that increase profit with minimal variability of such returns. However, none of the promising strategies in Table 2 possesses the dual attributes of having the highest mean yields coupled with minimum variance in its treatment class. One way of identifying suitable weed control strategies in such instances is to consider the sensitivity of various alternatives to fluctuations in economic and/or production environments. The 1995 results where control strategies with highest yields were not necessarily the most profitable imply that optimal weed control practices could be sensitive to changes in output prices and/or control costs.

The sensitivity analyses were conducted under two soybean price regimes: a low of \$203 and a high of \$295 $Mg^{-1}$ (\$5.50 and \$8.00 per bushel). These price levels also were complemented with two changes to weed control costs assuming that the overall effectiveness of the herbicides remained the same. The cost changes were a 50% decrease in weed control costs (e.g., by switching to a lower cost herbicide such as quizalofop) and a 50% increase in weed control costs (e.g., by switching to imazethapyr). The outcome of the sensitivity analysis is presented in Table 4. In 1995 when the best net return was obtained from herbicide-free treatments, results in Table 4 show that such low input weed control strategy will remain optimal under low soybean prices and/or high weed control costs. Although herbicide-based weed control strategies were favored in economic environments characterized by high soybean prices and/or low weed control costs, the herbicide-free rotary hoeing at any but the latest schedule consistently attained net revenues that were significantly above that of the check treatment and equivalent to or greater than the herbicide-based treatments.

In 1996 sethoxydim-based weed control provided the best returns, and this

TABLE 4. Sensitivity of net returns to changing soybean prices and weed control costs. Values within 5% of the maximum value within a column are highlighted in bold. Within each herbicide treatment for each year, net returns are highlighted if they are greater than the "No Rotary Hoeing" treatment by more than the monetary value of one LSD unit for yield (Table 2).

| | 1995 | | | | 1996 | | | |
|---|---|---|---|---|---|---|---|---|
| Weed Control & Target Date | Net Rev1 P↑C↑ | Net Rev2 P↑C↓ | Net Rev3 P↓C↑ | Net Rev4 P↓C↓ | Net Rev1 P↑C↑ | Net Rev2 P↑C↓ | Net Rev3 P↓C↑ | Net Rev4 P↓C↓ |
| | | | | - - - ($ ha⁻¹) - - - | | | | |
| A: Bentazon and: | | | | | | | | |
| No Rotary Hoeing | 173 | 216 | −11 | 31 | 195 | 238 | 1 | 44 |
| Rotary Hoeing 5 DAP | **333** | **387** | **93** | **147** | 182 | 236 | −13 | 41 |
| Rotary Hoeing 10 DAP | **238** | **291** | 28 | **81** | 236 | 291 | 24 | 79 |
| Rotary Hoeing 15 DAP | **339** | **393** | **97** | **151** | **301** | **355** | 68 | 123 |
| Rotary Hoeing 20 DAP | **285** | **338** | **60** | **114** | 290 | 345 | 61 | 116 |
| B. Sethoxydim and: | | | | | | | | |
| No Rotary Hoeing | 275 | 303 | 66 | 93 | 321 | 349 | 95 | 173 |
| Rotary Hoeing 5 DAP | **371** | **409** | **126** | **165** | 363 | 402 | 118 | 158 |
| Rotary Hoeing 10 DAP | 293 | 331 | 73 | 111 | 359 | 398 | 115 | 155 |
| Rotary Hoeing 15 DAP | 333 | 372 | 101 | 139 | 411 | **450** | 151 | 191 |
| Rotary Hoeing 20 DAP | 212 | 250 | 17 | 56 | **487** | **526** | **203** | **243** |
| C. No Herbicide with: | | | | | | | | |
| No Rotary Hoeing | 248 | 248 | 60 | 60 | 301 | 301 | 95 | 95 |
| Rotary Hoeing 5 DAP | **369** | **381** | **138** | **149** | 243 | 254 | 49 | 60 |
| Rotary Hoeing 10 DAP | **389** | **400** | **152** | **163** | 297 | 308 | 86 | 98 |
| Rotary Hoeing 15 DAP | **345** | **356** | **122** | **133** | 354 | 365 | 125 | 137 |
| Rotary Hoeing 20 DAP | 207 | 218 | 27 | 38 | **378** | **390** | **142** | **153** |

DAP = days after planting; Net Rev1 = net revenue or net return when the soybean price was set at $295 Mg⁻¹ ($8 per bushel) and the weed control cost was increased by 50% relative to the cost used in Table 2; Net Rev2 = net revenue or net return when the soybean price was set at $295 Mg⁻¹ and the weed control cost was reduced by 50%; Net Rev3 = net revenue or net return when the soybean price was set at $203 Mg⁻¹ ($5.50 per bushel) and the weed control cost was increased by 50%; Net Rev4 = net revenue or net return when the soybean price was set at $203 Mg⁻¹ and the weed control cost was reduced by 50%. (P= crop price, C = control cost, ↑= increase, ↓ = decrease.)

strategy was not sensitive to modest fluctuations in economic climate (Table 4). Again, this probably was due to the 4-fold greater foxtail densities in 1996 than 1995. Nevertheless, a late rotary hoeing in 1996 significantly enhanced net revenues across economic climates, and somewhat more so when prices were high and costs were low. These results indicate that weed control strategies involving some chemical usage, targeted at the predominant weed species, offer the best means of obtaining satisfactory returns under adverse economic and agronomic (weedy) environments. The results also suggest that a late rotary hoeing can increase net revenues substantially when the environment of the weed (e.g., microclimate) is adverse, which delays germination, and thereby enhances the efficacy of similarly delayed rotary hoeing.

## CONCLUSIONS

With the advent of computer-based models that can facilitate prediction of weed emergence, growth, and seedling height in real-time, effective weed management can be further enhanced, but only if these predictions are coupled with readily available information on the timing and mix of optimal weed control treatments. Results from the current study establish some link between the timing and efficacy of mechanical weeding for postplant weed control in soybeans. Unfortunately, no blanket optimal time windows (calendar dates) for conducting postplant rotary hoeing or inter-row cultivation can be announced, as these seem to depend upon a number of factors. However, use of accurately forecasted weed emergence values may alleviate this problem, at least partially, in that they appear to provide a consistent and biologically-based window for performing postplant mechanical weeding. The current analyses also show that while it is possible to obtain short term benefits from exclusive mechanical weed control, it is doubtful if such a practice would be continuously profitable due to the carryover effects of uncontrolled weed seed populations and to sensitivity to modest fluctuations in economic and/or production environments.

## REFERENCES

Anonymous. (1989-96). *Climatological Data: Minnesota.* National Oceanic and Atmospheric Administration, Asheville, North Carolina.

Bowman, G. (1997) *Steel in the Field: A Farmer's Guide to Weed Management Tools.* Sustainable Agriculture Network, Nat. Agric. Lib., Beltsville, MD. 128 p.

Buhler, D.D. and J.L. Gunsolus. (1996). Effect of date of preplant tillage and planting on weed populations and mechanical weed control in soybean (*Glycine max*). *Weed Science* 44:373-379.

Buhler, D.D., J.L. Gunsolus, and D.R. Ralston. (1992). Integrated weed management techniques to reduce herbicide inputs in soybean. *Agronomy Journal* 84:973-978.

Durgan, B.R., J.L. Gunsolus, R.L. Becker, and A.G. Dexter. (1996). *Cultural and Chemical Weed Control in Field Crops-1996.* BU-3157-F, Minnesota Extension Service, University of Minnesota, St. Paul.

Fernholz, C. (1990). How I control weeds without herbicides. *The New Farm.* March/April: 17-20.

Forcella, F. (1993). Seedling emergence model for velvetleaf. *Agronomy Journal* 85: 923-933.

Forcella, F. and K.R. Banken. (1996). Relationships among green foxtail (*Setaria viridis*) seedling development, growing degree days, and time of nicosulfuron application. *Weed Technology* 10:60-67.

Fuller, E., B. Lazarus, and L. Carrigan. (1994). *What to Grow in 1994: Crop Budget for Soil Area 4*. Minnesota Extension Service, University of Minnesota, St. Paul.

Gianessi, L.P. and C. Puffer. (1991). *Herbicide Use in the United States*. Quality of Environment Division, Resources for the Future, Washington, DC.

Gunsolus, J.L. (1990). Mechanical and cultural weed control in corn and soybean. *American Journal of Alternative Agriculture* 5:114-119.

Hartzler, R.G., B.D. Van Kooten, D.E. Stoltenberg, E.M. Hall, and R.S. Fawcett (1993). On-farm evaluation of mechanical and chemical weed management practices in corn (*Zea mays*). *Weed Technology* 7:1001-1004.

Lazarus, B. (1996). *Farm Machinery Economic Costs for 1996: Minnesota Estimates with Expanded Tillage Data for the U.S. and Canada*. Staff Paper P96-15, Department of Applied Economics, University of Minnesota, St. Paul.

Lovely, W.G., C.R. Weber, and D.W. Staniforth. (1958). Effectiveness of the rotary hoe for weed control in soybeans. *Agronomy Journal* 50:621-625.

Mohler, C.L., J.C. Frisch, and J. Mt. Pleasant. (1997). Evaluation of mechanical weed management programs for corn (*Zea mays*). *Weed Technology* 11:123-131.

Mt. Pleasant J., F.R. Burt, and J.C. Frisch (1994). Integrating mechanical and chemical weed management in corn (*Zea mays*). *Weed Technology* 8(2):217-223.

Mulder, T.A. and J.D. Doll. (1993). Integrating reduced herbicide use with mechanical weeding in corn (*Zea mays*). *Weed Technology* 7:382-389.

Olson, K.D., D.W. Nordquist, D.E. Talley, J. Christensen, E.J. Weness, and P.A. Fales. (1994). 1993 *Annual Report of the Southwestern Minnesota Farm Business Management Association*. Staff Paper P94-15, Department of Agricultural and Applied Economics, University of Minnesota, St. Paul.

Reese, C.R. and F. Forcella. (1997). *WeedCast: Forecasting Weed Emergence and Growth in Crop Environments*. <http://www.mrsars.usda.gov>.

Spurlock, S. R. and D.H. Laughlin. (1992). *Mississippi State Budget Generator User's Guide Version 3.0*. Mississippi Agricultural and Forestry Experiment Station. Agricultural Economics Technical Publication No. XX.

Stoller, E.W. and L.M. Wax. (1973). Periodicity of germination and emergence of some annual weeds. *Weed Science* 21:574-580.

Teasdale, J.R., E.C. Beste, and W.E. Potts. (1991). Response of weeds to tillage and cover crop residue. *Weed Science* 39:195-199.

Vencill, W.K. and P.A. Banks. (1994). Effects of tillage systems and weed management on weed populations in grain sorghum (*Sorghum bicolor*). *Weed Science* 42:541-547.

RECEIVED: 05/19/97
ACCEPTED: 12/12/97

# Multi-Year Evaluation
# of Model-Based Weed Control
# Under Variable Crop and Tillage Conditions

Melinda L. Hoffman
Douglas D. Buhler
Micheal D. K. Owen

**SUMMARY.** Selecting effective weed management options requires biological, ecological, and economic information. This study compared model-based to standard-herbicide weed control in a corn (*Zea mays* L.)/soybean [*Glycine max* (L.) Merr.] rotation that had a long-term history of different tillage and weed management practices. The model integrates weed population dynamics, herbicide efficacies, and economic information to evaluate preplant incorporated (PPI) or preemergence (PRE) weed control options based on weed seed bank size and postemergence (POST) strategies based on weed seedling densities. There were fewer weeds in standard-herbicide compared with model-based treatments. No-tillage had the greatest numbers of weeds 3 out of 4 years. Soybean yield was reduced the first year of the study in several treatments receiving model-based weed control and the third year in all model-based treatments. Corn yields were greatest in reduced tillage. Re-

Melinda L. Hoffman, Research Associate, Douglas D. Buhler, Research Agronomist, United States Department of Agriculture–Agricultural Research Service, National Soil Tilth Laboratory, 2150 Pammel Drive, Ames, IA 50011 USA. Micheal D. K. Owen, Professor, Department of Agronomy, Iowa State University, Ames, IA 50011 USA.

Address correspondence to: Melinda L. Hoffman, United States Department of Agriculture–Agricultural Research Service, National Soil Tilth Laboratory, 2150 Pammel Drive, Ames, IA 50011 USA (E-mail: hoffman@nstl.gov).

[Haworth co-indexing entry note]: "Multi-Year Evaluation of Model-Based Weed Control Under Variable Crop and Tillage Conditions." Hoffman, Melinda L., Douglas D. Buhler, and Michael D. K. Owen. Co-published simultaneously in *Journal of Crop Production* (Food Products Press, an imprint of The Haworth Press, Inc.) Vol. 2, No. 1 (#3), 1999, pp. 207-224; and: *Expanding the Context of Weed Management* (ed: Douglas D. Buhler) Food Products Press, an imprint of The Haworth Press, Inc., 1999, pp. 207-224. Single or multiple copies of this article are available for a fee from The Haworth Document Delivery Service [1-800-342-9678, 9:00 a.m. - 5:00 p.m. (EST). E-mail address: getinfo@haworthpressinc.com].

© 1999 by The Haworth Press, Inc. All rights reserved.

sults of using model recommendations to control weeds were mixed, with PRE recommendations being insensitive to a common cocklebur (*Xanthium strumarium* L.) infestation. Our conclusions agree with those of others that the nature of the weed pressure may be a prevailing influence on the outcome of using weed control recommendations of bioeconomic models. *[Article copies available for a fee from The Haworth Document Delivery Service: 1-800-342-9678. E-mail address: getinfo@haworthpressinc.com]*

**KEYWORDS.** Bioeconomic model, corn/soybean rotation, expert systems, weed population dynamics, weed seed banks

**ABBREVIATIONS.** PPI, preplant incorporated; PRE, preemergent; POST, postemergent; WAP, weeks after planting

## INTRODUCTION

There are numerous management options from which producers can choose to control weeds in corn and soybean. The variety of approaches listed by the Iowa Cooperative Extension Service includes 17 PRE and 24 POST herbicide treatments for corn, 20 PRE and 20 POST herbicide treatments for soybean, tillage, rotary hoeing, and interrow cultivation (Owen and Hartzler, 1996). Also facing producers are decreased profit margins and increased environmental accountability. Therefore, an inappropriate weed management decision leading to excessive herbicide use could have detrimental economic and environmental consequences. To select among the many weed control options available requires information on many subjects including weed biology, yield potential, treatment efficacy, economic factors, and environmental risk (Buhler et al., 1997). Recently, several computer-based decision aids that help managers integrate the wide array of information needed to make effective and efficient management decisions have become available.

One group of decision aids base weed control treatment selections on herbicide efficacy. Using product label information and efficacy ratings from university research, they recommend herbicides for maximum control of a given group of weeds. Included in this group are early weed control models such as WEEDIR (Kidder, Poser, and Miller, 1989), WEEDS (Linker, York, and Wilhite, 1990), SOYHERB (Renner and Black, 1991), CORNHERB (Kells and Black, 1993), and SELOMA (Stigliani and Resina, 1993).

Other weed management decision aids recommend control options based on herbicide cost and crop loss estimated from weed population dynamics

(Buhler et al., 1996; Mortensen and Coble, 1991). These models are generally more complex than herbicide-efficacy decision aids and are referred to as bioeconomic because their recommendations are based on optimum economic weed control (Swinton and King, 1994a). Within this group are two categories of models (Schweizer, Lybecker, and Wiles, 1998). One includes models such as HERB (Wilkerson, Modena, and Coble, 1991) that recommend POST weed control tactics based on weed seedling counts. Another category selects PPI or PRE weed control by predicting seedling numbers from numbers of weed seeds in the soil and uses seedling counts to make POST suggestions. The newest model of this type is GWM (General Weed Management) (Wiles et al., 1996). GWM evaluates weed control strategies in a manner similar to WEEDCAM (Lybecker, Schweizer, and King, 1991) and WEEDSIM (Swinton and King, 1994a) that preceded it. Its structure differs from the older models because it can be parameterized to provide weed control decision support for different crops and with various production practices (Wiles et al., 1996).

Although field evaluations of GWM are lacking, several bioeconomic models developed prior to GWM were tested in simulation and field studies. An expert system that was parameterized using weed biology, herbicide efficacy, yield loss, and economic return data from the literature was used to study velvetleaf (*Abutilon theophrasti* Medic.) control in a corn/soybean rotation (Lindquist et al., 1995). Results demonstrated that alternative weed management practices used with economic thresholds could increase economic returns and decrease herbicide use. A simulation study comparing weed control recommended by WEEDSIM to a prophylactic weed control strategy in a corn/soybean rotation found model recommendations were more profitable than the fixed strategy for POST but not PRE weed control (Swinton and King, 1994b). Less herbicide was used in corn but more was used in soybean. Weed control recommended by WEEDSIM was compared with herbicides standard for the region in a 4-year field test in Minnesota (Forcella et al., 1996). WEEDSIM recommendations used in continuous corn and rotated soybean/corn adequately controlled weeds, used less herbicide, decreased environmental risk, maintained crop yields, and increased gross margins. Another field study compared WEEDSIM-generated weed control with standard herbicides in Minnesota under higher weed densities. Compared with standard-herbicides, model-based weed control reduced amount of herbicide active ingredient used. Herbicide use reduction was approximately 20% greater in soybean (Buhler et al., 1997) compared with that in corn (Buhler et al., 1996). Frequency of herbicide application was not reduced in either study. Model-generated treatments resulted in weed control and crop yield similar to standard herbicides. Economic returns were not altered in corn, but using model recommendations in soybean increased net economic return to weed control 50% of the time.

Evaluations of GWM under a broader range of conditions would be useful to resolve some of the inconsistencies in previous research results. Our objective was to determine the effects of using GWM-recommended weed control in a soybean/corn rotation that included long-term tillage treatments and differing weed populations created by a history of various herbicide application methods.

## MATERIALS AND METHODS

A field experiment was conducted in 1994 through 1997 at the Iowa State University Agronomy and Agricultural Engineering Research Center in Boone County, IA. Soil type was a Clarion-Nicollet-Canisteo association consisting of silty clay loam soils (Typic Hapladolls), pH of 5.5, and 5% organic matter. Soybean was planted 3-cm deep in 0.76-m-wide rows at 388000 seeds ha$^{-1}$ in 1994 and at half that rate in 1996 due to mechanical malfunction (Table 1). Corn was planted 4-cm deep in 0.76-m-wide rows at 69000 seeds ha$^{-1}$ in 1995 and 1997. Crops were planted with a no-tillage planter adjusted to soil and residue conditions in each tillage system. Fertilizer applications were based on Iowa State University recommendations with the same rates applied to all plots.

The experiment was designed as a randomized complete block with four replications and a split-split-plot treatment arrangement. Whole plots consisted of pre-plant tillage systems, which were conventional-, reduced-, and no-tillage, reapplied to the same plots since 1988. Conventional-tillage plots were moldboard plowed in fall and disked and field cultivated in spring. Reduced-

TABLE 1. Date of preplant-incorporated (PPI) weed control, planting, pre-emergence (PRE) weed control, post-emergence (POST) weed control, interrow cultivation, and harvest at Boone County, IA in 1994, 1995, 1996, and 1997.

| Operation | 1994 | 1995 | 1996 | 1997 |
|---|---|---|---|---|
| PPI | May 12 | - - | May 22 | - - |
| Planting | May 13 | May 15 | May 31 | May 12 |
| PRE weed control | - - | May 17 | - - | May 15 |
| POST weed control | June 9 | June 12 | July 8 | June 9 |
| Interrow cultivation | June 20 | June 22 | July 18 | June 20 |
| Harvest | October 27 | October 24 | November 11 | October 10 |

tillage plots were chisel plowed in fall and field cultivated once in spring. No-tillage plots were left undisturbed until crop planting. The design of the experiment conducted in this field prior to establishing our study included 9.1 (12 rows) by 27 m subplots that were assigned different herbicide application methods. Herbicides were not applied or were either banded or broadcast applied to the same subplots for 6 years. Therefore, when our study began, weed populations in these subplots differed as a result of their herbicide application history. These subplots with differing herbicide application histories were divided into 4.6 (6 rows) by 27 m sub-subplots that received weed control recommended by GWM (Wiles et al., 1996) or standard-herbicides.

Model-based weed control recommendations included commonly used herbicides, rotary hoeing, and cultivation (Buhler et al., 1996). The highest ranking treatment generated by GWM was used (Table 2). PPI or PRE treatment recommendations were based on weed seed densities. Weed seedling counts were used to generate POST treatment recommendations. For corn, weed-free crop yield was estimated at 10 Mg ha$^{-1}$ and selling price was \$0.09 kg$^{-1}$. Weed-free crop yield for soybean was estimated at 3.35 Mg ha$^{-1}$ and selling price was \$0.20 kg$^{-1}$. Herbicide costs and efficacies (Owen and Hartzler, 1994) were obtained from Iowa State University.

The database on weed emergence and weed/crop interactions in GWM was generated from data collected about 320 km from our site in southern and western Minnesota (Swinton and King, 1994a). Because of this geographic discrepancy, a sensitivity analysis was conducted to determine if model output was consistent with expected results under local conditions (data not shown). The outcome of these simulations was deemed consistent with expected crop and weed behavior in central Iowa. In other field research in Iowa (Buhler and Forcella, 1996), weed control treatments recommended by GWM resulted in weed control and crop yields similar to standard herbicides in single year tests.

Standard-herbicide weed control consisted of locally popular treatments applied at the same time as PPI/PRE and POST model-based treatments. In soybean, standard-herbicide treatments were trifluralin [2,6-dinitro-$N,N$-dipropyl-4-(trifluoromethyl)benzenamine] at 1.1 kg a.i. ha$^{-1}$ applied PPI followed by imazethapyr {2-[4,5-dihydro-4-methyl-4-(1-methylethyl)-5-oxo-1$H$-imidazol-2-yl]-5-ethyl-3-pyridinecarboxylic acid} at 0.07 kg a.i. ha$^{-1}$ applied POST. Standard-herbicide treatments in corn were metolachlor [2-chloro-$N$-(2-ethyl-6-methylphenyl)-$N$-(2-methoxy-1-methylethyl)acetamide] at 3.4 kg a.i. ha$^{-1}$ applied PRE followed by dicamba (3,6-dichloro-2-methoxybenzoic acid) at 0.3 kg a.i. ha$^{-1}$ plus atrazine [6-chloro-$N$-ethyl-$N'$-(1-methylethyl)-1,3,5-triazine-2,4-diamine] at 0.6 kg a.i. ha$^{-1}$ applied POST.

PPI/PRE treatments were applied within 36 hours before or after crop planting (Table 1). Glyphosate [$N$-(phosphonomethyl)-glycine] at 0.4 kg a.e.

TABLE 2. Weed control strategies applied to model-based treatments at Boone County, IA in 1994, 1995, and 1996.[1]

| Tillage system | Herbicide application history | PRE/PPI treatment | POST treatment |
|---|---|---|---|
| **1994** | | | |
| Conventional | Band | None | Imazethapyr |
| | Broadcast | None | Cultivation |
| | None | Trifluralin | Cultivation |
| Reduced | Band | None | Thifensulfuron + sethoxydim |
| | Broadcast | None | Cultivation |
| | None | Trifluralin | Bentazon |
| No-tillage | Band | Metolachlor + glyphosate | Thifensulfuron + sethoxydim |
| | Broadcast | Metolachlor + glyphosate | Thifensulfuron |
| | None | Metolachlor + glyphosate | Thifensulfuron + sethoxydim |
| **1995** | | | |
| Conventional | Band | Cyanazine | 2,4-D |
| | Broadcast | Cyanazine | Nicosulfuron + bromoxynil |
| | None | Cyanazine | 2,4-D |
| Reduced | Band | Cyanazine | 2,4-D |
| | Broadcast | Cyanazine | Nicosulfuron + bromoxynil |
| | None | Cyanazine | Nicosulfuron + bromoxynil |
| No-tillage | Band | Cyanazine + glyphosate | Nicosulfuron + bromoxynil |
| | Broadcast | Cyanazine + glyphosate | Nicosulfuron + bromoxynil |
| | None | Cyanazine + glyphosate | Nicosulfuron + bromoxynil |
| **1996** | | | |
| Conventional | Band | Trifluralin | Cultivation |
| | Broadcast | Trifluralin | Thifensulfuron + sethoxydim |
| | None | Trifluralin | Thifensulfuron |
| Reduced | Band | Trifluralin | Sethoxydim + bentazon |
| | Broadcast | Trifluralin | Sethoxydim + bentazon |
| | None | Trifluralin | Imazethapyr |
| No-tillage | Band | Glyphosate | Imazethapyr |
| | Broadcast | Glyphosate | Imazethapyr |
| | None | Glyphosate | Imazethapyr |

| Tillage system | Herbicide application history | PRE/PPI treatment | POST treatment |
|---|---|---|---|
| **1997** | | | |
| Conventional | Band | Cyanazine | Dicamba |
| | Broadcast | Cyanazine | Cyanazine |
| | None | Cyanazine | Cyanazine |
| Reduced | Band | Cyanazine | Dicamba |
| | Broadcast | Cyanazine | Dicamba |
| | None | Cyanazine | Dicamba |
| No-tillage | Band | Cyanazine + glyphosate | Cyanazine |
| | Broadcast | Cyanazine + glyphosate | Cyanazine |
| | None | Cyanazine + glyphosate | Cyanazine |

[1] All herbicide applications followed Label-recommended rates.

ha$^{-1}$ was applied as a burn-down to no-tillage plots prior to planting corn and soybean. POST treatments were applied within 4 weeks of planting. Herbicides were broadcast using a tractor-mounted pressurized sprayer delivering 190 L ha$^{-1}$. The sprayer was equipped with flat-fan (TeeJet 11003, Spraying Systems Co., Wheaton, IL) nozzles spaced 0.38 m apart. In all plots, interrow cultivation followed application of POST treatments by about 10 days.

To determine seed bank composition, 15 soil cores (5-cm diam. by 10-cm deep) were collected in a M-shaped pattern from each plot prior to planting. Samples were composited by experimental unit and stored at $-5°C$ to prevent germination of seeds before extraction. Seeds were extracted from 100 g subsamples using a flotation/centrifugation method (Buhler and Maxwell, 1993). Seeds were air dried for 12 hours and stored in envelopes until they were identified by species and counted with the aid of a dissecting microscope. Seeds resisting pressure with forceps were counted as viable (Roberts and Ricketts, 1979). Counts were expressed as number per mass of soil. To express number of seeds of selected taxa as a proportion of the seed bank, relative abundance for each plot was calculated as: (number of seeds of an individual weed taxa/total number of weed seeds in the seed bank).

Weed seedling densities were assessed 3 weeks after planting (WAP) and used to generate recommendations by the model for POST weed control. Weed seedlings were counted by species in six 0.1 m$^2$ quadrats centered over one of the four middle rows in each plot. Weeds were also counted by species at 8 WAP in six 0.5 m$^2$ quadrats centered over one of the four middle rows in each plot. Weed densities were converted to plants m$^{-2}$.

Crops were machine harvested at maturity. Weight and moisture of grain in the center 3 rows of corn and the center 2 rows of soybean in each plot were determined. Grain yields were corrected to 15.5% moisture for corn and 13.0% moisture for soybean.

All variables were analyzed using a split-split-plot analysis of variance model (SAS, 1988). Data were tested to determine if transformations were necessary and none were needed. Because there was a large amount of data and many significant effects and interactions, only those means differing significantly were presented. Means were separated by Fisher's Protected LSD Test at the 5% level of significance using the appropriate error term.

## RESULTS AND DISCUSSION

Five taxa represented the most prevalent weeds each year. Foxtail species were dominated by giant foxtail (*Setaria faberi* Herrm.) but green foxtail [*S. viridis* (L.) Beauv.] and yellow foxtail [*S. glauca* (L.) Beauv.] were also present. Forb species included common lambsquarters (*Chenopodium album* L.), Pennsylvania smartweed (*Polygonum pensylvanicum* L.), redroot pigweed (*Amaranthus retroflexus* L.), and velvetleaf. Densities of individual species were such that detecting treatment differences was unlikely so treatment effects on total number of weeds and weed seeds were quantified. Although foxtails were usually not dominant in number, they were the weed group most consistently affected by treatments. Other species, including common dandelion (*Taraxacum officinale* Weber.) and common cocklebur, comprised less than 5% of the weed population so they were not initially quantified. Common cocklebur density increased rapidly during the study such that it was analyzed in 1996 and 1997.

Analysis of seed number, weed density, and crop yield data indicated year was significant and there were numerous interactions with year so these data are presented individually for each year. Year did not influence relative abundance of foxtails in the soil seed bank so these data are presented averaged over years. Weed density at 3 WAP, used to evaluate PRE weed control treatments and to generate model-based POST weed control, and those taken at 8 WAP to evaluate weed control treatments applied POST are also presented separately due to interactions.

### The Soil Weed Seed Bank

The soil seed bank was sampled once a year before treatments were applied in spring, so 1994 samples were taken before model-based and standard-herbicide weed control was initiated. In 1995, there was a tillage system by weed control interaction such that there were fewer weed seeds in standard-herbicide compared with model-based weed control treatments in re-duced-and no-tillage plots (data not shown). Numbers of weed seeds were reduced 40% in 1996 and nearly 50% in 1997 by standard-herbicide compared with model-based treatments averaged over tillage system and herbicide application history. In 1994 through 1996, there were 40% fewer weed seeds in plots where herbicides had been applied during the 6 years prior to estab-lishing this study compared with plots receiving none. Compared with other tillage systems, no-tillage consistently had more weed seeds. In no-tillage, numbers of weed seeds in 1 kg of soil exceeded 1500 in 1994 and 1995 but decreased to 100 in 1997. Weed seed numbers generally declined over the course of the study.

Averaged over the four years of this study, relative abundance of foxtail seeds decreased in reduced tillage and in plots having a herbicide application history that included band and broadcast applied herbicides compared with other treatments (Table 3). Averaged over 1995, 1996, and 1997, standard-herbicide weed control decreased relative abundance of foxtails to 23% compared with 30% in model-based treatments (data not shown).

## Weed Control

At 3 WAP, a tillage system by herbicide application history interaction affected weed seedlings in 1994 and 1995 such that weed densities differed among tillage systems in plots where herbicides had not been applied (data not shown). Average number of weeds at 3 WAP in conventional-and reduced-tillage plots was 40% of that in no-tillage plots in 1994. In 1995, average

TABLE 3. Abundance of foxtail seeds relative to total number of weed seeds in the soil seed bank was influenced throughout the course of this experiment by a tillage system by herbicide application history interaction.

| | Relative abundance of foxtail seeds | | |
| | Herbicide application history[1] | | |
| Tillage system | Band | Broadcast | None |
|---|---|---|---|
| | | % | |
| Conventional | 34 a | 33 a | 46 a |
| Reduced | 20 b | 16 b | 38 a |
| No-tillage | 35 a | 31 a | 33 a |

[1]Means followed by the same lower-case letter are not significantly different according to Fisher's LSD (P = 0.05) test.

number of weeds in conventional-and no-tillage plots was 40% of that in reduced-tillage plots. PPI/PRE applied weed control consisting of trifluralin in conventional-and reduced-tillage plots and metolachlor in no-tillage plots (Table 2) likely contributed to the effect in 1994. In 1995, cyanazine {2-[[4-chloro-6-(ethylamino)-1,3,5-triazin-2-yl]amino]-2-methyl-propanenitrile} was applied PRE to all treatments so the effect of tillage on weed density was not confounded by differing weed control treatments. Among herbicide application histories, numbers of weeds in 1994 and 1995 were decreased by reduced-tillage in plots where herbicides had been applied. There was a similar trend in no-tillage in 1994 but not in 1995.

In 1994 and 1995, weed control by tillage system and by herbicide application history interactions affected weed densities at 3 WAP (data not shown). Averaged over herbicide application history, standard-herbicide weed control in 1994 reduced weed numbers 75% in conventional-tillage plots and 50% in reduced-tillage plots compared with model-based weed control treatments. Weed numbers did not differ between model-based and standard-herbicide weed control treatments in no-tillage plots in 1994 and in conventional-and no-tillage plots in 1995. In 1995, numbers of weeds at 3 WAP were reduced 40% in reduced-tillage plots where standard-herbicide compared with model-based weed control was applied. Averaged over tillage system, numbers of weeds were reduced by standard-herbicide compared with model-based treatments in 1994 if herbicides had been banded and in 1995 if herbicides had been broadcast.

In 1994 at 8 WAP, the tillage system by herbicide application history by weed control interaction was significant because numbers of weeds following POST applied weed control differed in some model-based compared with standard-herbicide treatments (Figure 1). Model-based weed control consisting of nothing applied PRE and cultivation POST (Table 2) did not control weeds as well as standard-herbicides in conventional-and reduced-tillage where herbicides had been broadcast in previous years. Where herbicides were not applied previously and tillage was reduced, bentazon [3-(1-methylethyl)-(1$H$)-2,1,3-benzothiadiazin-4(3$H$)-one-2,2-dioxide] applied POST in model-based treatments did not control weeds as well as the standard-herbicide. Weed control in no-tillage with POST applied thifensulfuron {[[[[(4-methoxy-6-methyl-1,3,5-triazin-2-yl)amino]carbonyl]amino]sulfonyl]-2-thiophenecarboxylic acid} and sethoxydim {2-[1-(ethoxyimino)butyl]-5-[2-(ethylthio) propyl]-3-hydroxy-2-cyclohexen-1-one} that followed metolachlor applied PRE did not differ from standard-herbicides.

In 1995, numbers of weeds at 8 WAP did not differ between model-based and standard-herbicide treatments (data not shown). POST weed control in model-based treatments included 2,4-D (2,4-dichlorophenoxy acetic acid) and nicosulfuron {2-[[[[(4,6-dimethoxy-2-pyrimidinyl)-amino]-carbonyl]

FIGURE 1. In 1994 at 8 WAP, a tillage system by herbicide application history by weed control treatment interaction affected total number of weeds in conventional (CON), reduced (RED), and no-tillage (NOT) plots. Letters above bars indicate significant differences according to Fisher's LSD (P = 0.05) test. BND, band; BCT, broadcast; NON, none.

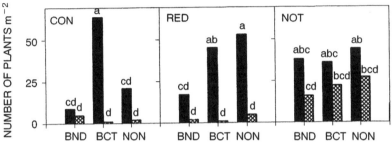

HERBICIDE APPLICATION HISTORY

amino]sulfonyl]-$N,N$-dimethyl-3-pyridinecarboxamide} plus bromoxynil (3,5-dibromo-4-hydroxybenzonitrile) (Table 2).

In 1996 and 1997, treatments affected weed densities at 8 WAP following POST applied treatments such that there were fewer weeds in conventional- and reduced-compared with no-tillage (Figure 2). At 8 WAP in 1997, weed densities averaged 64 and 34 plants $m^{-2}$ in model-based and standard-herbicide weed control treatments, respectively (data not shown).

As a consequence of the increase that prompted us to begin recording its density in 1996, common cocklebur densities differed among treatments in 1997. Because the soil sampling method used did not reliably detect common cocklebur, the model recommended treatments that were insensitive to this species. The result was an almost 20-fold increase in common cocklebur densities at 3 WAP in reduced-tillage plots receiving model-based weed control compared with other treatments (data not shown). Reduced-tillage practices, such as disking, created a more favorable environment for common cocklebur than no-tillage (Sims and Guethle, 1992). Also shown was that control of common cocklebur in no-tillage and conventionally-tilled plots does not differ under conditions with adequate rainfall such as ours (Mills and Witt, 1989; Mills, Witt, and Barrett, 1989). The large populations of other forbs that drove the model to recommend dicamba as the POST treatment in reduced-tillage was fortuitous. This was also an appropriate choice to inhibit common cocklebur and thus avert crop yield loss.

As a result of treatments, there were trends in relative abundance of fox-

FIGURE 2. The main effect of tillage system influenced total number of weeds at 8 WAP in 1996 and 1997. Letters above bars indicate significant differences according to Fisher's LSD (P = 0.05) test. CON, conventional; RED, reduced; NOT, no-tillage.

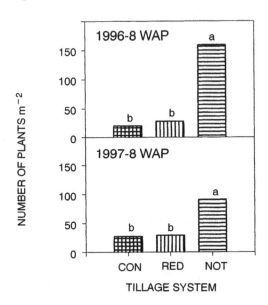

tails. In 1994, relative abundance of foxtails was 30% in reduced-tillage plots receiving standard herbicides compared with an average of 60% in other treatments (data not shown). The trend repeated in 1995 and included conventional- and no-tillage where relative abundance of foxtails was 50% in standard-herbicide compared with 75% in model-based treatments. In 1995 and 1996, relative abundance of foxtails was less if herbicides had been applied historically compared with plots that had received none (data not shown). In 1997, foxtail relative abundance was greatest among model-based treatments in no-tillage and least in reduced-tillage (Figure 3). Relative abundance of foxtails did not differ among standard-herbicide treatments in 1997. Differences in relative abundance of foxtails at 8 WAP were trivial and are not presented.

### Crop Yield

Soybean yield was reduced in some model-based compared with standard herbicide treatments. Among plots with different herbicide application histories, yield in model-based treatments in 1994 was reduced if herbicides had

FIGURE 3. At 3 WAP in 1997, a tillage system by herbicide application history by weed control treatment interaction affected relative abundance of foxtails in conventional (CON), reduced (RED), and no-tillage (NOT) plots. Letters above bars indicate significant differences according to Fisher's LSD (P = 0.05) test. BND, band; BCT, broadcast; NON, none.

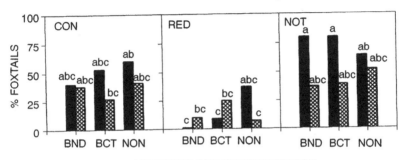

been broadcast or not applied compared with banded in conventional tillage (Table 4). This yield reduction likely resulted from differential POST weed control, which was cultivation in plots where herbicides had been broadcast or not applied compared with imazethapyr in plots where herbicides had been banded in previous years (Table 2). Compared to standard herbicides, model-based weed control reduced yield in conventional-tillage plots where herbicide application histories were broadcast and none and POST weed control was cultivation (Table 4). In reduced-tillage plots, yields were reduced where herbicide application history was none and bentazon was applied POST. These bentazon treated plots were among those with large numbers of uncontrolled weeds at 8 WAP (Figure 1) and had the largest population of weeds among treatments at 3 WAP the following year (data not shown). In 1996, average yield in model-based weed control treatments was consistently less than that in standard-herbicide treatments (data not shown).

Corn yield was influenced by tillage systems used in this study. Compared with conventional- and reduced-tillage, no-tillage reduced yield both years corn was grown (Table 5). Unlike a similar experiment that found yield reduced 2 out of 3 years by model-based compared with standard-herbicide weed control in continuous corn (Hoffman, Buhler, and Owen 1997), weed control treatments did not affect corn yield in this soybean/corn rotation study.

TABLE 4. Soybean yield at Boone County, IA was influenced by a tillage system by herbicide application history by weed control interaction in 1994.

| Tillage system | Herbicide application history | Soybean yield[1] | |
|---|---|---|---|
| | | Model-based weed control | Standard-herbicide weed control |
| | | kg ha$^{-1}$ | |
| Conventional | Band | 4248 a | 3000 abc |
| | Broadcast | 1029 c | 4013 ab |
| | None | 1334 c | 3978 ab |
| Reduced | Band | 2092 bc | 3868 ab |
| | Broadcast | 2070 bc | 4072 ab |
| | None | 1459 c | 4028 ab |
| No-tillage | Band | 2709 abc | 3380 ab |
| | Broadcast | 2450 bc | 3171 ab |
| | None | 3504 ab | 2766 abc |

[1]Means followed by the same lower-case letter are not significantly different according to Fisher's LSD (P = 0.05) test.

## CONCLUSIONS

Treatment effects on weeds and yields in this study utilizing a soybean/ corn rotation differed from those in a continuous corn study (Hoffman, Buhler, and Owen, 1997). Although the two studies can not be statistically compared, qualitative comparisons are of interest since the experiments were located on adjacent areas with identical soil types and similar treatment histories. In continuous corn, effects of model-based weed control included increased numbers of weed seeds in the seed bank, domination of the weed community by foxtails after the third year, and decreased corn yield 2 out of 3 years. By the end of the study, model-based weed control did not differ among treatments. Compared with the continuous corn study, the weed community in the soybean/corn rotation was more diverse with foxtails comprising about 25% of the seed bank. There were differences throughout the study in model-based weed control applied to plots with different treatment histo-

TABLE 5. Corn yield at Boone County, IA was influenced by tillage system in 1995 and 1997 and a tillage system by weed control interaction in 1997.

| | Corn yield | | | | | |
|---|---|---|---|---|---|---|
| | 1995 | | | 1997 | | |
| Tillage system | Model-based weed control | Standard-herbicide weed control | Mean | Model-based weed control | Standard-herbicide weed control | Mean |
| | kg ha$^{-1}$ | | | | | |
| Conventional | 10240 ns[1] | 10250 | 10245 a[2] | 6980 a | 7210 ab | 7095 b |
| Reduced | 10397 | 10537 | 10467 a | 7313 a | 8040 a | 7677 a |
| No-tillage | 9158 | 8945 | 9052 b | 7660 a | 7084 b | 7372 b |

[1]The tillage system by weed control interaction was not significant, therefore individual tillage system by weed control means were not compared by LSD.
[2]Means within a column followed by the same lower-case letter are not significantly different according to Fisher's LSD (P = 0.05) test.

ries. Weed control affected soybean yield in some model-based treatments, but corn yield was not affected. Using a more diverse cropping system increased diversity among weeds and treatments and resulted in greater opportunity for model recommendations to perform at a level similar to that of standard-herbicides.

The objective of this research was to evaluate model-based weed control under an extended range of weed populations and production conditions. We were particularly interested in the response of weed populations over years. Given the potential impact of weed seed production by uncontrolled weeds in threshold-based management (Bauer and Mortensen 1992), multi-year responses are important when using thresholds.

The varying weed populations in plots with different herbicide application histories allowed us to determined the extent to which effects of herbicide application method interacted with weed control strategies included in our study. During the first year, previous herbicide use history had a profound effect on weed control treatments recommended by the model. However, not all treatments controlled weeds well enough to maintain crop yields. This poor weed control was likely the result of either inappropriate weed threshold levels or inflated control efficacy ratings.

Differences among model-generated treatments declined by the second year of this study. This was primarily due to seed production from uncontrolled weeds eliminating differences in weed densities due to previous treatment history. As reported by Burnside et al. (1986), seed production from a single year can increase weed densities to previous levels following several years of weed-free conditions. While the model did not reduce herbicide use under these conditions, model-generated treatments resulted in corn yields similar to the standard treatment. When corn was grown in the fourth year of the study, there was little variation among model-generated treatments, and corn yields were equal to or greater than the standard-herbicide treatment.

As pointed out in previous studies (Buhler et al., 1996, 1997; Forcella et al., 1996), the performance of the model was influenced by the nature of the weed populations. When weed populations were high, the model recommended treatments that controlled weeds and maintained crop yields, especially in corn. At lower and more variable densities occurring early in this study, the model suggested that little or no herbicide was needed. Unfortunately, these strategies did not control weeds sufficiently to maintain crop yields. Based on the results of this research, GWM parameters need to be adjusted to make it more effective under these conditions.

This study is unique in that it documented the rapid escape of common cocklebur from model-based weed control. Many large-seeded weed species such as common cocklebur occur in corn and soybean fields. Such weed species are not included in the existing GWM database. Although the program easily accepts information for new species, information on the biology and competitiveness of many weeds is lacking. As our results demonstrate, ignoring their presence can result in rapidly increasing infestations. However, in many cases, adding them to the database requires using unsubstantiated parameters. Therefore, more weed biology information are needed in order for model-based weed control to have long-term success.

## REFERENCES

Bauer, T.A., and D.A. Mortensen. (1992). A comparison of economic and economic optimum thresholds for two annual weeds in soybean. *Weed Technology* 6: 228-235.

Buhler, D.D., and F. Forcella. (1996). Field evaluations of bioeconomic weed management models. *Proceedings of the North Central Weed Science Society* 51: 170-171.

Buhler, D.D., R.P. King, S.M. Swinton, J.L. Gunsolus, and F. Forcella. (1996). Field evaluation of a bioeconomic model for weed management in corn (*Zea mays*). *Weed Science* 44: 915-923.

Buhler, D.D., R.P. King, S.M. Swinton, J.L. Gunsolus, and F. Forcella. (1997). Field evaluation of a bioeconomic model for weed management in soybean (*Glycine max*). *Weed Science* 45: 158-165.

Buhler, D.D., and B.D. Maxwell. (1993). Seed separation and enumeration from soil using $K_2CO_3$–centrifugation and image analysis. *Weed Science* 41: 298-302.

Burnside, O.C., R.S. Moomaw, F.W. Roeth, G.A. Wicks, and R.G. Wilson. (1986).

Weed seed demise in soil in weed-free corn (*Zea mays*) production across Nebraska. *Weed Science* 34: 248-251.

Forcella, F., R.P. King, S.M. Swinton, D.D. Buhler, and J.L. Gunsolus. (1996). Multi-year validation of a decision aid for integrated weed management in row crops. *Weed Science* 44: 650-661.

Hoffman, M.L., D.D. Bugler, and M.D.K. Owen. (1997). Effects of cropping systems and management strategies recommended by a bioeconomic model on weed control in corn. *Proceedings of the North Central Weed Science Society* (in press).

Kells, J.J., and J.R. Black. (1993). CORNHERB–Herbicide options program for weed control in corn: An integrated decision support computer program (Version 2.2). *Agricultural Experiment Station Publication, Michigan State University,* East Lansing, MI.

Kidder, D., B. Poser, and D. Miller. (1989). WEEDIR: Weed control directory (Version 3.0). *University of Minnesota Extension Service Publication AG-CS-2163,* St. Paul, MN.

Lindquist, J.L., B.D. Maxwell, D.D. Buhler, and J.L. Gunsolus. (1995). Modeling the population dynamics and economics of velvetleaf (*Abutilon theophrasti*) control in a corn (*Zea mays*)-soybean (*Glycine max*) rotation. *Weed Science* 43: 269-275.

Linker, H.M., A.C. York, and D.R. Wilhite, Jr. (1990). WEEDS–A system for developing a computer-based herbicide recommendation program. *Weed Technology* 4: 380-385.

Lybecker, D.W., E.E. Schweizer, and R.P. King. (1991). Weed management decisions in corn based on bioeconomic modeling. *Weed Science* 39: 124-129.

Mills, J.A., and W.W. Witt. (1989). Effect of tillage systems on the efficacy and phytotoxicity of imazaquin and imazethapyr in soybean (*Glycine max*). *Weed Science* 37: 233-238.

Mills, J.A., W.W. Witt, and M. Barrett. (1989). Effects of tillage on the efficacy and persistence of clomazone in soybean (*Glycine max*). *Weed Science* 37: 217-222.

Mortensen, D.A., and J.D. Coble. (1991). Two approaches to weed control decision-aid software. *Weed Technology* 5: 445-452.

Owen, M.D.K., and R.G. Hartzler. (1994). Herbicide manual for agricultural professionals. *Iowa State University Extension Service Publication Pm-92,* Ames, IA.

Owen, M.D.K., and R.G. Hartzler. (1996). Weed management guide for 1997. Ames, Iowa: *Iowa State University Cooperative Extension Service Publication Pm-601,* Ames, IA.

Renner, K.A., and J.R. Black. (1991). SOYHERB–A computer program for soybean herbicide decision making. *Agronomy Journal* 83: 921-925.

Roberts, H.A., and M.E. Ricketts. (1979). Quantitative relationship between the weed flora after cultivation and the seed population in the soil. *Weed Research* 19: 269-275.

(SAS) Statistical Analysis System. (1988). *SAS User's Guide.* Release 6.03 edition. Cary, NC: SAS Institute, Inc.

Schweizer, E.E., D.W. Lybecker, and L.J. Wiles. (1998). Important biological information needed for bioeconomic weed management models. In *Integrated Weed and Soil Management,* eds. J.L. Hatfield, D.D. Buhler, and B.A. Stewart. Chelsea, MI: Ann Arbor Press, pp.1-24.

Sims, B.D., and D.R. Guethle. (1992). Herbicide programs in no-tillage and conventional-tillage soybeans (*Glycine max*) double cropped after wheat (*Triticum aestivum*). *Weed Science* 40: 255-263.

Stigliani, L., and C. Resina. (1993). SELOMA: Expert system for weed management in herbicide intensive crops. *Weed Technology* 7: 550-559.

Swinton, S.M., and R.P. King. (1994a). A bioeconomic model for weed management in corn and soybean. *Agricultural Systems* 44: 313-335.

Swinton, S.M., and R.P. King. (1994b). The value of weed population information in a dynamic setting: The case of weed control. *American Journal of Agricultural Economics* 75: 36-46.

Wiles, L.J., R.P. King, E.E. Schweizer, D.W. Lybecker, and S.M. Swinton. (1996). GWM: General weed management model. *Agricultural Systems* 50:355-376.

Wilkerson, G.G., S.A. Modena, and H.D. Coble. (1991). HERB: Decision model for postemergence weed control in soybean. *Agronomy Journal* 83:413-41.

RECEIVED: 01/28/98
ACCEPTED: 03/03/98

# Knowledge-Based Decision Support Strategies: Linking Spatial and Temporal Components Within Site-Specific Weed Management

Gregg A. Johnson
David R. Huggins

**SUMMARY.** Research and application of site-specific farming has focused on technological advancements that enable site-specific field operations. The current emphasis on technological tools of site-specific farming, rather than the development of new crop and pest management systems that integrate and optimize the use of these new tools, limits our abilities to realize the full potential and promise of site-specific management. The objective of this paper is to introduce a conceptual framework that integrates spatial and temporal data, information, knowledge, and wisdom into a knowledge-based decision support strategy (KBDSS). Our goal is to stimulate researchers, engineers, farmers, and other agri-business personnel to refocus their approach to site-specific resource management. Moreover, this knowledge-based decision support strategy must embrace questions of economic, environmental, and social sustainability as well as create an opportunity for people to share information and experiences in addressing emerging issues in crop production. Current strategies emphasize data collection and infor-

Gregg A. Johnson, University of Minnesota, Southern Experiment Station, Waseca, MN 56093 (E-mail: Johns510@tc.umn.edu).
David R. Huggins, USDA/ARS, 215 Johnson Hall, Pullman, WA 99164-6421.
Address correspondence to Gregg A. Johnson at the above address.

[Haworth co-indexing entry note]: "Knowledge-Based Decision Support Strategies: Linking Spatial and Temporal Components Within Site-Specific Weed Management." Johnson, Gregg A., and David R. Huggins. Co-published simultaneously in *Journal of Crop Production* (Food Products Press, an imprint of The Haworth Press, Inc.) Vol. 2, No. 1 (#3), 1999, pp. 225-238; and: *Expanding the Context of Weed Management* (ed: Douglas D. Buhler) Food Products Press, an imprint of The Haworth Press, Inc., 1999, pp. 225-238. Single or multiple copies of this article are available for a fee from The Haworth Document Delivery Service [1-800-342-9678, 9:00 a.m. - 5:00 p.m. (EST). E-mail address: getinfo@haworthpressinc.com].

© 1999 by The Haworth Press, Inc. All rights reserved.

*225*

mation management followed by an immediate action. We argue that implementation of decisions derived using only data and information circumvents knowledge and wisdom, often leading to inappropriate resource management. A successful KBDSS will incorporate a better understanding of interdependency among factors affecting or affected by a site-specific decision. We believe that the KBDSS outlined in this paper is a first step in improving integration of weed biology, enhancing the utility of bioeconomic models, shaping farmer/advisor relationships, and recognizing the importance of long-term learning and experience-building within the landscape. *[Article copies available for a fee from The Haworth Document Delivery Service: 1-800-342-9678. E-mail address: getinfo@haworthpressinc.com]*

**KEYWORDS.** Modelling, weed biology

## INTRODUCTION

Site-specific agriculture has been defined as "an information and technology based agricultural management system to identify, analyze, and manage spatial and temporal variability within fields for optimum profitability, sustainability, and protection of the environment" (Robert, Rust, and Larson, 1994). This holistic and forward-thinking definition is not, regrettably, the current perception or application of site-specific agriculture. Instead, research and application of site-specific farming has focused on technological advancements that enable site-specific field operations such as yield monitors, variable rate controllers, global positioning systems (GPS), and mapping software. The current emphasis on technological tools of site-specific farming, rather than the development of new crop and pest management strategies that integrate and optimize the use of these new tools, limits our abilities to realize the full potential and promise of site-specific management. It is critical that site-specific agriculture transcend the technology and embrace the concept of knowledge building and wisdom.

New advances in technology allow us to explore new ideas and act on those ideas in ways never before imagined. Clearly, the ability to accurately obtain and analyze spatial information on crop, pest, and landscape elements creates new opportunities for enhanced dialogue, learning, and decision building. Moreover, new advances in technology also allow us to implement spatially heterogeneous decisions. However, the technology must not be at the center of our thinking; it must be peripheral to the philosophy.

The development of knowledge-based decision support strategies (KBDSS) must embrace questions of economic, environmental, and social sustainability. The system should create opportunities for people to share information and experiences with others in addressing emerging issues in crop production. Our conceptualized decision support strategy, suggested to refocus and redefine site-specific resource management, should not be confused with decision support systems that Greer et al. (1994) define as computer programs that encode expert knowledge to assist the user in making management decisions. Rather, we are proposing a framework for "systems" thinking that may or may not include computer-based decision support systems as defined above.

Our conceptual framework for a KBDSS is applicable to the overall management of farm resources as well as specific resource components such as weed, insect, disease, soil, and water management. Examples of integrated weed management in a site-specific context are used to explore the fundamental elements and application of our proposed KBDSS. Our rationale is that the consequence of weed populations being spatially aggregated across agricultural landscape have only recently been explored in integrated weed management systems (Cardina, Sparrow, and McCoy, 1995; Johnson, Mortensen, and Gotway, 1996b; Mortensen, Dieleman, and Johnson, 1998). We argue that an understanding of spatial weed distribution has tremendous consequences for future scouting routines, development of more refined yield loss predictions, and an understanding of the processes that cause changes in spatial and temporal distribution of weeds (Wiles, Wilkerson, and Gold, 1992; Johnson, Mortensen, and Gotway, 1996b; Cardina, Johnson, and Sparrow, 1997). Therefore, spatial and temporal components are required in a KBDSS to facilitate enhanced functionality, adaptability, and site-specificity in the development of integrated weed management strategies.

## FUNDAMENTALS OF A KNOWLEDGE-BASED DECISION SUPPORT STRATEGY

Knowledge-based decision support strategies integrate data, information, knowledge, and wisdom to provide sound decision support strategies. Currently, data is often confused with information and information with knowledge. Arriving at a judicious decision is the goal of a decision support strategy, one that requires more than just knowledge. Understanding the interdependency of each element affecting the decision-making process will allow us to identify those elements that are lacking or when decisions are implemented at various scales.

Data, information, knowledge, and wisdom must all be incorporated into a KBDSS with the primary goal being optimal resource management (Figure 1). The entire KBDSS concept is based on a dynamic learning process through which individual elements are enhanced and expanded over time. Data collection is simply an entry point into the cycle. Data put into context creates

FIGURE 1. Proposed conceptual framework for a KBDSS incorporating data, information, knowledge, and wisdom.

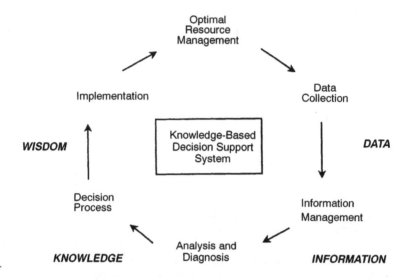

information that, coupled with analytic and diagnostic tools, results in knowledge. Knowledge that incorporates risks, biases, experiences, and insights (wisdom) creates a decision making process and implements a course of action that, over time and several iterations of the cycle, results in optimal resource management. To fully comprehend the value of a KBDSS, further development of what defines data, information, knowledge, and wisdom is critical. It is also critical to understand how each of these components contributes to site-specific management and how they will facilitate agriculture's entrance into the information and computer age.

### Data

Davis and Botkin (1994) state that data provide the basic building blocks of information, but that data is nothing more than numbers, images, words, or sounds. In its purest form, data does not mean anything or have any direct value. How we process, store, or manipulate data determines its value (Davis and Botkin, 1994). By itself, data is not information; it is simply a means by which we obtain information. Since data is considered the building block of information, it is important to recognize that data can be inaccurate, biased, or irrelevant. Therefore, careful thought should be given to data quality.

Many useful methods of data collection are emerging with site-specific management including field survey and sampling, sensor technology, and remote sensing. Field scouting is the most common method of data collection for weed management. Field scouts typically collect qualitative measures related to distinct features about weed populations in a field. Qualitative features are usually categorical in nature and can be very subjective, e.g., weed severity may be characterized as low, medium, or high. Under certain circumstances, scouts may obtain quantitative data comprising discrete numbers, usually weed density per unit area. This method can, however, be very time consuming. To enhance the rapidity at which data is acquired, many are looking towards new technology to provide data collection tools. Remote sensing, for example, is being used to collect color images or infrared/red reflectance patterns (Lass, Carson, and Callihan, 1996).

The entire premise of site-specific weed management is to selectively apply a given weed management strategy based on an understanding of spatial patterns of weeds and the surrounding landscape. Weeds are clearly not homogeneously distributed across the landscape, nor are landscape variables that interact with weed populations. The need for and type of weed management strategy therefore requires that data be put in a spatial context (Cardina, Johnson, and Sparrow, 1997). To add spatial context to data, it is necessary to know the location of each data collection point or area in the field. Data can be spatially referenced using a systematic grid system or relative position assessment, or through differentially corrected global positioning systems (DGPS) technology. Design of a weed sampling procedure has received little attention, especially as it relates to spatial weed distribution. Sampling for weed density and/or spatial distribution is a complex issue and beyond the scope of this paper. The reader is referred to Johnson et al. (1996a), Gold, Bay, and Wilkerson (1996), and Cardina, Johnson, and Sparrow (1997) for more information on this subject. Regardless of how the data is collected, it is still only data, i.e., it means little without further manipulation.

## *Information*

Information is data that has been arranged into meaningful patterns (Davis and Botkin, 1994). The concept of patterns is especially relevant in site-specific farming strategies whereby spatially referenced weed data can be used to graphically display the position of quantitative or qualitative weed features across a landscape. For example, Lass, Carson, and Callihan (1996) acquired high resolution multispectral digital images of vegetation in semiarid rangeland in Idaho. This data was then used to detect yellow starthistle (*Centaurea solstitialis* L.) and common St. Johnswort (*Hypericum perforatum* L.) based on specific spectral signatures of these species. The transition of raw digital images to the detection and mapping of starthistle and common St. Johnswort represents the development of information.

The method and time frame in which data is put into context is important from an information management perspective. Spatially referenced data collected by remote sensing or direct field scouting is termed historic, i.e., the transformation of data to information requires some type of post-processing (Mortensen, Dieleman, and Johnson, 1998). This can take several hours to several days, depending on the data source and mapping software. Conversely, intermittent herbicide application technology uses a real-time approach where the transformation of data to information (and from information to action) occurs in a very short time period. For example, intermittent herbicide application equipment uses sensor technology to obtain infrared/red reflectance values. These values, obtained only a short distance ahead of the spray boom, are then sent to an on-board computer that calculates a normalized difference index (NDI). When the NDI exceeds a given threshold, it is assumed that a plant is present in a non-plant background. Information is derived from the data almost instantaneously.

The most appropriate method for data collection and interpretation depends on how the information is going to be used and in what time frame. On-board sensor technology can obtain and process data very quickly and makes use of the information for making instantaneous, real-time decisions. However, data used to create information in this context is usually directed to a specific outcome, usually very limited in scope, that does not require real-time development of knowledge or wisdom. The potential for poor decision making exists if the technology is extrapolated to circumstances other than originally designed without re-thinking the situation and including applicable new knowledge and wisdom. Conversely, information based on historic data can be very broad in scope and used for many different applications. For example, historic data on weed density and diversity can be used to ascertain past management successes or failures thereby providing a basis for enhanced decision making in light this new information. The ability to expand the scope and context of data thereby creating new information improves the utility of KBDSS and naturally leads to integration of knowledge.

### Knowledge

Rogers (1995) suggests that knowledge is information arranged in a way that will enhance learning. For example, it is not enough to simply show someone a map of weed populations; there must be a structure in place whereby the information can be customized to facilitate interpretation and productive use of the available information. Transforming information into knowledge through learning is at the heart of a decision support strategy. Rogers and Skyrme (1994) state that supporting knowledge processes with technology "needs an approach based not on thinking machines but on thinking humans." In order to learn from the available data, we must have methods in place to analyze information thereby leading to the accurate diagnosis of a given situation. The use of bioeconomic models is one example of ongoing efforts to organize and analyze the available data to determine best management practices (Mortensen and Coble, 1991). Knowledge should also be systematic, e.g., understanding that changes in one factor may have a significant impact on other factors. This aspect of knowledge must be recognized and managed for effective implementation of management strategies across all scales of influence. For example, an aggressive tillage strategy may be needed to control certain perennial weed species. However, these tillage strategies may lead to increased soil erosion and undesirable transport of nutrients. The result of a field-scale solution to a specific weed problem may, therefore, create an even greater problem on a watershed scale.

Knowledge starts with accurate and reliable information that can be rigorously examined and analyzed. Introducing a spatial component to this process adds another dimension to learning activities. Understanding temporal dynamics of the management system through appropriate analysis and diagnostic procedures helps to sharpen our learning skills. The goal is to have a platform on which to begin asking questions using the available information to optimize the integration of crop, weed, and landscape management tactics. Moreover, this dynamic learning process leads to a better understanding of what changes are needed to meet economic, environmental, and social needs on a landscape, regional, and national scale. To fully meet these challenges and facilitate socioeconomic changes, we must integrate wisdom into the conceptual framework of a knowledge-based decision support strategy.

## *Wisdom*

Wisdom is a dynamic concept whereby existing knowledge is incorporated into new patterns through a process of reflection and further learning. It is the sum total of the learning and experience residing in an individual, community, or group (Rogers, 1995). The process of adding knowledge to information results in choices. It is wisdom that creates decisions. However, wisdom is best served through interactions among and between groups rather than the individual. Mortensen, Dieleman, and Johnson (1998) suggested that site-specific management strategies offer the possibility for farmers and consultants to build upon individual expertise through the exchange of ideas. Risk, either perceived or real, is oftentimes the limiting factor in adoption of new ideas and strategies. All segments of society, ranging from individuals to groups and communities, have different risks and biases. Moreover, risks and biases are oftentimes poorly communicated between these segments. It is this exchange of ideas and experiences that empower all segments of society to make decisions that reduce overall environmental, economic, and societal risk.

### Implementation

The overall goal of a decision support strategy is to implement judicious decisions. Decisions that are derived from adding wisdom to knowledge are of little use unless there is a clear path to implementation of that decision. The correct path depends on whether a decision is made in light of spatial (heterogeneous) or non-spatial (homogeneous) outcomes. New technology is, or will soon be, available to assist farmers and other agricultural professionals in implementing decisions based on spatial outcomes. It is important that farmers do not feel compelled to use technology in the implementation phase of a decision support strategy simply because the information, knowledge, and/or wisdom incorporates spatial elements. In fact, understanding the spatial and temporal nature of weed populations may actually lead to homogeneous implementation of a decision. This may be due to unfamiliarity with the technology or risk associated with adoption of new technology. When thinking about taking an action based on spatial components, we may consider options such as intermittent herbicide application or tillage, variable rate herbicide application, and construction of management zones on which individual management strategies are employed.

## CURRENT DIRECTION

Currently, most site-specific and traditional weed management programs involve data collection and information management followed immediately by an action (Figure 2). Moreover, there is little if any recognition of spatial elements of both weed and landscape elements. Implementation of a decision derived primarily from information circumvents knowledge and wisdom. For example, a consultant may obtain qualitative or quantitative data on weed diversity and/or density across a given field. This *data* is then used as a basis for determining the need for and type of weed control strategy to employ (*information and action*). The need for control is based on perceived grower risks and attitudes while the type of control usually involves the selection of one or more herbicides and tillage combinations to insure a given level of control. This perception-based decision making does not encompass knowledge, as defined in our model, and can therefore lead to significant economic risk on the part of consultants who must anticipate grower needs and risks. The result is conservative decisions that reduce overall risk for not only the farmer but also the business associated with consulting activities. The same case can be made for decisions not directly made in consultation with outside

FIGURE 2. Current site-specific and traditional weed management programs involve data collection and information management followed immediately by an action. Implementation of a decision derived primarily from information circumvents knowledge and wisdom.

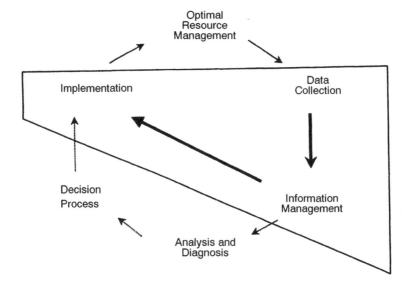

professionals. For example, the grower may fully understand his/her economic, environmental, and societal risks and biases (wisdom) but may not have access to certain tools or information necessary to adopt new management strategies (knowledge).

Without an understanding of how to communicate knowledge and wisdom on the part of both the grower and consultant, overall risk may actually increase and the ability of producers to adapt to changing political, economic, and social climates may be significantly impaired. Increased risk due to a lack of structure for integrating knowledge and wisdom results in these conservative and perception-based decisions. The KBDSS we present is the first step in developing a structure for integrating knowledge and wisdom, along with an appreciation of spatial structure, into the overall management of agricultural landscapes. This structure can facilitate and enhance interaction between the grower and consultant.

## TAKING THE NEXT STEP: INCORPORATING KNOWLEDGE INTO A DECISION SUPPORT STRATEGY

Weed management decision-making is a complex endeavor requiring integration of weed biology, environmental risks, labor needs, crop yield potential, efficacy of a given control measure, and economics (Buhler et al., 1996). One way that growers and consultants can manage the integration of these complex factors is through the use of bioeconomic models. Maxwell (1996) states that management-oriented bioeconomic models strive to increase our understanding of how weed biology and management strategies interact, assessing policy decisions, or aiding producers or consultants in making weed management decisions. Bioeconomic weed management models have been shown to adequately control weeds, reduce herbicide use/amount, decrease environmental risk, and lower weed management costs (Forcella et al., 1996; Buhler et al., 1996; Buhler et al., 1997).

Bioeconomic models require *data* in the form of weed composition and density, anticipated crop value, and cost of various weed management strategies (Figure 1). Economic thresholds, based on the crop loss function, are then used to create *information* from the aforementioned data. Other information such as environmental consequences of a given management strategy can also be derived from herbicide and soils properties data as well as depth to water table (Martin, Mortensen, and Harvill, 1996). *Knowledge* is gained though analysis of dynamic multi-species and multiple-control tactics to identify weed management strategies that are more profitable than those currently used (Swinton and King, 1994; Martin, Mortensen, and Harvill, 1996). Additional knowledge can be gained by including information such as risk and timeliness issues. For example, Swinton and King (1994) extended the bioeconomic model WEEDSIM to include simulations of whole-field weed management thereby allowing re-evaluation of bioeconomic model recommendations in light of field time constraints due to competing tasks and/or inclement weather. This simulation model uses data on number of cropped acres, equipment capacity, labor, and climatic conditions to adjust gross income determined in the bioeconomic models. This adjustment creates information and knowledge that producers and consultants can use to individualize weed management directions.

Bioeconomic models also create an environment for enhanced learning by understanding the impact of one management factor on others. For example, a grower may consider a given management strategy based on previous experiences, efficiency, and risk perception. In the current mode, it is very difficult to determine if this is the most optimal strategy given economic and environmental constraints. Bioeconomic models allow users to explore alternative weed management options and compare them to the current strategy with respect to profitability, efficiency, and environmental fate. Because these models are interactive, they can help to facilitate greater interaction between farmers and consultants through a common format in which all information is clearly stated, and available options ranked based on a number of different factors.

The integration of spatial and temporal components in weed management is the next step in furthering the knowledge-building process. Knowledge of weed species diversity and severity (quantitative or qualitative) by location allows for site-specific integration of weed management tools that are effective for a given weed species complex. For example, a farmer may find that grass species dominate a given area in the field while large-seed broadleaf species dominate others. Tillage can have a profound impact on weed species density and diversity whereby small-seeded grass species tend to do well in no-till while large-seeded broadleaf species are adapted to more aggressive tillage. This information can then be used as a basis for altering tillage practices across the field based on an understanding of species location and diversity. Moreover, tracking changes in weed density and diversity over time permits the farmer/consultant to adjust management strategies that address species shifts as a result of a given management strategy. Site-specific implementation of a given management strategy will, over time, result in the appearance of weed species adapted to that environment. Understanding and anticipating these shifts allows the farmer to take quick and directed action that will reduce risk of economic yield loss.

Geographic information systems (GIS) have also been used to create information and knowledge based on spatial data. Mitchell, Pike, and Mitasova (1996) demonstrated the use of a GIS to locate potential hazard areas for pesticide movement on a watershed. In this example, a GIS was used to successfully locate hazard areas for pesticide movement, using this information to provide land managers with additional knowledge that could be used to effectively reduce the risk for adverse environmental consequences. Clearly, changes in weed, crop, and soil management techniques can have a significant effect on the degree to which herbicides impact the environment. Improving management techniques to minimize environmental effects can be difficult, especially when considering that herbicide fate is dependent on heterogeneous site characteristics (Mitchell, Pike, and Mitasova, 1996). A GIS approach could also be used to evaluate alternative management practices on water quality. Reducing surface and groundwater contamination by herbicides is critical and much emphasis has been placed on predicting the fate of herbicides under various management strategies. However, Mitchell, Pike, and Mitasova (1996) suggest that the difficulty with improving management techniques is that they are dependent on site characteristics. For example, changes in slope and soil physical and chemical properties across a given field or landscape can have a significant impact on environmental fate of herbicides.

## INTEGRATING WISDOM

Wisdom is a difficult concept to incorporate in the decision support strategy outlined in this paper. Nevertheless, it is one of the most important aspects of a decision support strategy. We are not suggesting a complicated methodology by which to derive analytical measures of wisdom. Rather, the idea of wisdom should center on creating options by which wisdom can be incorporated and used in the decision making process. Wisdom for site-specific management tends to be diffuse, residing with practitioners who have first-hand knowledge of the resource, rather than centralized at more distant locations. Therefore, individuals and communities that rely on the resource must be recognized as an important source of localized wisdom required for site-specific management and not circumvented by the technology. Open communication of risks and biases, as it relates to all persons involved in the decision-making process, is paramount to reducing overall risk and allowing growers to consider more integrated management strategies. Without open communication and interaction, risk increases and leads to conservative decision making based primarily on perception of what is the best strategy. Assessments of risk should also encompass time and labor management consideration on an individual basis.

As previously discussed, wisdom is the sum total of the learning and experience residing in an individual (Rogers, 1995). Wisdom is augmented through experiences with other individuals and groups. As one moves through the decision support strategy, as outlined in this paper, new experiences and opportunities for individual and community learning will occur. It is this evolutionary process that leads to greater wisdom and is therefore key in resolving difficulties in adoption of integrated weed and crop management strategies.

## CONCLUSIONS

Current applications of site-specific farming technology often circumvent and fail to incorporate knowledge and wisdom appropriate for a site-specific scale. This circumstance leads to risk conserving decisions that may actually increase the risk of failure or misapplication of these new technologies. Concepts of site-specific weed management must be explored and integrated into knowledge based decision support strategies. Site-specific farming provides an opportunity for knowledge-building with the goal of understanding field-scale variability in a way that facilitates more refined management of the landscape. Spatial and temporal information and knowledge about weeds and the surrounding landscape allows for the study and understanding of interactions and how these interactions can or cannot be managed. We believe that the knowledge-based decision support strategy outlined in this paper is a first step in improving integration of weed biology, enhancing the utility of bioeconomic models, shaping farmer/advisor relationships, and recognizing the importance of long-term learning and experience-building within the landscape.

Anderson and Nielsen (1996) state that knowledge of weed ecology is the foundation for integrated weed management systems. Bioeconomic weed management models are tremendous vehicles for integrating weed biology and ecology in long and short-term decision making. Recognizing the nature and consequences of spatial heterogeneity of weeds and the landscape will further enhance the utility of these models. Bioeconomic models also serve to enhance communication between farmers and advisors, thereby introducing wisdom into the design and implementation of integrated weed management strategies.

The development of computer technologies linked to electronic controllers has introduced new possibilities for spatial management of farm resources and immersed farm managers into the information and computer age. However, application of site-specific farming will require greater understanding of the processes that govern spatial heterogeneity at site-specific scales as well as the capability to collect, process, and apply data appropriate for the scale of interest.

## REFERENCES

Anderson, R.L. and D.C. Nielsen. (1996). Emergence pattern of five weeds in the central Great Plains. *Weed Technology* 10:744-749.

Buhler, D.D., R.P. King, S.M. Swinton, J.L. Gunsolus, and F. Forcella. (1996). Field evaluation of a bioeconomic model for weed management in corn. *Weed Science* 44:915-923.

Buhler, D.D., R.P. King, S.M. Swinton, J.L. Gunsolus, and F. Forcella. (1997). Field evaluation of a bioeconomic model for weed management in soybean. *Weed Science* 45:158-165.

Cardina, J., G.A. Johnson, and D.H. Sparrow. (1997). The nature and consequences of weed spatial distribution. *Weed Science* 45:364-373.

Cardina, J., D.H. Sparrow, and E.L. McCoy. (1995). Analysis of spatial distribution of common lambsquarters in no-till soybean. *Weed Science* 43:258-268.

Davis, S. and J. Botkin. (1994). The coming of knowledge-based business. *Harvard Business Review* Sept-Oct. 165-170.

Forcella F., R.P. King, S.M. Swinton, D.D. Buhler, and J.L. Gunsolus. (1996). Multi-year validation of a decision aid for integrated weed management in row crops. *Weed Science* 44:650-661.

Gold, H.J., J. Bay, and G.G. Wilkerson. (1996). Scouting for weeds, based on the negative binomial distribution. *Weed Science* 4:504-510.

Greer, J.E., S. Falk, K.J. Greer, and M.J. Bentham. (1994). Explaining and justifying recommendations in an agricultural decision support system. *Computers and Electronics in Agriculture* 11:195-214.

Johnson, G.A., D.A. Mortensen, L.J. Young, and A.R. Martin. (1996a). Parametric sequential sampling based on multistage estimation of the negative binomial parameter k. *Weed Science* 44:555-559.

Johnson, G.A., D.A. Mortensen, and C.A. Gotway. (1996b). Spatial and temporal analysis of weed seedling populations using geostatistics. *Weed Science* 44:704-710.

Lass, L.W., H.W. Carson, and R.H. Callihan. (1996). Detection of yellow starthistle and common St. johnswort with multispectral digital imagery. *Weed Technology* 10:466-474.

Martin, A.R., D.A. Mortensen, and T.E. Harvill. (1996). WeedSoft, a bioeconomic decision aid. *Proceedings of the North Central Weed Science Society* 51:172

Maxwell, B.D. (1996). The structure and application of bioeconomic models. *Proceedings of the North Central Weed Science Society* 51:169-170.

Mitchell, K.M., D.R. Pike, and H. Mitasova. (1996). Using a geographic information system for herbicide management. *Weed Technology* 10:856-864.

Mortensen, D.A., and H.D. Coble. (1991). Two approaches to weed control decision-aid software. *Weed Technology* 5:445-452.

Mortensen, D.A., J.A. Dieleman, and G.A. Johnson. (1998). Weed spatial variation and weed management. In *Integrated Weed and Soil Management,* eds. J.L. Hatfield, D.D. Buhler, and B.A. Stewart. Chelsea, MI: Ann Arbor Press, pp. 293-310.

Robert, P.C., R.H. Rust, and W.E. Larson. (1994). Preface. In *Site-Specific Management for Agricultural Systems,* eds. P.C. Robert, R.H. Rust, and W.E. Larson. Madison, WI: ASA, CSSA, SSSA.

Rogers, D.M.A. (1995). The challenge of 5th generation R&D. (1996). *Research Technology Management* 39:33-42.

Rogers, D.M.A. and D.J. Skyrme. (1994). Leveraging knowledge through socio-technological fusion. *Journal of Technology Studies.* Fall 1994. 1-7.

Swinton, S.M. and R.P. King. (1994). A bioeconomic model for weed management in corn and soybean. *Agricultural Systems* 44:313-335.

Wiles, L.J., G.G. Wilkerson, and H.J. Gold. (1992). Value of information about weed distribution for improving postemergence control decisions. *Crop Protection* 11:547-554.

RECEIVED: 06/13/97
ACCEPTED: 04/02/98

# Development of Weed IPM: Levels of Integration for Weed Management

J. Cardina
T. M. Webster
C. P. Herms
E. E. Regnier

SUMMARY. Integrated pest management (IPM) for weeds is considered to be in the early stages of development, especially when compared with insect and disease management. The need to develop IPM principles for weed management will increase as weed population shifts, species adaptation, and environmental impacts of weed control methods threaten the sustainability of agricultural systems. In this paper, we propose a framework for discussion and development of the weed component of IPM. We describe five levels of weed IPM that encompass progressively larger spatial scales and longer time periods, with corresponding changes in goals and complexity of supporting technology and information. The first level consists of the use of a

J. Cardina, T. M. Webster, and C. P. Herms, Department of Horticulture and Crop Science, Ohio State University, Wooster, OH 44691. E. E. Regnier, Department of Horticulture and Crop Science, Ohio State University, Columbus, OH 43210.

Address correspondence to: J. Cardina, Department of Horticulture and Crop Science, Ohio State University, Wooster, OH 44691-4096 (E-mail: cardina.2@osu.edu).

The authors thank Casey Hoy and Denise Sparrow for helpful discussions, and Diane Miller and Frank Hall for reviewing versions of this manuscript. Salaries and research support were provided by state and federal funds appropriated to the Ohio Agricultural Research and Development Center, The Ohio State University. Manuscript No. 35-98.

[Haworth co-indexing entry note]: "Development of Weed IPM: Levels of Integration for Weed Management." Cardina, J. et al. Co-published simultaneously in *Journal of Crop Production* (Food Products Press, an imprint of The Haworth Press, Inc.) Vol. 2, No. 1 (#3), 1999, pp. 239-267; and: *Expanding the Context of Weed Management* (ed: Douglas D. Buhler) Food Products Press, an imprint of The Haworth Press, Inc., 1999, pp. 239-267. Single or multiple copies of this article are available for a fee from The Haworth Document Delivery Service [1-800-342-9678, 9:00 a.m. - 5:00 p.m. (EST). E-mail address: getinfo@haworthpressinc.com].

© 1999 by The Haworth Press, Inc. All rights reserved.

single weed management tool at the field level, such as the use of a herbicide as the sole means for weed control. Level II integration incorporates tactical use of multiple tools for weed management, including combinations of herbicides, crop rotation, and mechanical and biological controls, as well as other cultural practices designed to reduce the damage caused by weeds. Level III represents the threshold of much current weed science research, and involves the design of entire cropping systems at the farm and landscape level as a strategy to resist weed invasion, tolerate weed presence, and decrease weed population survival and persistence. Levels IV and V are more speculative because the ecological basis for them is in development. Level IV addresses the management of habitats and landscapes at the ecosystem level to address large scale problems related to weed spread, dispersal, and invasion. At Level V, national trade and environmental policies are considered with respect to their impact on weed populations at a global scale. Approaches that might be taken at the various levels to management of herbicide resistance are described. Research has tended to focus on the first two levels of integration; however, development of IPM principles for weeds will require that attention be given to all levels of integration so that agricultural systems can be designed to more effectively deter and withstand the inevitable presence of weeds. *[Article copies available for a fee from The Haworth Document Delivery Service: 1-800-342-9678. E-mail address: getinfo@haworthpressinc.com]*

**KEYWORDS.** Herbicide resistance, integrated pest management, spatial scales

## INTRODUCTION

Integrated pest management (IPM) was developed as an approach to reduce and make rational use of pesticides (Bottrell, 1979; Frisbie and Smith, 1989; Prokopy, 1994). IPM has advanced from simply monitoring specific pest populations and applying pesticides where needed when populations exceed an economic threshold level to a more comprehensive approach that considers conserving or enhancing environmental quality and user profits. For some crops and pest disciplines, IPM theory and practice are highly developed and include regional monitoring of pests and weather conditions, complex predictive models, and combinations of genetic, cultural, biological control, and biotechnology resources (NRC, 1996). However, for weeds, IPM has been characterized as being in the early stages of development (Hall, 1995). Some of the concepts and approaches of IPM can be borrowed from other pest disciplines and applied to weed management. However, other fundamental IPM concepts, such as tolerable pest injury levels, and tactics,

such as the use of genetic resistance or maintaining populations of benefi-
cials, have been difficult to apply to weed management. There may be IPM
principles unique to weed science that are yet to be developed. Weed scientists will be required to go beyond current technologies to consider how the
many complex biological interactions among weeds and their ecosystem
might be exploited in weed management.

Weed scientists integrate information and technologies from basic and
applied disciplines to manage populations of plants that interfere with human
activity. Integration of diverse technology and information is essential for
weed management because the agricultural ecosystems in which weeds reside are complex. One result of this complexity is that efforts to manage a
particular weed in a field have implications for many other components of the
ecosystem, including other weeds in that field, weeds in adjacent fields, and
other organisms that are positively or negatively affected by the presence or
absence of those weeds. Many species in the food chain are affected by the
change in primary production brought about by weed control, including some
species beyond the field where control occurs (Crossley et al., 1984). Conversely, weed populations are affected directly or indirectly by virtually all
crop, soil, or other pest management practices, as well as any manipulation of
the landscape or global environment. In other words, there are complex
interactions between weeds and their environment at spatial and temporal
scales beyond which most weed control measures are implemented. Understanding and exploiting these interactions may reveal new opportunities for
managing weeds and reducing the negative consequences associated with
current weed management practices. The type and complexity of technology
and information directed to weed management must be appropriate so that
the results for weed populations and agricultural ecosystems are effective and
compatible with society's need for a sustainable agriculture.

The concept of progressive levels of integration of pest management practices was outlined by Prokopy and colleagues with respect to IPM in apples
(*Malus* spp.) in the northeastern U.S. (Prokopy et al., 1990a; Prokopy, Johnson, and O'Brien, 1990b; Prokopy 1994). They described various levels of
IPM in terms of the number and complexity of management tactics applied
from a field level to an entire production system, with increasing levels of
involvement by stakeholders. Their concept of levels of integration was
limited to technologies and social involvement. More recently, principles of
integration in IPM have been developed around appropriate combinations of
preventive and therapeutic practices (Pedigo, 1995), especially those applicable to non-weed pests. Elmore (1996) suggested a hierarchy of six levels of
weed management ranging from "weed control" to "integrated crop management" that describe increasing numbers of practices directed toward
weeds.

In this paper we attempt to extend these concepts to include integration of information and technology over time and space. We will describe levels of integration characterized by different management approaches that vary in the types of technology and knowledge systems that are used to address weed problems, as well as the spatial and temporal scales at which weed problems are studied and managed. The levels of integration are outlined and described briefly in Table 1. Because herbicide resistance poses such an important problem for current and future weed management, we will use this issue to exemplify how the goals at each level of integration might be applied to management of herbicide resistant weeds.

## *LEVEL I: WEED CONTROL*

The most basic level of integration in weed management involves the control of one or more weeds with a single tool or technology (Elmore, 1996). The tool may be chemical, mechanical, or biological, but the main objective of weed control is to reduce the impact of the target weed, usually by killing it. The greater the mortality, or percent control, the more effective the given tool is regarded. The spatial scale of interest at Level I is relatively narrow, i.e., a single weed or a group of susceptible species in a production field. The time scale of interest ranges from a few days to a growing season.

The need for research effort and grower attention at this level has been justified by weed biology studies showing that weed densities of 1 plant $m^{-2}$, a realistic occurrence in growers' fields (Stoller et al., 1987), reduce crop yields from 10 to 75% (Bloomberg, Kirkpatrick, and Wax, 1982; Kirkpatrick, Wax, and Stoller, 1983; Webster et al., 1994; Wyse, Young, and Jones, 1986). Without control, weeds would cause an estimated $2 billion in yearly losses in both corn (*Zea mays*) and soybean (*Glycine max*) crops throughout the corn belt states (Illinois, Indiana, Iowa, Missouri, and Ohio) (Bridges, 1992). These potential losses help to explain why herbicides account for approximately 70% of the pesticides applied annually in the U.S. (Coble, 1994). Level I has depended on an enormous data base on herbicide efficacy that has been generated through university and industry research over the past 50 years (Timmons, 1970). The information that flows from university research trials to growers in the form of weed control guides is primarily aimed at this level of integration. Results of herbicide evaluation studies are formatted to provide yearly data on label changes and new products. This information has been widely adopted and has served many growers well. Therefore, Level I continues to be a highly focused research thrust of a multi-billion dollar herbicide industry that has produced biologically active chemicals to effectively control most weeds of importance in major cropping systems. For considerably fewer weeds, biological controls (i.e., insects or pathogenic organisms) have been developed (Julien, 1992), but the overall research effort has been relatively small compared to that devoted to discovery and development of chemical herbicides.

The limitations of this level result from the relative lack of true integration. Regardless of which technology is used, research and grower efforts have sometimes emphasized the technology–often a single tool–over the intended ecological consequence. It has been argued that too much emphasis in weed control has been on killing weeds instead of crop protection (Vandeman, 1994). By focusing on killing weeds, the Level I approach overlooks the possibility that positive interactions among biological components of agroecosystems may act to keep weed populations in check. For example, increased complexity of rotations might contribute to reducing weed populations; however, few studies have been conducted where effects of rotation were not confounded with effects of herbicides (Liebman and Dyck, 1993). As a result, the possible effects of crop rotations, in and of themselves, are virtually unknown with respect to management of weed populations.

Lack of integration is a limitation for long-term effective and environmentally responsible weed management regardless of which technology is used. Herbicides provide the most obvious historical example. Overuse of any tool or technology at the expense of appropriate integration with other approaches would be expected to have analogous consequences. In the case of herbicides, broader scale limitations of the Level I approach have become increasingly evident over time. Initially, considerable research (and marketing) effort was devoted to learning how to use chemical tools, and only later was attention given to their potential environmental effects. The social and economic impact of these tools on rural communities has received relatively little attention. Perhaps the greatest shortcoming of Level I is that, in spite of more than four decades of herbicide use, weeds are still a problem. The 100 or so herbicides registered for use on field crops have allowed for enormous increases in efficiency and reliability of weed control. Nevertheless, some weeds are frequently more of a problem now than when herbicides were first introduced (e.g., giant foxtail, *Setaria faberi*, in corn, giant ragweed, *Ambrosia trifida*, in soybeans, Florida beggarweed, *Desmodium tortuosum*, in peanuts, *Arachis hypogaea*, morningglories, *Ipomoea* spp., in cotton, *Gossypium hirsutum*) (Loux and Berry, 1991; Webster and Coble, 1997). In addition, significant species shifts have occurred, especially in reduced tillage systems (Buhler et al., 1994; Tripplett and Lytle, 1972), so that some weeds, including hard-to-control perennials (e.g., hemp dogbane, *Apocynum cannabinum*, poison ivy, *Toxicodendron radicans*, sumac, *Ruhus* spp., brambles, *Rubus* spp., etc.), that were rarely a concern before, are now the target of chemical control efforts in field crops. Furthermore, the value that was given to percent control rather than crop protection, as well as the aesthetic value of weed-free fields,

TABLE 1. Conceptual outline of advancing levels of integration for weed management in IPM and the corresponding spatial scale, methods, time frame, knowledge base, and research questions of interest.

| Level of Integration | Goal/Focus | Spatial Scale | Method/ Approach | Temporal Scale | Conceptual Basis | Types of Questions |
|---|---|---|---|---|---|---|
| I | weed control | plant–field | individual tools or technologies | days–weeks | data | What kills weed X? |
| II | weed management | plant–field– farm | tactical use of tools | growing season | integrate information; knowledge-based models | Can we coordinate use of multiple tools to maintain damage below economic injury level? |
| III | cropping system design | farm– landscape | cropping systems; strategic use of tactics | years | complex knowledge; long-term management models | Can seedbanks be re-duced? Can we limit spread of weeds across landscape? What is the best cropping system design? Can herbicide carryover, movement be checked? Can farm-level management of herbicide resistance be realized? |

| | | | | | |
|---|---|---|---|---|---|
| IV | landscape and regional pest management | landscape–region | regional coordination of habitat, landscape management | years | complex experimental and experiential knowledge | What are downstream effects of management practices? How are communities affected? Regional management of resistant weeds and invasive plants. How to structure agriculture over a region to minimize the impact of weeds? |
| V | agro-ecoregion management | regional–global | national policies | historical | very complex interactions of historical knowledge with global economic or climate models | Regional/global trends in pest populations with respect to historical changes in weather patterns or atmospheric $CO_2$ levels. Large scale migration of multiple pests interacting with global trade patterns. |

has put enormous selection pressure on weed populations. As a result, growers are now concerned about herbicide resistant and cross-resistant biotypes appearing in their fields only a few years after the adoption of a confusing suite of herbicide products. These problems have arisen not because of the tools themselves (i.e., herbicides), but because of the narrowly focused manner in which they have been used in Level I for weed control rather than for crop protection.

The primary means of managing herbicide resistance at this level are (1) use of an alternative herbicide to which the weed is not resistant; (2) tank mixing herbicides of different modes of action; and (3) rotating among herbicides with different modes of action. Because the focus remains on herbicides as the primary means of weed control, the solution to a resistance problem is simply to change or mix herbicides (Gressel and Segel, 1990). The occurrence of herbicide resistance has increased and now includes many weed species with resistance to common herbicides with varying modes of action (Powles et al., 1997). Therefore, it is not always possible to find effective alternatives or companion products that do not pose risks of resistance development. The potential for over application and inappropriate tank mixing of herbicides is high when management of herbicide resistance is limited to use of alternative herbicides.

## LEVEL II: WEED MANAGEMENT

At the next level of integration is the use of more than one method for reducing weeds below acceptable levels of infestation (VanGessel, 1996). Describing similar goals, Buchanan (1976), Elmore (1996), Swanton and co-workers (1991; 1996), and Liebman and Gallandt (1997) have all pointed to the need for an array of technology options including crop rotation, mechanical controls, herbicides, and crop competition. Recent examples of the use of multiple technologies are described in Table 2. A common theme at Level II is reduction in herbicide use for the sake of lowering the cost and environmental impact of herbicide-dependent weed control programs. In row crops, reduced herbicide use is often offset by inter-row cultivation or other mechanical control methods. The net economic returns of Level II systems compared with herbicide-based systems have generally favored the multiple technology approach (Buhler et al., 1996, 1997). The reduction in herbicide use has generally been used to imply a lower environmental impact, although the total environmental and economic costs of additional tillage and cultivation are difficult to compute.

The spatial scale at this level of integration ranges from that of a field to an entire farm, and the temporal scale ranges from one to several growing seasons within a crop rotation. Thus, the goal is to reduce population levels

TABLE 2. Examples of combinations of technologies and tools used at Level II integration.

| Technologies Used | Comparisons / Objectives | Conclusions | Reference |
|---|---|---|---|
| stale seedbed, tillage, herbicides | control weeds | Main benefit is timely planting and stand establishment. | Shaw, 1996 |
| ridge-tillage, cultivation, herbicides | control weeds | Integrating cultivation and herbicides controls a broader spectrum of weeds than cultivation or herbicides alone. | Klein et al., 1996 |
| herbicides, in-row cultivator, rotary hoe, crop rotation, seedbank model | reduce seedbank, control weeds | Moderate inputs needed to reduce and maintain the seedbank; high herbicide inputs are not justified for seedbank depletion or economics. | VanGessel et al., 1996 |
| crop rotation, tillage systems, herbicides | determine interactions of rotation, tillage, and intensity of weed control on economics and soil erosion | Developed an economical 3-year crop rotation that controlled weeds, reduced erosion, and reduced risk. | Young et al., 1996 |
| cultivation, cover crops, herbicides, planting patterns, crop rotation, competitive crop cultivars | improve agroecosystem health | Adoption of multiple technologies will affect water quality, soil quality, and crop productivity. | Swanton and Murphy, 1996 |
| crop rotation, cultivation, herbicides | control weeds, reduce seedbank, maintain yields | Low herbicide inputs with cultivation were as effective as full-rate herbicide inputs. | Mulugeta and Stoltenberg, 1997 |
| "sanding," hand removal, water management, herbicides | inhibition of dodder emergence in cranberry | Sanding reduces initial population, lessening need for postemergence control. | Sandler et al., 1997 |

247

over time rather than simply to control a given weed in the short-term to grow a single crop. Because the spatial scale is at the field level, the weed seedbank becomes important at Level II, and with it the recognition that numerous species pose a threat to crop production and must be managed. The application of multiple tools helps to assure that shifts to alternative competitive species either do not occur, or can be managed using various tools applied at different times within a growing season and throughout the rotation, with greater potential to impact multiple species at critical phases in their life cycles (Hass and Streibig, 1982).

Level II integration makes use of information from studies on individual components (i.e., herbicides, cultivation, etc.). This approach recognizes that single-tactic management has often been ineffective in long-term weed management and that reliance on a single tactic has resulted in shifts among weed species or to herbicide resistant biotypes (Gressel and Segel, 1990; Wrubel and Gressel, 1994). Integrating multiple biological, mechanical, and chemical agents favors a decline in seedbanks and reduction in populations. The need for multiple tools is as important for chemical-dependent management systems as it is for biological control. In successful examples of classical biological control of weeds, more than one organism acts in concert with other stresses to suppress the weed population (Charudattan, 1986; Maddox and Mayfield, 1979). For instance, musk thistle (*Carduus nutans*) control has been successful with a combination of insects that attack the flowerhead (*Rhinocyllus conicus*) and the vegetative rosette (*Trichosirocalus horridus*) (McDonald, 1993). Because the focus is on population management, the Level II approach can make use of technologies that by themselves are not considered to be effective, but contribute to weed suppression as part of an integrated system.

Seedbank, emergence, and bioeconomic models have become important knowledge-based tools for decision making at this level of integration (Forcella, 1993; Forcella et al., 1997; Mohler, 1993). The use of models represents the integration of a knowledge-based tool to guide management tactics. Models add knowledge to decision-making because they evaluate biological properties of the system, such as seedbanks, with reference to recent research, such as studies on emergence. In addition, some models consider multi-year effects of seed return from sub-threshold populations of weeds (Bauer and Mortensen, 1992). Such considerations would not be possible without population dynamics models. In practice, bioeconomic models have been used mostly for choosing which herbicide is most appropriate, and possibly the rate and time of application. A bioeconomic model for weed management in corn selected mostly triazine-based herbicide programs, plus one or no inter-row cultivation treatments, resulting in net economic returns that were not different from "standard" herbicide treatments (Buhler et al., 1996). In soybeans, a bioeconomic model selected herbicide and cultivation combinations that were different from the "standard" treatment and increased economic returns 50% of the time (Buhler et al., 1997).

Field mapping and site-specific technologies are relatively recent approaches to weed management. Patch sprayers and other site-specific herbicide application methods that increase the efficiency of herbicide application are among the current adaptations of this technology. The goal of precision herbicide applicators is to detect the weed, select an appropriate control (i.e., herbicide), and apply it at the proper rate where the weed is present (Sudduth, Hummel, and Birrell, 1997). However, site-specific weed management does not need to be limited to herbicide control, since the spatial distribution of many other factors that modify and regulate weed populations (such as cultivars, plant density, or companion crops) might be arranged more efficiently. Herbicide resistant crops and competitive crop cultivars are examples of the application of gene technology in crop production as a tool for improving the efficiency of weed control.

Level II integration represents progress beyond basic technology testing, and is founded on the fundamental IPM concept of the use of multiple and diverse technologies (Bottrell, 1979). The multiple tool approach can exploit synergistic or cumulative effects that are not evident when a single technology is used. By basing management on several different methods, the risk of failure is lessened if one approach fails. There seems to be no technological limit to the number of tools that can be directed toward weeds, including allelopathic cover crops, biological control agents, herbicides banded at reduced rates, inter-row cultivation, competitive herbicide-tolerant crop cultivars, and other approaches within crop rotations (Liebman and Gallandt, 1997). Use of these tactics will possibly come to represent the standard agronomic practice. Economic considerations might be the limiting factor for the number and complexity of approaches that a grower is willing to invest for weed management. However, secondary benefits of components such as cover crops, reduced tillage, and rotations may contribute to overall crop productivity but are difficult to quantify.

Herbicide resistance management at Level II is attempted through the use of multiple tools to reduce selection pressure by a single mode of action herbicide. Rotating herbicide mode of action is often coupled with crop rotation or secondary tillage as a means of suppressing resistant weeds. Other approaches, such as altering planting date or late season herbicide applications to hinder weed seed production, along with herbicide mode of action rotation, could also be effective in delaying selection or limiting the spread of herbicide resistance. For example, ALS-resistant kochia (*Kochia scoparia*) emerges earlier in the season than susceptible biotypes (Dyer, Chee, and Fay, 1993); therefore, altering time of field operations might help to manage resistance in this species.

One drawback of this level is that if Level I can be described as "tools," Level II might be described simply as "more tools." In many articles describing a multiple tool approach, the experimental treatments have involved comparisons such as herbicides alone versus herbicides plus cultivation (Buhler, Gunsolus, and Ralston, 1993; Burnside et al., 1994; Poston, Murdock, and Toler, 1992). Among the few cases where combinations of three technologies have been studied, crop or companion crop competition has been a common feature (Malik, Swanton, and Michaels, 1993; Shilling et al., 1995; Teasdale, 1995). However, where competitive crop cultivars, alternative planting patterns, or reduced herbicide rates have been used, the focus has been restricted to the effect of the multiple tools on weed control and crop yields. More could be done to address biological interactions or fundamental approaches to cropping system design that might help in the development of more efficient management systems.

The second difficulty with Level II is the lack of clear association with IPM and the many alternate terms that have been used to describe weed management systems that are, in fact, IPM. The use of alternate designations for weed IPM research or management systems separates weeds from other pests and thereby weed science from other pest disciplines that traditionally have greater opportunity for IPM funding (Coble, 1994). Governments recognize IPM as an established area for funding. Separating weeds from IPM by a confusion of terms only assures that support will go to other disciplines at the expense of weed science research and extension programs. Furthermore, this separation gives the impression that weed scientists are not interested in IPM, but are stuck at Level I thinking. Regardless of terminology, much excellent research has been conducted at Level II integration, much of it motivated by a desire to meet the economic and biological needs of growers as well as the environmental needs of society. For example, Swanton and Murphy (1996) have made an effort to put Level II weed management into the context of "agroecosystem health," as measured by indicators of quality, productivity, and efficiency.

We believe that there are also higher levels of integration at which weed management can be addressed. Because they involve larger spatial and temporal scales, they are difficult to study. The linkage of weed management at Levels I and II with the broader levels described below is an important challenge for weed science.

## LEVEL III: CROPPING SYSTEM DESIGN

The third level of integration represents the threshold of innovation in weed science and addresses fundamental questions of cropping system–and perhaps agricultural ecosystem–design. Instead of a technology focus, which emphasizes how to use the tools of weed management, this level focuses on how to design and manage the cropping system itself. The tools developed at Levels I and II will obviously play an important role, but Level III recognizes that cropping systems–the conglomerate of management practices directed to crop production–result in physical, chemical, and biological characteristics that can influence weed populations. These characteristics may not be apparent if we focus solely on the tools of weed management. For example, in cropping systems where particular crop and cover crop residues are returned to the soil, biological interactions might occur that would result in microbial communities and a soil chemical environment that would favor some weed species over others. The only way to study such complex interactions is in long-term cropping system experiments. Shifts in weed species composition of seedbanks in long-term tillage experiments suggest that cropping systems may have such effects (Cardina, Regnier, and Harrison, 1991; Egley and Williams, 1990; Oryokot, Murphy, and Swanton, 1997).

Making use of Level III integration will require integration of knowledge from more subject areas than any single individual can possess, and will require use of new tools to synthesize information. This will also require integration of experiential and indigenous knowledge with experimentally generated information and knowledge. For example, there is abundant and accurate non-experimental knowledge among people who have explored and observed weed populations over time, and who often have a sense of relationships among species or between soil characteristics and species adaptation. Cultures with sophisticated agricultural systems have existed for centuries without digital electronics (Rhoades, 1997). Biological systems are extremely complex, and knowledge about them comes from many sources, all of which should be considered and utilized in management systems.

The spatial scale at Level III ranges from the farm to the local landscape. Weeds are not uniform across fields, so it makes little sense to manage weeds–or crops–uniformly across fields. Cropping systems should be designed with the landscape characteristics in mind. This concept is in practice in agricultural areas with a rolling terrain, where hillsides are pastured and bottom land is devoted to row crops. It is at this level of integration that limiting weed seed spread to vulnerable parts of the farm or landscape settings becomes important. It is also at this level that "downstream" effects of management become important considerations, not just to follow the pending regulations, but for stewardship of the land.

At Level III, weed management could be viewed in terms of strategies that derive from cropping system design, rather than as a collection of technologies or tactics. All practices relating to crop production are considered with respect to how they might contribute to managing weed populations. In other words, the cropping system itself is designed as a weed management strategy.

The time scale extends beyond a rotation cycle to many years, with the expectation that species composition will change and populations fluctuate, but the overall difficulty of weed management will decrease, or at least remain constant. A goal at this level is to achieve what Jordan and Jannink (1997) referred to as "genostasis" in the weed population by depriving weeds of sufficient genetic variation to allow them to adapt to the management system. At Level III integration, a grower would consider the *susceptibility* (the probability of invasion) of the cropping system to weeds and the *vulnerability* (the degree of damage caused by weeds) for each crop in the rotation. A grower would design the cropping system to allow for successful management where invasion and damage have already occurred, such as rotating to a crop where effective control of the particular weed can be accomplished. The system should also enhance the ability of susceptible crops to *tolerate* future weed outbreaks and be more competitive, so as to reduce weed survival and regeneration. Weed tolerance may be accomplished by increasing the carrying capacity of the system so weeds have less impact. Therefore, important research questions at this level involve developing ecological principles that underlie the impact of cropping systems on weed populations.

*1. Reduce cropping system susceptibility to invasion* by weeds that have the ability to spread, compete, dominate, and persist to an extent that extraordinary measures are required to grow crops in an economical and ecologically responsible manner. There has been little research to develop cropping systems that resist weed invasion, because the design and funding of such studies are more difficult than for Level I and II studies. Many interacting principles could be explored to develop weed resistant cropping systems; three of these are described below.

*a.* One ecological principle behind the development of cropping systems that resist weed invasion and establishment is that weeds occupy ecological niches not utilized by crop plants. Therefore, the cropping system must be designed to maximize resource capture by the crop or other desirable species. Resources are stratified above and below ground and are dispersed in horizontal space as well as in time. Therefore, desirable species need to be included in the cropping system to maximize resource capture over space and time. For example, a multi-species combination of cover crops with different life cycles, rooting depths, and canopy architectures, could sequester resources above and below ground and minimize the potential for weed invasion (Burke and Grime, 1996; Thebaud et al., 1996). These cover crop communities could be selected to provide habitat for weed seed predators that would reduce seed survival of invading species (Reader, 1991). They could also elicit allelopathic chemicals that would inhibit seedling emergence and establishment of invaders (Bradow and Connick, 1990; Putnam, DeFrank, and Barnes, 1983).

*b.* Another principle is that weed invasion is an ecological response to disturbance, which stimulates germination of dormant seeds and provides a habitat for invasion of species from other locations within or outside of the field or farm (Lavorel, O'Neill, and Gardner, 1994). To design cropping systems that resist weeds, disturbances that allow for colonization by invaders must be minimized. It will be impossible to eliminate disturbance in agroecosystems, which are by their nature disturbed systems, but the intensity and duration of disturbances can be reduced, or at least be timed so that they are not synchronized with major periods of weed seed dispersal or emergence. For example, it might be possible to link disturbance events to multi-species cover crop life cycles in a way that would shift seedbanks to communities of desirable or non-competitive plant species.

*c.* A third principle that could be applied to cropping systems is that, in spite of being sessile organisms, the survival of weed populations depends on dispersal of propagules to suitable niches. Weed spread is a result of processes of dispersal interacting with disturbance. Assuming that dispersal comes from outside a given field or farm, efforts to resist invasion must first avoid or block dispersal of propagules into susceptible fields. This requires that managers focus on a broader spatial scale to determine where threats of dispersal are, and to avoid movement therefrom. Management of troublesome species outside the field or farm may be justified to avoid spread to growing areas. For example, many perennial weeds tend to grow harmlessly outside of crop fields, until disturbance by herbicides provides an open and suitable niche for invasion by seed dispersal or creeping rhizomes or roots. Management of such species at a landscape level might effectively reduce potential for dispersal and repetition of the cycle of disturbance and invasion.

*2. Reduce cropping system vulnerability* to weeds; in other words, reduce the ability of weeds to affect crop yield and quality. This involves development of competitive crop varieties that suppress weed growth, and that yield well in spite of the presence of weeds. Competitive crop cultivars have been studied for over 20 years as an approach to lessening the impact of weeds on crop productivity (Malik, Swanton, and Michaels, 1993; Wicks et al., 1994). Some progress has been made in the development of competitive crops, with emphasis on the ability of varieties to suppress weed growth and development (Bussan et al., 1997; Monks and Oliver, 1988). The need for crops that produce well in the presence of other plants will be necessary for tolerance of weeds, and for crop production in the presence of some cover crops. These might be perennial cover crops or living mulches (such as crownvetch, *Coronilla varia*), or cover crops that senesce after the main crop has become established (such as subterranean clover, *Trifolium subterraneum*) (Enache and Ilnicki, 1990).

An alternate approach to reducing cropping system vulnerability to damage caused by weeds would be to design systems with increased carrying capacity. The idea here is that resources available for plant growth are not fixed. If the equilibrium population density can be increased, then more weeds can be tolerated without sacrificing crop productivity. The high level of variability over years and locations in crop yield loss and economic threshold studies suggests that the carrying capacity varies with environmental conditions. It is common practice to alter the carrying capacity of agricultural systems with irrigation or fertilizers. But it may also be possible to modify the carrying capacity by system design. For example, a cropping system that fixes large quantities of carbon in the soil would favor nutrient cycling, support microbial communities that release nitrogen over a long period of crop growth, provide higher levels of water and nutrient holding capacity, and allow for increased rooting volume. To develop systems that favor carbon fixation would require creative cropping system designs. In and of themselves, these systems would not be expected to "control" weeds, but they would decrease the damage caused by weeds and thereby reduce the need for additional weed management inputs.

*3. Reduce weed population survival and persistence.* To be "successful," an individual weed has to provide only a single like individual to replace itself in the next generation. The survival strategy of most annual weeds is to produce a large quantity of seeds with inherently variable dormancy, germination, and growth response characteristics, to ensure survival of progeny. This strategy tolerates the loss of many individuals to guarantee survival of the population. One form of loss is density-dependent mortality, which occurs when initial population levels exceed that which can be supported by available resources. Non-density-dependent losses that constrain weed populations include predation, disease, and of course, many cultural, biological, and chemical weed management practices. At Level III integration, cropping systems would be designed to maximize natural constraints on weed populations. The idea here is to determine the sources of natural weed mortality and to design systems in which these forces operate effectively to help reduce the weed population.

An example of a natural constraint that has not been exploited is seed mortality. Although there are reports of seeds surviving for centuries (under unnatural conditions) or decades (under natural conditions), this represents only a small fraction of the seed population (Fenner, 1995). In fact, many seeds die before they are able to germinate, and many seedlings die before they are able to emerge. Several studies have reported that 80% or more of seeds in the seedbank are dead, or die between fall and spring (Forcella, 1992; Forcella et al., 1992; Gross, 1990; Gross and Renner, 1989). Up to half of the weed seedlings that make their way toward the soil surface fail to emerge (Sanchez del Arco, Torner, and Quintanilla, 1995). Even for species like velvetleaf (*Abutilon theophrasti*), whose seeds have reportedly survived 50 years (Warwick and Black, 1988), recent work shows that only 30% survive from fall to spring (Lindquist et al., 1995). If natural causes of mortality were understood, management systems could be designed to enhance the level of seed mortality that occurs in the seedbank throughout the year. For example, Lueschen et al. (1993) showed that the rate of velvetleaf seed loss was more than three times as fast in continuous corn or oats (*Avena sativa*) than in continuous alfalfa (*Medicago sativa*). These findings suggest that cropping systems could be designed to maximize the effect of soil macro- and micro-organisms on weed seed survival.

Standard cultural, mechanical, and chemical tools will still play important roles at Level III, because it is unrealistic to think that systems can be designed to completely eliminate weeds. Therefore, some level of directed therapeutic herbicide use is compatible with this level of integration. It is unlikely that herbicide resistance would develop in such a system, because of the complexity of approaches, and because the system itself would be designed to protect against such adaptation. A more likely scenario is that herbicide resistant biotypes might be introduced from outside by various dispersal mechanisms. However, the expectation is that selection pressure for development and dominance by resistant biotypes would be very low.

## LEVEL IV: LANDSCAPE AND REGIONAL MANAGEMENT

There are additional levels of integration at which weed populations could be studied and managed, especially those associated with spatial scales of geographic dimensions. It is at these levels that regional habitat management and land use patterns are considered with respect to movement and invasion of weeds. This is beyond the scope of a single farm focused manager, and requires community support and wide scale regional cooperation to make IPM work. Because of the lack of research at these levels, our discussion will be somewhat speculative.

The spatial scale at Level IV ranges from landscape to regional, and includes the current distribution of weeds as well as their potential distribution. Because of this large scale of interest, new technologies and methodologies for sampling, detecting, and mapping weeds will be required. It is at this level of integration that new spatial models will be developed to help understand the basis for current weed distribution patterns and to anticipate future spread. It is also at this level that the social, environmental, and economic impact of weeds and weed management practices must be considered.

The ecological basis for a landscape perspective of weed populations rests on the observation that variations in soil type and drainage associated with topography result in corridors along which species can easily spread and barriers across which spread is more difficult (Cousens and Mortimer, 1995). This results in a mosaic of habitats and a corresponding mosaic of weed communities. The positions of some larger scale landscape features, such as patches of a particular soil type or drainage pattern, may be considered relatively stable over time. Terrain characteristics determine the distribution of soil water and temperature in fields, and thus the distribution of soil and microclimate resources that influence seed dormancy status, seedling recruitment rate, and possibly other population processes within the landscape. Landscape characteristics also dictate land use, such as the type of farming, and therefore the types of weeds associated with particular management practices. Richardson and Bond (1991) concluded that soil environmental stresses were predominant factors determining the extent of invasion by a species. Therefore, the position and stability of landscape features over time and space might be used to model the spatial distribution of weed populations over several scales of interest.

The proximity of fields to seed sources, such as waterways, road ditches, pastures, conservation reserve land, and wooded areas, influences not only the patchiness of weed species, but also the magnitude and direction of species invasion across a landscape (Bunce and Hallam, 1993). Presumably, local environmental conditions differ with proximity to landscape features, and cause differences in processes that control weed seedling establishment, survival, and spread. Field boundary populations of *Bromus sterilis* were found to be the major seed source for field reinfestations in spite of herbicide applications within and outside the field (Theaker, Boatman, and Froud-Williams, 1995). The diversity of weed species is generally greater along field edges than in field interiors, because of differences in the light environment, proximity to alternative seed sources, movement of animals and machinery, and variable herbicide dosage (Cavers and Benoit, 1989). Ratio of the field perimeter to total field area is likely to be important in determining weed invasibility. Therefore, population processes probably exhibit considerable spatial heterogeneity that is directly related to the nature of the landscape. However, few weed population studies have examined environmental interactions at the landscape and regional scales.

An example of Level IV integration is the regional effort to manage leafy spurge (*Euphorbia esula*) with combinations of conventional and biological controls (USDA-APHIS, 1992). Considerable cooperation among agencies, states, and communities has made this possible. Another example is the regional effort to contain witchweed (*Striga asiatica*) in 11 counties in the Carolinas (Sand, Hampstead, and Manley, 1990). Efforts to manage the invasion of exotic weeds in Florida require a spatial and temporal perspective that is well beyond that of previous levels of integration (Hall, Currey, and Orsenigo, 1997; White and Westbrooks, 1997). Qualitative and quantitative information derived from various knowledge sources, such as simulation results, historical data, and heuristic information, must be integrated. A key technology for investigation of weeds at the landscape and regional levels is geographical information systems.

The need for Level IV integration can be understood by considering the history of some of the more important weeds in the U.S. cornbelt. Velvetleaf was introduced to the U.S. as a fiber crop in pre-colonial days (Spencer, 1984) and became a major problem in soybean and corn. Giant foxtail has become the most troublesome annual grass weed in corn and soybeans, but was not an important weed before about 1950 (Pohl, 1951). The Level I approach to velvetleaf and giant foxtail management was to develop and test new herbicides in an attempt to find more effective chemical controls. A Level IV approach would have been to develop a regional effort to contain these weeds and to limit their future spread before they expanded to the entire Midwest and portions of PA, NY, VA, and southern Canada, as well as CA. Such an effort might still be relevant to constrain the spread of new or existing species. To accomplish this, knowledge of weed biology and adaptation could be used to identify areas where cultural practices could be used to reduce the potential spread and impact of this weed. It would be important to consider what characteristics of cropping systems make them susceptible to velvetleaf, as described for Level III. There remain many cropping systems where velvetleaf is not a problem weed (such as most vegetable and fruit production systems as well as turf and pastures), even within the main region of its current distribution. Mapping the historical distribution of velvetleaf and associated crops, landscape characteristics, and management methods would be an important first step. This would illustrate that the spread of velvetleaf has been greatly enhanced by the expansion of soybean acreage, the large scale use of acetanilide herbicides with little activity on this weed, and the use of conventional rather than reduced tillage systems. Maps could also be used to explore the potential influences of soil properties, landscape characteristics, presence of seed predators, soil pathogens (verticilium), foliar pathogens (several), and other site characteristics on velvetleaf distribution. From this information, a regional management plan could be developed, much like that used to reduce the impact of the boll weevil in cotton (Pencoe and Phillips, 1987). The regional plan would take into consideration the location of farms relative to the weed's distribution. Controlling velvetleaf on a farm is made more difficult by uncontrolled infestations on adjacent farms, due to the many modes of seed dispersal. If the farm is at the edge of the regional infestation, or outside of it, the spatial implications for control and invasion protection measures will be quite different.

If the weed is herbicide resistant, it becomes even more important to consider the regional distribution of the weed with respect to the regional distribution of control practices. Now that growers are required to keep records on herbicide applications, these data could be made available to map the distributions of herbicide use and herbicide resistant weeds. A grower's proximity to resistant weed outbreaks, surrounding management practices, and routes of weed dispersal would be valuable information to support management decisions needed to protect farms from spread of resistant biotypes. A coordinated effort to monitor and map historical trends in resistant weed populations with respect to landscape features and field management could help support regional management programs.

## LEVEL V: ECO-REGION MANAGEMENT

There is probably at least one more level of integration that functions at spatial scales of agricultural eco-regions and beyond. This is the scale at which national agricultural policy and world food trade policies might affect dispersal of weeds as contaminants in agricultural products. Even though weed populations may not function at a global level, easing of trade restrictions is a concern for importation of exotic invasive species. Global concerns also relate to exploration for biological control agents in native habitats to help manage weeds that have escaped without associated pests. At this stage of integration, the impact of global warming on weed populations becomes an issue (Patterson, 1995), and extremely complex models are needed to help describe these changes with respect to anticipated dispersal patterns. Efforts at Level V will probably focus on methods that will preserve weed, non-weed plant, and animal population stability in an effort to minimize selection pressure by humans.

Goals for managing weeds at a regional or global scale are as yet unclear. Weeds and other non-crop plants play essential roles in ecosystems as pioneer species, erosion protection, nutrient cycling, wildlife habitat, and sources of valuable genetic material, to name a few (Crossley et al., 1984). Therefore, it is important to consider how to manage global weed populations to balance these needs with needs of crop production. Plants that are considered to be weeds in one region of the world may be economic crops or important sources of germplasm elsewhere.

Management of plant populations with the skills and resources available to weed science could be brought to bear on the issue of global plant biodiversity. In agricultural production, the importance of maintaining biodiversity in weed populations is not easy to determine, regardless of whether the scale of interest is a single field or the collective area devoted to food production. The changes that weed management practices have had on species composition are receiving increased attention with concerns over endangered as well as invading species (Froud-Williams, 1997). Herbicides, and weed management in general, are less of a threat to global plant biodiversity than are habitat disturbance and colonization by invasive plants. However, there are several cases of near extinction of species formerly considered to be agronomically important weeds (Wilson, Boatman, and Edwards, 1990), mostly due to improvements in seed cleaning. Over a 20-yr period, most common weeds in Danish fields declined and two species were nearly eliminated due to intensive herbicide use (Andreasen, Stryhn, and Streibig, 1996). In other cases, increased herbicide use has allowed for the invasion of weeds (Theaker, Boatman, and Froud-Williams, 1995). Therefore, common practices, when carried out on many farms, can have a large impact on regional–and possibly global–plant communities.

The time scale at Level V integration can range from years to several decades to consider how management affects long-term changes in weed populations. The focus here is on abundance and species composition as well as biodiversity issues. An insight into long-term changes in weed populations can be obtained by reading early studies on weeds. In the first Ohio weed book (Selby, 1897), for example, today's most important annual grass weed (giant foxtail) did not appear. Velvetleaf and giant ragweed, the two most important broadleaf weeds (Loux and Berry, 1991), were considered occasional roadside weeds. Selby gave considerable attention to the weeds he considered to be most troublesome, the most important of which was red sorrell (*Rumex acetosella*). In other words, the agriculture of the late 1800's, based on forages, lacking adequate seed cleaning technology and information on the importance of liming to adjust soil pH, selected for an acid-tolerant contaminant of alfalfa seed. With the Level V approach to integration, weed scientists might consider how current trends in agriculture, with large scale use of herbicide tolerant crops, monoculture production systems, and reduced tillage, together with global climate changes, are selecting for tomorrow's weeds. If Selby's (1897) book is any guide, those weeds, which may be increasingly difficult to manage, may be on our roadsides and in our fence rows. Non-weedy plant populations exposed to selection pressure in annual cropping systems can evolve multiple-trait phenotypes similar to adapted weed populations within relatively few years (Jordan, 1989). There remains a need for botanical and genetic studies of potentially weedy species.

The importance of herbicide resistance at Level V has a global dimension. Of the 100-plus reported cases of weeds that have developed resistance (Powles et al., 1997), many are relatively localized. Resistant biotypes have spread rapidly to adjacent arable fields and non-crop lands by natural and human-aided dispersal (Thill and Mallory-Smith, 1997). Annual/Italian ryegrass (*Lolium* spp.) with multiple resistance is widespread throughout Australia. Movement of this, or any other, herbicide resistant weed around the world to contaminate soil seedbanks in major food production regions represents an important threat to agriculture where weed management relies on herbicides with limited modes of action. Widespread use of glyphosate tolerant soybeans, corn, and other crops will increase the potential for resistance development in weeds by increasing the scale of the selection pressure on susceptible weed populations.

A related issue relevant to Level V is the global trade in genetically modified herbicide tolerant crop plants that have potential to introgress with weedy relatives. Gene transfer is more likely for crops with close weedy relatives than for crops whose domestication has resulted in ecological and reproductive isolation (Ellstrand and Hoffman, 1990). Transgene movement between genetically modified *Brassica* crops and weedy species is possible and inevitable (Snow and Palma, 1997). If commodity shipments drop transgenic crop seeds where related wild relatives are a problem, spread from a nascent focus can be rapid (Thill and Mallory-Smith, 1997). The potential ecological consequences of the escape and spread of resulting progeny have implications for weed management (herbicide tolerant weeds) as well as wildlife (toxicity and allergenic effects) (Rogers and Parkes, 1995). The shipping of herbicide-resistant crop seeds as a global trade commodity has become a national political issue in Canada, Japan, Mexico, and the U.S., but resistance management specialists have apparently been left out of major decisions that require a Level V perspective of weed populations (Gressel, 1997).

## CONCLUSIONS

As we consider the progression of integration from weed control to eco-region management, several trends become evident, and more questions than answers arise regarding the direction of current and future weed IPM research. Whereas IPM theory and practice are dominant features of the training in other pest disciplines, there is a large knowledge gap in IPM theory related to weeds. Moreover, at the higher levels described in this paper, with associated larger spatial and temporal scales of interest, there is little information available that would provide the basis for establishment of IPM concepts.

The most perplexing observation from this appraisal of levels of integration is the increasing difficulty in defining appropriate goals of weed management, particularly at the more advanced levels of integration. The goal of weed eradication has generally been considered unrealistic and probably unwise, although for some weeds in some habitats such a goal might be acceptable (Sand, Hampstead, and Manley, 1990). In arable fields, efforts to eradicate a given weed will inevitably result in replacement by another weed–from large, species-rich seedbanks or from propagules dispersed from other cropped or non-crop fields. The goal of weed suppression, maintaining populations at low levels, is consistent with a general IPM approach, but acceptable population levels and associated levels of seed return are difficult to define, or defend, to growers. A goal of protection of agricultural and other ecosystems from invasion by potentially harmful species is one that might find more general acceptance. Methods for detection remain in development, but determining who is responsible for monitoring and designing landscapes to limit invasion is a difficult practical question. Once an invasive species has been introduced, eradication might be an appropriate goal. The goal of maintaining biodiversity in seedbanks or in entire ecosystems might work to protect against invasive species and to limit the potential for development of resistance to control measures. For example, Jordan and Jannink (1997) suggested that spatial and temporal diversification of agricultural systems at a landscape scale might impede weed adaptation and lead to genostasis within weed populations. Weed science research at Level IV, in cooperation with landscape ecologists, will be needed to determine how to achieve this diversification.

The various levels of integration are all necessary for a comprehensive approach to development of IPM principles for weed management. The different goals associated with the levels of integration will necessarily vary, depending on the weeds of interest, their habitat, and the agricultural ecosystem in which they are found. The past and current emphasis on Level I is logical because this level deals with concrete steps an individual can take to solve immediate needs. This is favored by funding sources and current approaches to education. The lack of research and education at higher levels of integration is unfortunate but not surprising. However, students who enter the discipline of weed science with a desire to make a contribution to agriculture might consider the relative long-term impact of the direction of their studies and research relative to their contribution to the advancing levels of integration.

## REFERENCES

Andreasen, C., H. Stryhn, and J. C. Streibig. (1996). Decline of the flora in Danish arable fields. *J. Appl. Ecol.* 33:619-626.

Bauer, T. R., and D. A. Mortensen. (1992). A comparison of economic and economic optimum thresholds for two annual weeds in soybeans. *Weed Technol.* 6:228-235.

Bloomberg, J. R., B. L. Kirkpatrick, and L. M. Wax. (1982). Competition of common cocklebur (*Xanthium pensylvanicum*) with soybean (*Glycine max*). *Weed Sci.* 26:556-559.

Bottrell, D. G. (1979). *Integrated Pest Management.* Council on Environmental Quality, U.S. Government Printing Office, Washington, DC.

Bradow, J. M., and W. J. Connick, Jr. (1990). Volatile seed germination inhibitors from plant residues. *J. Chem. Ecol.* 16:645-666.

Bridges, D. C. (1992). *Crop Losses Due to Weeds in the United States–1992.* Weed Science Society of America, Champaign, Illinois. 403 pp.

Buchanan, G. A. (1976). Management of the weed pests of cotton (*Gossypium hirsutum*). In *Proc. U.S.-U.S.S.R. Symposium: The integrated control of the arthropod, disease and weed pests of cotton, grain sorghum and deciduous fruit,* pp. 168-294, Lubbock, TX.

Buhler, D. D., D. E. Stoltenberg, R. L. Becker, and J. L. Gunsolus. (1994). Perennial weed populations after 14 years of variable tillage and cropping patterns. *Weed Sci.* 42:205-209.

Buhler, D. D., J. L. Gunsolus, and D. F. Ralston. (1993). Common cocklebur (*Xanthium strumarium*) control in soybean (*Glycine max*) with reduced bentazon rates and cultivation. *Weed Sci.* 41:447-453.

Buhler, D. D., R. P. King, S. M. Swinton, J. L. Gunsolus, and F. Forcella. (1996). Field evaluation of a bioeconomic model for weed management in soybean (*Glycine max*). *Weed Sci.* 45:158-165.

Buhler, D. D., R. P. King, S. M. Swinton, J. L. Gunsolus, and F. Forcella. (1997). Field evaluation of a bioeconomic model for weed management in corn (*Zea mays*). *Weed Sci.* 44:915-923.

Bunce, R. G. H., and C. J. Hallam. (1993). The ecological significance of linear features in agricultural landscapes in Britain. In R. G. H. Bunce, L. Ryszkowski, and M. G. Paoletti (eds.). *Landscape Ecology and Agroecosystems,* pp. 11-19, Lewis Publishers, Boca Raton, FL.

Burke, M. J. W., and J. P. Grime. (1996). An experimental study of plant community invasibility. *Ecology* 77:776-790.

Burnside, O. C., W. H. Ahrens, J. J. Holder, M. J. Wiens, M. M. Johnson, and E. A. Ristau. (1994). Efficacy and economics of various mechanical plus chemical weed control systems in dry beans (*Phaseolus vulgaris*). *Weed Technol.* 8:238-244.

Bussan, A. J., O. C. Burnside, J. H. Orf, E. A. Ristau, and K. J. Puettmann. (1997). Field evaluation of soybean (*Glycine max*) genotypes for weed competitiveness. *Weed Sci.* 45:31-37.

Cardina, J., E. Regnier, and K. Harrison. (1991). Long-term tillage effects on seed banks in three Ohio soils. *Weed Sci.* 39:186-194.

Cavers, P. B., and D. L. Benoit. (1989). Seed banks in arable land. In M. A. Leck, V. T. Parker, and R. L. Simpson (eds.). *Ecology of Soil Seed Banks,* pp. 309-328, Academic Press, San Diego.

Charudattan, R. (1986). Integrated control of waterhyacinth (*Eichornia crassipes*) with a pathogen, insects, and herbicides. *Weed Sci.* 34 (suppl.):26-30.

Coble, H. D. (1994). Future directions and needs for weed science research. *Weed Technol.* 8:410-412.

Cousens, R., and M. Mortimer. (1995). *Dynamics of Weed Populations.* Cambridge University Press, Great Britain. 332 pp.

Crossley, D. A., G. J. House, R. M. Snider, R. J. Snider, and B. R. Stinner. (1984). The positive interactions in agroecosystems. In R. Lowrance, B. R. Stinner, and G. J. House (eds.). *Agricultural Ecosystems: Unifying Concepts,* pp. 73-81, John Wiley and Sons, New York.

Dyer, W. E., P. W. Chee, and P. K. Fay. (1993). Rapid germination of sulfonylurea-resistant *Kochia scoparia* L. accessions is associated with elevated seed levels of branched chain amino acids. *Weed Sci.* 41:18-22.

Egley, G. H., and R. D. Williams. (1990). Decline of weed seeds and seedling emergence over five years as affected by soil disturbance. *Weed Sci.* 38:504-510.

Ellstrand, N. C., and C. A. Hoffman. (1990). Hybridization as an avenue of escape for engineered genes. *BioScience* 40:438-442.

Elmore, C. L. (1996). A reintroduction to integrated weed management. *Weed Sci.* 44:409-412.

Enache, A. J., and R. D. Ilnicki. (1990). Weed control by subterranean clover (*Trifolium subterraneum*) used as a living mulch. *Weed Technol.* 4:534-538.

Fenner, M. (1995). Ecology of seed banks. In J. Kigel and G. Galili (eds.). *Seed Development and Germination*, pp. 507-528, Marcel Dekker, New York.

Forcella, F. (1992). Prediction of weed seedling densities from buried seed reserves. *Weed Res.* 32:29-38.

Forcella, F. (1993). Seedling emergence model for velvetleaf. *Agron. J.* 85:929-933.

Forcella, F., R. G. Wilson, J. Dekker, R. J. Kremer, J. Cardina, R. L. Anderson, D. Alm, K. A. Renner, R. G. Harvey, S. Clay, and D. D. Buhler. (1997). Weed seed bank emergence across the corn belt. *Weed Sci.* 45:67-76.

Forcella, F., R. G. Wilson, K. A. Renner, J. Dekker, R. G. Harvey, D. A. Alm, D. D. Buhler, and J. Cardina. (1992). Weed seedbanks of the U.S. corn belt: magnitude, variation, emergence and application. *Weed Sci.* 40:636-644.

Frisbie, R. E., and J. W. Smith, Jr. (1989). Biologically intensive integrated pest management: the future. In J. J. Menn and A. L. Steinhauer (eds.). *Progress and Perspectives for the 21st Century*, pp. 151-164, Proceedings of Entomological Society of America Centennial National Symposium, Washington, DC.

Froud-Williams, R. J. (1997). Invasive weeds: implications for biodiversity. *BCPC Symp. Proc.* 69:41-52.

Gressel, J. 1997. Can herbicide resistant oilseed rapes from commodity shipments potentially introgress with local *Brassica* weeds, endangering agriculture in importing countries? *Resistant Pest Mgt.* 9:2-5.

Gressel, J., and L. A. Segel. (1990). Modelling the effectiveness of herbicide rotations and mixtures as strategies to delay or preclude resistance. *Weed Tech.* 4:186-198.

Gross, K. L. (1990). A comparison of methods for estimating seed numbers in the soil. *J. Ecol.* 78:1079-1093.

Gross, K. L., and K. A. Renner. (1989). A new method for estimating seed numbers in the soil. *Weed Sci.* 37:836-839.

Hall, D., W. L. Currey, and J. R. Orsenigo. (1997). Weeds from other places, the Florida beachhead is established. *WSSA Abstracts* 37:134.

Hall, R. (1995). Challenges and prospects of integrated pest management. In R. Reuveni (ed.). *Novel Approaches to Integrated Pest Management*, pp. 1-19, Lewis Publishers, Boca Raton.

Hass, H., and J. C. Streibig. (1982). Changing patterns of weed distribution as a result of herbicide use and other agronomic factors. In H. M. LeBaron and J. Gressel (eds.). *Herbicide Resistance in Plants*, pp. 57-80, John Wiley & Sons, New York.

Jordan, N. (1989). Predicted evolutionary response to selection for tolerance of soybean (*Glycine max*) and intraspecific competition in a nonweed population of poorjoe (*Diodia teres*). *Weed Sci.* 37:451-457.

Jordan, N. L., and J. L. Jannink. (1997). Assessing the practical importance of weed evolution: a research agenda. *Weed Res.* 37:237-246.

Julien, M. H. (1992). *Biological Control of Weeds: A World Catalogue of Agents and Their Target Weeds.* CAB International, Wallingford, U.K. 186 pp.

Kirkpatrick, B. L., L. M. Wax, and E. W. Stoller. (1983). Competition of jimsonweed with soybean. *Agron. J.* 75:833-836.

Klein, R. N., G. A. Wicks, and R. G. Wilson. (1996). Ridge-till, an integrated weed management system. *Weed Sci.* 44:417-422.

Lavorel, S., R. V. O'Neill, and R. H. Gardner. (1994). Spatio-temporal dispersal strategies and annual plant species coexistence in a structured landscape. *Oikos* 71:75-88.

Liebman, M., and E. Dyck. (1993). Crop rotation and intercropping strategies for weed management. *Ecol. Appl.* 3;92-122.

Liebman, M., and E. R. Gallandt. (1997). Many little hammers: ecological management of crop-weed interactions. In L. E. Jackson (ed.). *Ecology in Agriculture,* pp. 291-343, Academic Press, New York.

Lindquist, J. L., B. D. Maxwell, D. D. Buhler, and J. L. Gunsolus. (1995). Velvetleaf (*Abutilon theophrasti*) recruitment, survival, seed production, and interference in soybean (*Glycine max*). *Weed Sci.* 43:226-232.

Loux, M. M., and M. A. Berry. (1991). Use of a grower survey for estimating weed problems. *Weed Technol.* 5:460-466.

Lueschen, W. E., R. N. Andersen, T. R. Hoverstad, and B. K. Kanne. (1993). Seventeen years of cropping systems and tillage affect velvetleaf (*Abutilon theophrasti*) seed longevity. *Weed Sci.* 41:82-86.

Maddox, D. M., and A. Mayfield. (1979). Biology and life history of *Amynothrips andersoni,* a thrip for the biological control of alligatorweed. *Ann. Entomol. Soc. Am.* 72:136-140.

Malik, V. S., C. J. Swanton, and T. E. Michaels. (1993). Interaction of white bean (*Phaseolus vulgaris*) cultivars, row spacing, and seeding density with annual weeds. *Weed Sci.* 41:62-68.

McDonald, R. (1993). *Biological Control of Musk Thistle Using Introduced Weevils.* North Carolina Department of Agriculture Publication, Raleigh.

Mohler, C. L. (1993). A model of the effects of tillage on emergence of weed seedlings. *Ecol. Appl.* 3:53-73.

Monks, D. W., and L. R. Oliver. (1988). Interactions between soybean (*Glycine max*) cultivars and selected weeds. *Weed Sci.* 36:770-776.

Mulugeta, D., and D. E. Stoltenberg. (1997). Weed and seedbank management with integrated methods as influenced by tillage. *Weed Sci.* 45:706-715.

NRC (National Research Council). (1996). *Ecologically Based Pest Management: New Solutions for a New Century.* Committee on Pest and Pathogen Control through Management of Biological Control Agents and Enhanced Cycles and Natural Processes, Board on Agriculture, Washington, DC. National Academy Press.

Oryokot, J. O. E., S. D. Murphy, and C. J. Swanton. (1997). Effect of tillage and corn on pigweed (*Amaranthus* spp.) seedling emergence and density. *Weed Sci.* 45:120-126.

Patterson, D. T. (1995). Weeds in a changing climate. *Weed Sci.* 43:685-701.

Pedigo, L. P. (1995). Closing the gap between IPM theory and practice. *J. Agric. Entomol.* 12:171-181.

Pencoe, N. L., and J. R. Phillips. (1987). The cotton boll weevil: legend, myth, reality. *J. Entomol. Sci. Suppl.* 1:30-51.

Pohl, R. W. (1951). The genus *Setaria in Iowa*. *IA St. J. Sci.* 25:501-508.

Poston, D. H., E. C. Murdock, and J. E. Toler. (1992). Cost-effective weed control in soybean (*Glycine max*) with cultivation and banded herbicide applications. *Weed Technol.* 6:990-995.

Powles, S. B., C. Preston, I. B. Bryan, and A. R. Jutsum. (1997). Herbicide resistance: impact and management. *Adv. Agron.* 58:57-93.

Prokopy, R. J. (1994). Integration in orchard pest and habitat management: a review. *Agric. Ecosyst. Environ.* 50:1-10.

Prokopy, R. J., M. Christie, S. A. Johnson, and M. T. O'Brien. (1990a). Transitional step toward second-stage integrated management of arthropod pests of apple in Massachusetts orchards. *J. Econ. Entomol.* 83:2405-2410.

Prokopy, R. J., S. A. Johnson, and M. T. O'Brien. (1990b). Second-stage integrated management of apple arthropod pests. *Entomol. Exp. Appl.* 54:9-19.

Putnam, A. R., J. DeFrank, and J. P. Barnes. (1983). Exploitation of allelopathy for weed control in annual and perennial cropping systems. *J. Chem. Ecol.* 9:1001-1010.

Reader, R. J. (1991). Control of seedling emergence by ground cover: a potential mechanism involving seed predation. *Can. J. Bot.* 69:2084-2087.

Rhoades, R. E. (1997). *Pathways Toward Sustainable Mountain Agriculture in the 21st Century.* Kathmandu: International Centre for Integrated Mountain Development. 161 p.

Richardson, D. M., and W. J. Bond. (1991). Determinants of plant distribution evidence from pine invasions. *Am. Nat.* 137:639-668.

Rogers, H. J., and H. C. Parkes. (1995). Transgenic plants and the environment. *J. Exp. Bot.* 46:467-488.

Sanchez del Arco, M. J., C. Torner, and C. Fernandez Quintanilla. (1995). Seed dynamics in populations of *Avena sterilis* spp. *ludoviciana*. *Weed Res.* 35:477-487.

Sand, P. F., N. C. Hampstead, and J. D. Manley. (1990). The witchweed eradication program survey, regulatory and control. *Monogr. Ser. Weed Sci. Soc. Am.* 5:141-150.

Sandler, H. A., M. J. Else, and M. Sutherland. (1997). Application for inhibition of swamp dodder (*Cuscuta gronovii*) seedling emergence and survival on cranberry (*Vaccinium macrocarpon*) bogs. *Weed Technol.* 11:318-323.

Selby, A. D. (1897). A first Ohio weed manual. *Oh. Agric. Exp. Stn. Bul.* 83: 249-400.

Shaw, D. R. (1996). Development of stale seedbed weed control programs for southern row crops. *Weed Sci.* 44:413-416.

Shilling, D. G., B. J. Brecke, C. Hiebsch, and G. MacDonald. (1995). Effect of soybean (*Glycine max*) cultivar, tillage and rye (*Secale cereale*) mulch on sicklepod (*Senna obtusifolia*). *Weed Technol.* 9:339-342.

Snow, A. A., and P. M. Palma. (1997). Commercialization of transgenic plants: potential ecological risks. *BioScience* 47:86-96.

Spencer, N. R. (1984). Velvetleaf, *Abutilon theophrasti* (Malvaceae), history and economic impact in the United States. *Econ. Bot.* 38:407-416.

Stoller, E. W., S. K. Harrison, L. M. Wax, E. E. Regnier, and E. D. Nafziger. (1987). Weed interference in soybeans (*Glycine max*). *Rev. Weed Sci.* 3:155-181.

Sudduth, K. A., J. W. Hummel, and S. J. Birrell. (1997). Sensors for site-specific management. In J. J. Pierce and E. J. Sadler (eds.). *The State of Site Specific Management for Agriculture*, pp. 183-210, ASA, Madison, WI.

Swanton, C. J., and S. D. Murphy. (1996). Weed science beyond the weeds: the role of integrated weed management (IWM) in agroecosystem health. *Weed Sci.* 44:437-445.

Swanton, C. J., and S. F. Weise. (1991). Integrated weed management: the rationale and approach. *Weed Technol.* 5:657-663.

Teasdale, J. R. (1995). Influence of narrow row / high population corn (*Zea mays*) on weed control and light transmittance. *Weed Technol.* 9:113-118.

Theaker, A. J., N. D. Boatman, and R. J. Froud-Williams. (1995). Variation in *Bromus sterilis* on farmland: evidence for the origin of field infestations. *J. Appl. Ecol.* 32:47-55.

Thebaud, C., A. C. Finzi, L. Affre, M. Debussche, and J. Escarre. (1996). Assessing why two introduced *Conyza* differ in their ability to invade Mediterranean old fields. *Ecology* 77:791-804.

Thill, D. C., and C. A. Mallory-Smith. (1997). The nature and consequence of weed spread in cropping systems. *Weed Sci.* 45:337-342.

Timmons, F. L. (1970). A history of weed control in the United States and Canada. *Weed Sci.* 18:294-307.

Tripplett, G. B, and G. D. Lytle. 1972. Control and ecology of weeds in continuous corn grown without tillage. *Weed Sci.* 20:453-457.

USDA-APHIS. (1992). *Biological Control of Leafy Spurge.* USDA-APHIS Program Aid No. 1435.

Vandeman, A. M. (1994). *Adoption of Integrated Pest Management in U.S. Agriculture.* Agric. Info. Bull. No. 707. USDA-ERS. 26 p.

VanGessel, M. J. (1996). Successes of integrated weed management–a symposium. *Weed Sci.* 44:408.

VanGessel, M. J., E. E. Schweizer, D. W. Lybecker, and P. Westra. (1996). Integrated weed management systems for irrigated corn (*Zea mays*) production in Colorado–a case study. *Weed Sci.* 44:423-428.

Warwick, S. I., and L. D. Black. (1988). The biology of Canadian weeds. 90. *Abutilon theophrasti. Can. J. Plant Sci.* 68:1069-1085.

Webster, T. M., and H. D. Coble. (1997). Changes in the weed species composition of the southern United States: 1974 to 1995. *Weed Technol.* 11:308-317.

Webster, T. M., M. M. Loux, E. E. Regnier, and S. K. Harrison. (1994). Giant ragweed (*Ambrosia trifida*) canopy architecture and interference studies in soybean (*Glycine max*). *Weed Technol.* 8:559-564.

White, P. S., and R. G. Westbrooks. (1997). Exotic pest plants in the United States: an ecological explosion in slow motion. *WSSA Abstracts* 37:136.

Wicks, G. A., P. T. Nordquist, G. E. Hanson, and J. W. Schmidt. (1994). Influence of winter wheat (*Triticum aestivum*) cultivars on weed control in sorghum (*Sorghum bicolor*). *Weed Sci.* 42:27-34.

Wilson, P. J., N. D. Boatman, and P. J. Edwards. (1990). Strategies for conservation of endangered arable weeds in Great Britian. In *European Weed Research Society Symposium 'Integrated Weed Management in Cereals,'* pp. 93-101, Helsinki.

Wrubel, R. P., and J. Gressel. (1994). Are herbicide mixtures useful for delaying the rapid evolution of resistance? A case study. *Weed Technol.* 8:635-648.

Wyse, D. L., F. L. Young, and R. L. Jones. (1986). Influence of Jerusalem artichoke (*Helianthus tuberosus*) density and duration of interference on soybean (*Glycine max*) growth and yield. *Weed Sci.* 34:243-247.

Young, F. L., A. G. Ogg, D. C. Thill, D. L. Young, and R. I. Papendick. (1996). Weed management for crop production in the northwest wheat (*Triticum aestivum*) region. *Weed Sci.* 44:429-436.

RECEIVED: 01/28/98
ACCEPTED: 03/24/98

# Index

Abutilon theophrasti. See Velvetleaf
Acetanilide herbicides, 257
Actinomycetes, 100
Aegilops cylindrica (jointed
    goatgrass), 128
Aeschynomene virginica (northern
    jointvetch), 127
Agricultural policy, national, 258
Agro-ecosystems
    complexity of, 241
    weed management of, 241,245
        ecological principles of, 96-97
Alachlor, 169-170
Alfalfa (Medicago sativa)
    as smother crop, 78
    weed economic thresholds for,
        34,37
    weed management in, 47,53-54,259
Allelochemical-resistant crop plants,
    86
Allelochemicals
    as biomass component, 80
    of cover crops, 252
Allelopathy
    of cover crops, 78-86,106
        genetics of, 81-86
        relationship to weed seed size,
            109,111
    definition of, 79
    of microbial metabolites, 131
Alopecurus myosuroides (blackgrass),
    37,102-103
Alternaria cassiaeis, 127
Alternaria crasse, 128
Amaranth, Powell (Amaranthus
    powellii), 169-171
Amaranthus. See Pigweed
Amaranthus retroflexus. See Pigweed,
    redroot

Ambrosia artemisiifolia (common
    ragweed), 16
Ambrosia trifida (giant ragweed),
    243,259
American Society of Agricultural
    Engineers, 196
Ammonium nitrate fertilizers,
    comparison with green
    manures, 107-109,110
Ammonium sulfate, 105
Anther smut, 127
Anthracnose, 127
Aphids, economic thresholds for, 41
Apocynum cannabinum (hemp
    dogbane), 243
Apples, integrated pest management
    of, 241
Arachis hypogaea (peanut), 127,243
Arthropods, economic thresholds for,
    32,33
    population ecology principles of,
        38-44
Atrazine, 211
Avena fatua. See Oats, wild
Avena sativa (oats), 112-113


Bacteria. See also Soil
    microorganisms
    pathogenic, insect-assisted
        transmission of, 128
    weed-suppressive, effect on weed
        competitiveness, 132
Bacterial diseases
    of seedlings, 127
    of seeds, 126,127
Barley
    as smother crop, 80
    weed competitive ability of, 15,62

© 1999 by The Haworth Press, Inc. All rights reserved.

weed interference with
    effect of fertilizers on, 105
    effect on yield, 14
    weed thresholds for, 12
Barnyardgrass (*Echinochlea
    causgalli*), thresholds for, 16
    economic thresholds, 37,47,48-50
    no-seed thresholds, 47-48,53-54
*bas* gene, of *Streptomyces*, 85-86
Beans
    dry (*Phaseolus vulgaris*), 16,68
    faba (*Vicia faba*), 14
    green manuring of, 109,110
Beetles, ground (*Harpulus*), 102
*Bemisia* (whiteflies), 41
Benomyl, 130
Bentazon, 171,175,176-177,199,200,
    216
Bias, communication of, in
    decision-making systems,
    231-233,236
Bindweed, field (*Convolvulus
    arvensis*), 128
Biodiversity
    of ecosystems, 261
    of weeds, 258-259
        in field edge vs. field interiors,
        256
        of seed banks, 143-144,154,261
        site-specific management of,
        235
        as weed management technique,
        154
Bioeconomic models, use in weed
    management, 231,248-249
    definition of, 208-209
    General Weed Management
        (GWM) model, 209,210-222
        effect on crop yield,
        211,218-220,221,222
        postemergence strategies of,
        210,211, 213,214,216-217
        preplant incorporated/
        preemergence strategies of,
        210,211,212,213,214,216

relationship to tillage systems,
    210-221
    effect on weed seed banks,
    214-215,220
    use in risk management, 183-184
    use in weed management
        decision-making,
        233-234,237
    of weed thresholds, 17-18,19
    whole-farm, 184
Biological control agents,
    127-128,242-243
    exploration for, 258
    indigenous, use in weed control,
        132-133
    insects as, 128,131
    use of multiple organisms as, 248
    synergistic interaction with
        herbicides, 128-129
BioMal (biocontrol agent), 127
Biomass
    allelochemical components of, 80
    of cover crops, 80,83
    of weeds
        as competitive index basis, 15
        effect of crop genotype on,
        69-70
        crop stand relationship of, 66-67
        early increase in, as weed
        competitive trait, 61-62
        effect of tillage methods on, 68
        for weed population assessment,
        45
Biotechnology, effect on weed
    competitive crops, 70-71
*Bipolaris sorghicolam*, 128
Blackgrass (*Alopecurus myosuroides*),
    37,102-103
Blight, 126,127
Bollworms, economic thresholds for,
    41
Boswell Company, 48,51-52
Bramble (*Rubus*), 243
Branching, as weed competitive
    characteristic, 62
Brassicaceae

allelochemicals of, 85
germplasm of, allelopathic
potential of, 82
*Brassica oleracea*, effect of soil
properties on, 103
*Brassica rapus* (rapeseed), 107
Broadleaf weed species
crop competition with, 62
threshold densities of, 12
Brome, 103-104,128,256
Bromoxynil, 212,216-217
*Bromus japonicus*, 128
*Bromus madritensis*, 103-104
*Bromus rigidus*, 103
*Bromus sterilis*, 256
*Bromus tectorum*, 128
Buckwheat (*Fagopyrum esculentum*),
78

Calcium, accumulation by weeds,
104-105
California
no-seed threshold use in, 51-53
velvetleaf and giant foxtail control
in, 257
Canada
herbicide-resistant crop seed trade
issue in, 260
velvetleaf and giant foxtail control
in, 257
Canopy closure, rapid, 62
Canopy establishment, as weed
competitive trait, breeding
for, 69
Carbofuran, 131
Carbon fixation of, by cropping
systems, 254
in no-till systems, 132
Carboxin, 130
*Carduus nutans* (musk thistle), 248
Carrying capacity, of cropping
systems, 253-254
Caryopses, of *Setaria*, fungal
colonization of, 101,130-131
*Cassia obtusifolia* (sickle pod), 127

Catchweed bedstraw (*Galium
aparine*), 12,102-103
*Centaurea solstitialis* (yellow
starthistle), 229-230
*Centauria maculosas* (spotted
knapweed), 128
Cereal crops. *See also* Barley; Oats;
Wheat
weed economic thresholds for,
12-13,37,45
weed size hierarchies in, 15
*Chenopodium album. See*
Lambsquarters
Chickweed (*Stellaria media*),
14,102-103
*Cirsium arvense* (Canada thistle), 128
Cloning, 70,84
Clover
berseem (*Trifolium alexandrinum*),
53
crimson (*Trifolium incarnatum*), as
green manure, 107-109
red (*Trifolium pratense*), as green
manure, 109,110,112-113
subterranean (*Trifolium
subterraneum*), 253
Cocklebur, common (*Xanthium
strumarium*), control of
bioeconomic model-based,
214,217,222
by eradication, 151
with herbicides, 171,172
comparison with inter-row
cultivation, 171-172
efficiency frontiers of, 175,177
with inter-row cultivation,
171-172
postemergence,
171-172,175,176
risk efficiency of, 175,176
Collego (biocontrol agent), 127
*Colletotrichum coccodes*, 127-128
*Colletotrichum gloeosporiooides* f. sp.
*Aeschynomeme*, 127
Competitive indices, of weeds, 14-16
Competitiveness

applied research in, 67-69
basic research in, 65-67
crop characteristics of, 61-63
economic factors in, 64,65
farmers' utilization of, 71-72
history of, 61-63
as integrated weed management
    component, 10,242,247
lack of development of, 63-64
mechanistic models of, 20-21
plant breeding for, 59-76
    biotechnology use in, 70-71
    genetic variability in,
        61-62,63-64,67-68
    standard methods for, 69-70
production practices for, 60
effect of soil-applied herbicides on,
    17
effect of soil bacteria on, 132
stimulation for development of,
    64-65
Compost and composting, 105,106
Computer-based decision aids,
    183-185
Conservation tillage, weed species
    shifts in, 131-132
*Convolvulus arvensis* (field
    bindweed), 128
Corn (*Zea mays*)
    allelopathic suppression of, 80
    glyphosate-tolerant, 260
    weed competitiveness of, 64
    weed management of, 47,50,53-54
        bioeconomic models of, 248
        with green manuring, 107-109
        with herbicides, 190
        with multiple weed control
            techniques, 168
    weed economic threshold levels
        for, 12,16,18-19,32,34,37
    yield loss, weed-related, 18-19,104
Corn earworm, economic thresholds
    for, 41
CORNHERB (weed management
    model), 208
Corn seed maggot (*Delia paltura*), 111

Corn/soybean rotation, bioeconomic
        model-based vs.
        standard-herbicide treatments
        in, 209-222
    crop yield results, 211,218-220,
        221,222
    postemergence strategies, 210,211,
        213,214,216-217
    preplant incorporated/
        preemergence strategies,
        210,211,212,213,214,216
    relationship to tillage systems,
        210-221
    weed seed bank effects of,
        214-215,220
*Coronilla varia* (crownvetch), 253
Cotton
    handweeding of, 51-52
    hoeing of, 48
Cover crops
    allelopathy of, 78-86,109,111
    deleterious effects on crops, 111
    efficacy of, 84-85
    as emergency forage, 106
    genetic-based development of, 81-86
    multi-species, 252,253
    for nutrient recycling, 106
    for plow pan breakage, 106
    selectivity of, 82,84-85
    for soil erosion control, 106
    for soil improvement, 97
    for water retention, 106
    weed interference by, 79-81
    as weed management technique,
        79-81,106,247,249
    weeds as, 154
Crickets, field (*Gryllus*), 102
Critical period, of weed interference,
    182-183
Crop breeding. *See* Plant breeding
Crop equivalents, 15
Cropping system design. *See also*
        No-till cropping systems;
        Reduced-till cropping
        systems

carrying capacity of, 253-254
conceptual basis of, 244
ecological principles of, 252-255
goal/focus of, 244,250-251
as level II of integrated weed
    management, 244,250-255
methods of, 244
multiple, weed competitiveness of,
    68
spatial scale of, 244,251
strategies for, 251,252-255
temporal scale of, 244,252
Cropping system studies, of soil
    management for weed
    control, 101,111-113
Crop residues. *See also* Cover crops;
    Green manures
definition of, 80
in no-till systems, 101-102
Crop rotation. *See also* Corn/soybean
    rotation
limited crop choices for, 4
of smother crops, 78
as soil management technique,
    111-113,114
as weed management technique,
    10,145,158,243,246,247
Crop stand, relationship to weed
    biomass, 66-67
Crop yield loss. *See* Yield loss
Cross-resistant biotypes, 246
Crownvetch (*Coronilla varia*), 253
Cultivated land, weed seed banks of,
    131
Cultivation, as weed management
    technique, 247. *See also*
    Inter-row cultivation
cost of, 173
field work time of, 179
risk efficiency of, 176
Cyanazine, 212,213,216


Dandelion (*Taraxacum officinale*),
    128,214
Decay, of seeds, 98,99,126,130,132-133

Decision aids, in weed management,
    183-185
Decision making, in weed
    management
bioeconomic model use in,
    233-234,237
efficiency frontiers in, 174
knowledge-based. *See*
    Knowledge-based decision
    support strategies (KBDSS)
timeliness factors in, 176-183
    field working days, 178-181,185
    time density equivalents,
        182-183
in weed seed bank management,
    149-150,151-152
traditional approach in, 232-233
Dehydrogenase, 132
*Delia paltura* (corn seed maggot), 111
Density
of crop plants, effect on weed
    competitive ability, 60,62
of weeds
    consistency of weed control and,
        172
    effect on crop yield,
        11-12,100,242
    duration of weed interference
        and, 16
    field work time requirements
        and, 178,179
    seed production and, 221-222
    weed emergence time and, 16-17
    site-specific management of, 235
Density equivalents, of
    multiple-species weed
    infestations, 16
Density response, 68
Deoxyribonucleic acid (DNA)
    analysis, of allelopathic
    potential, 84
*Desmodium tortuosum* (Florida
    beggarweed), 243
DeVine (biocontrol agent), 127
Dicamba, 211,213,217

Digital imaging, use in site-specific
    weed management, 229-230
Discoloration, of seeds, 126
Dispersal, of weed seeds, 253,259-260
Diversity. *See* Biodiversity
Domestication, effect on allelopathy,
    83
Dormancy
    of arthropods, 42
    of weed seeds, 40,42,142,147,148,
        150-151
Drainage, 145,255-256


Earthworms, 102
*Echinochloa crus-galli. See*
    Barnyardgrass
Ecological niches, of weeds,
    141,252,253
Economic optimum thresholds. *See*
    Thresholds, economic
    optimum
Economic thresholds. *See* Thresholds,
    economic
Ecophysiological models, of weed
    thresholds, 20
Eco-region integrated weed
    management, 245,258-260
Ecosystems. *See also*
    Agro-ecosystems
    biodiversity of, 261
    roles of weeds in, 258
Efficiency frontiers, 174-175,177
Embryo death, as seed loss cause,
    98,99
Emergence time
    in barnyardgrass/sugarbeet system,
        48-50
    early, as weed competitive trait,
        61,62-63
    modeling of, for mechanical weed
        control timing, 189-205
        crop yields, 199,200
        economics of, 195-196
        net economic returns,
            195-196,199-204

sensitivity analysis, 202-204
    weed densities, 196-199,202
    relationship to consistency of weed
        control, 172
    relationship to crop yield loss,
        16-17
    relationship to weed thresholds, 22
    timing of duration of, 150-151
Environmental impact, of herbicides
    geographic information system
        management of, 235
    risk indices for, 184-185
    water quality, 68
Environmental interactions, of weeds,
    241
*Epilobium* (willowherb), 41
Eradication, of weeds, 151,260-261
*Eriochloa villosa* (woolly cupgrass),
    154
Ethephon, 131
*Euphorbia esula* (leafy spurge), 256
Experimental error, in weed seed bank
    studies, 158-159
Explanatory models, of weed
    thresholds, 20
Extinction, of
    agronomically-important
    weeds, 259


*Fagopyrum esculentum* (buckwheat),
    78
Fecundity per generation,
    weed-arthropod comparison
    of, 39,42
Fertility programs, effect on seed
    banks, 158
Fertilizers, 10. *See also* Manures
    differential application of, 71
    interspecific interference effects of,
        104-105
    for soil improvement, 97
    weed seed bank effects of, 145
Field bindweed (*Convolvulus
    arvensis*), 128
Field mapping, 249

Field scouting, 172,229,230
Field survey and sampling, 228-229
Field working days, 178-181,184,185
First International Weed Control
    Congress, 34
Fitness, relationship to genetic
    variance, 143
Flax (*Linum usitatissimum*), 14,15
Florida, exotic weed control in, 256
Florida beggarweed (*Desmodium*
    *tortuosum*), 243
Food trade policies, 258
Foxtail
    giant (*Setaria faberi*), 243,257
        bioeconomic model-based
            control of, 214,215,217-218,
            220
        emergence time-related
            mechanical weeding of,
            195,197-199,202,203
        herbicide control of, 169-171
        interference by, 79
        mechanical control of, 170-171
    green (*Setaria viridis*), 104
        crop yield interference by, 14
        emergence time-related
            mechanical weeding of,
            195,196-199,202,203
    seed banks of, 144
    yellow (*Setaria lutescens*), 104
        bioeconomic model-based control
            of, 214,215,217-218,220
        emergence time-related mechanical
            weeding of, 195,196-197,
            202,203
Foxtail millet (*Setaria italica*), 78
Frost injury, mulch-related, 79
Fungal diseases, of seedlings and
    seeds, 126,127
Fungi
    as biological control agents,
        synergism with herbicides,
        131
    pathogenic, insect-assisted
        transmission of, 128

plant suppressive metabolites of,
    126
saprophytic, insect-assisted
    transmission of, 128
as seed decay cause, 126,130
*Setaria* caryopses colonization by,
    101,130-131
Fungicides, 99,127
*Fusarium*, 128
*Fusarum oxysporum*, 131


*Galeopsis tetrahit. See* Hempnettle
*Galium aparine* (catchweed bedstraw),
    12,102-103
Gene cloning, 70,84
General Weed Management (GWM)
    bioeconomic model,
    209,210-222
    effect on crop yield, 211,218-220,
        221,222
    postemergence strategies of,
        210,211,213,214,216-217
    preplant incorporated/
        preemergence strategies of,
        210,211,212,213,214,216
    effect on weed seed banks,
        214-215,220
Generation time, weed-arthropod
    comparison of, 39,40-41
Genetics. *See also* Plant breeding
    of allelopathy of cover crops, 81-86
Genetic variance
    relationship to fitness, 143
    in weed competitiveness,
        61-62,63-64,67-68
Gene transfer, 70,85,260
Genostasis, 252
Geographic information systems,
    71,235,257
Germination, of weed seeds
    changes in ability for, 146
    fatal, 152
    by heterogenous populations, 144

microbial inhibition of,
    126,129-131
microsites for, 157
in seed banks, 98,99,146,148
effect of soil structure on, 132
timing of, 142
Germplasm
    of cover crops, allelopathic
        potential of, 81-83,85
    of weed-competitive crop species,
        67-68,69-70
    of weeds, 258
Global management, of weeds,
    245,258-260
Global positioning systems, 71,226
    differentially-corrected, 229
Global warming, 258
Glucofosinate herbicides, 85-86
*Glycine max. See* Soybeans
Glyphosate, 211,212,213
*Gossypium hirsutum. See* Cotton
Grains. *See also* Barley; Oats; Wheat
    biocontrol agent weed control for,
        127
Grasses
    annual, crop competition with, 62
    in conservation tillage systems,
        131-132
    threshold densities of, 12
    weed seed banks of, 131
Green manures
    deleterious effects on crops, 111
    as nutrient source, 106-109
    for soil erosion control, 106
    for soil improvement, 97
    for weed management,
        106,107-109
Grewal, Mark, 48,51-52
Groundsel (*Senecio vulgaris*), 41,105
Growth, of weeds
    crop suppression of, 61
    effect of microorganisms on,
        126,127
    rates of, 104-105

Hairy vetch (*Vicia villosa*), as green
    manure, 112-113
Hand-harvested crops, weed
    eradication from, 151
Handweeding, 247
    cost of, 47-48,51-52
*Harpulus* (ground beetles), 102
Height, of plants, as weed competitive
    trait, 62,69
*Helianthus annus* (sunflower), 12,37
*Helicoverpa zea* Bodie, 41
Hemp dogbane (*Apocynum
    cannabinum*), 243
        Hempnettle (*Galeopsis
        tetrahit*), 47
    economic thresholds for, 37
    no-seed thresholds for, 53-54
HERB decision model, 19
Herbicide-resistant crops,
    67,70,72,85-86,249,255
Herbicide-resistant weeds, 65,78,246
    dispersal of, 259-260
    historical trends in, 258
    management of, 154,246,249,
        257-258
    selection for, 153-154
Herbicides, 78-79. *See also* names of
        specific herbicides
    alternatives to, 2
    biologically-effective, 22
    decision model for, 19
    differential application of, 71
    efficacy of, 63,78,123-124
    environmental effects of
        geographic information system
            management of, 235
        risk indices for, 184-185
        water quality, 68
    evaluation studies of, 242
    factor-adjusted, 22
    farmers' preference for, 168-169
    gene transfer with
        herbicide-resistant crops, 260
    for horticultural crops, 13
    increased use of, 123-124

intermittent application of, sensor
  technqiues for, 230
mechanical weed control versus,
  168-169
myco-, 127
overapplication of, 19
as percentage of total annual
  pesticide use, 242
postemergence application of
  cost of, 173
  field work time required for, 179
  risk efficiency of, 175
  timing of, 182
precision application of, 249
records of use of, 258
reduced use of, 176-178,246
registration of, 65,78,243
restriction of, 123-124
risk-efficiency of, 174
single-pass application of, 168
site-specific application of, 249
soil-applied, 17
  cost of, 173
  field work time required for, 179
  reductions in, 176-178
as species shift cause, 149
sub-lethal doses of, 17
synergistic interactions of, 128-129
tax on, 168-169
time required for application of,
  176-178,179
weed competitiveness effects of, 17
weed seed bank effects of, 145,158
as weed seed mortality cause, 152
as weed species extinction cause,
  259
yield variability and, 168-169
Hoeing. *See also* Rotary hoeing
  cost of, 47-48
*Hordeum glaucum*, paraquat-resistance
  of, 46
*Hordeum vulgare. See* Barley
Horticultural crops, weed threshold
  densities for, 13
Hydroxamic acid, 81,85

*Hypericum perforatum* (common
  St. Johnswort), 229-230
Hypertrophy diseases, 126

Imazethapyr, 202,211,212
Insects. *See also* Arthropods
  beneficial, effect of tillage timing
    on, 155
  as biological control agents, 128
    synergistic interaction with
      microorganisms, 128,131
  as pests, effect of cover crops on,
    111
Integrated pest management, 2-3
  development of, 240
  ecological principles of, 96-97
  relationship to weed management,
    250
Integrated weed management. *See also*
    Weed management; specific
    techniques of integrated
    weed management
  comparison with weed control, 168
  components of, 60
  definition of, 168
  goals of, 113-114
  levels of integration in. See
    Integration levels, of
    integrated weed management
  timeliness factors in, 176-183
    field working days, 178-181,185
  weed competitiveness evaluation
    in, 68
  as weed threshold supplement, 22
  effect on weed seed production, 18
Integration levels, of integrated weed
    management, 239-264
  definitions of, 239-240
  level I (weed control),
    239-240,242-244,246
  level II (weed management),
    240,244,246-250
  level III (cropping system design),
    240,244,250-255

level IV (landscape and regional
management),
240,245,255-258
level V (eco-region management),
240,245,258-260
Interference, 61,79-81. *See also*
Allelopathy;
Competitiveness
additivity in, 13-14
components of, 79,80-81
by cover crops, 79-81
critical period of, 182
crop yield effects of, 21,66
definition of, 79
effect of fertilizers on, 104-105
indirect sources of, 79
by multiple weed species, 13-14
plant breeding strategies for, 66
Inter-row cultivation, 249
combined with herbicide
treatments, 169-170
comparison with full-labeled
herbicide application,
175-176
risk efficiency of, 174
weed emergence-based timing of,
193-194
effect on net economic returns,
195-196,199-204
effect on weed densities,
196-199,202
relationship to herbicide use,
195,196,199-203
sensitivity analysis of, 202-204
Invasion, by weeds
cropping systems' resistance to,
252-253
economic threshold concept of,
50-51
protection from, 261
Iowa Cooperative Extension Service,
208
Iowa State University Agronomy and
Agricultural Engineering
Research Center, 210

*Ipomoea* (morning glory), 51-52,243
Irrigation, 105,145
Isothiocynate, 107


J.G. Boswell Company, 48,51-52
Japan, herbicide-resistant crop seed
trade issue in, 260
Jimsonweed (*Datura stramonium*),
128
Johnsongrass (*Sorghum halepense*),
128
Jointed goatgrass (*Aegilops
cylindrica*), 128
Jointvetch, northern (*Aeschynomene
virginica*), 127


Knowledge-based decision support
strategies (KBDSS), 225-258
fundamentals of, 227-232
goal of, 232
knowledge incorporation into,
233-235
spatial context of, 229,230,231,232,
235,236
temporal context of, 230,231,232,
235,236
wisdom incorporation into,
235-236
*Kochia*, ALS-resistant, 249


Lambsquarters (*Chenobium albium*)
bioeconomic model-based control
of, 214
biological control of, 128
establishment of, effect of soil
properties on, 103
germination of, fertilizer-induced,
104
effect of green manures on,
107-109
herbicide control of, 169-170
mechanical control of, 170-171
emergence timing for, 196,199

as yield loss cause, prediction of, 21

Landscape and regional habitat weed management, 245,255-258

Landscape variables, in site-specific weed management, 229

*Latuca serriola* (prickly lettuce), 41

Leaching, of soil, 78,96

Leaf area, of weeds
for crop yield loss assessment, 17
for weed population assessment, 45

Leaf area index, as weed competitive trait, 61-62,69

Leaf expansion rate, as weed competitive trait, 69

Leaf galls, 127

Leafminers (*Liriomyza*), 41

Leaf necrosis, 127

Leaf size, as weed competitve trait, 62

Leafy spurge (*Euphorbia esula*), 256

Legumes, as green manure, 107

Lentils, 127

Lepidoptera, economic thresholds for, 41

Lettuce, prickly (*Lactuca serriola*), 41

Light, soil penetration depth of, 103

*Linum usitatissimum* (flax), 14,15

Liriomyza (leafminers), 41

*Lolium* (annual/Italian ryegrass), 259-260

Longevity, of populations
of populations, weed-arthropod comparison of, 39,40
of seeds, 149,152

Lygus bugs, 41


Magnesium, accumulation by weeds, 104-105

Maine Potato Ecosystem Project, 111-113,114

*Malus. See* Apples

*Malva pusilla* (round leaf mallow), 127

Manures
animal, 105-106,112

green. *See* Green manures

Manzoni, Louis, 52-53

Mapping
of fields, 249
in regional management of weeds, 257,258
software for, 226,230

Maturity, of crops
in multiple cropping systems, 68
as weed competitive trait, 64,69

Maximum economic return per unit of land, 152

Mean-variance economic analysis model, of integrated weed management systems, 172-173

Mechanical weed control, 246,248. *See also* Inter-crop cultivation; Rotary hoeing
field work time required for, 176,178
herbicide use versus, 168-169
in row crops, emergence modeling for, 189-205
economics of, 195-196,202
as weed seed mortality cause, 152

Mechanistic models, of weed thresholds, 20-21

Mechanistic research, in weed competition research, 65

*Medicago sativa. See* Alfalfa

Memory, genetic, 40

Meristem galls, 127

Methyl bromide, 52-53

Metoachlor, 211,212,216

Metribuzin, 113,169-170

Mexico, herbicide-resistant crop seed trade issue in, 260

Mice, field (*Peromyscus maniculatus*), 102

Microorganisms. *See also* Soil microorganisms
effect on agro-ecosystems, 124,125
diversity of, 124-125
as seed loss cause, 98,99

Millet
  foxtail (*Setaria italica*), 78
  wild proso (*Panicum miliaceum*),
    154
*Mimosa pigra*, 99,130-131
Mississippi State Budget Generator,
  195
Mites, economic thresholds for, 41
Mobility, weed-arthropod comparison
  of, 39,41
Modeling, use in weed management,
  207-224,248-249. *See also*
  Emergence time modeling;
  Simulation models
  of crop yield loss, 11-12
  in weed competition research, 65
  in weed seed bank management,
    149,150-151
  of weed thresholds, 20,44-45
Molecular genetic analysis, of
  allelopathic potential, 84
Molecular markers
  of allelopathic potential, 84
  of weed competitiveness, 70
Morning glory (*Ipomea*), 51-52,243
*Morrenia odorata* (stranglervine), 127
Mulch
  cover-crop, 80
  definition of, 80
  living, 253
  wheat straw, 79,102
Mustard
  white (*Brassica hirta*), 105
  wild (*Sinapis arvensis*), 109,110
Mycorrhizal spores, 100

Natural selection, 143
Near-isogenic lines, 84
Necrosis, of seeds, 126
Nematodes, economic thresholds for,
  32
New York State, velvetleaf and giant
  foxtail control in, 257
Nicosulfuron, 212,216-217

*Niesthrea louisianica* (scentless plant
  bugs), 128
Nightshade
  eastern black (*Solanum
    ptycanthum*), 21-22,127-128
  no-seed thresholds for, 51-52
Nitrogen
  accumulation by weeds, 104-105
  mineralization by green manure,
    107
  sources of, 107-108
Nitrogen fertilizers, effect on
  interspecific interference,
    104-105
Normalized difference index (NDI),
  230
North Carolina, witchweed
  management in, 256
No-till cropping systems
  fungal colonization of seeds in, 131
  residue zones of, seed decay in,
    123-133
  surface-residing weeds in, 132
  weed management in, 64-65
    with microbial biocontrol
      agents, 101
    with seed predators, 102
    effect on soil properties,
      101-102
  weed seed distribution in, 132,153
Nutrients. *See also* Carbon fixation;
    Nitrogen; Phosphorus;
    Potassium
  from cover crops, 106-107

Oats *Avena sativa*, as green manure,
  112-113
Oats, wild (*Avena fatua*)
  competition with winter wheat, 21
  competitive index for, 15
  economic thresholds for, 37
  germination of
    fertilizer-induced, 104
    inhibition of, 130
  seeds of, 131

zero threshold for, 46
Onions (*Allium*), weed economic
    thresholds for, 34
*Oryza sativa. See* Rice
Ovary galls, 127

PALWEED-WHEAT model, 45
*Panicum miliaceum* (wild proso
    millet), 154
Paraquat, 113
Paraquat-resistance, of *Hordeum
    glaucum*, 46
Parasite-related rot, 126
Parasites, microbial, 126
Parasitic weeds, 151
*pat* gene, of *Streptomyces*, 85-86
Pea (*Pisium*)
    as green manure, 112-113
    long-vined and short-vined, 105
Peanuts (*Arachis hypogaea*), 127,243
Pedigree method, use in weed
    competitive crop breeding,
    69
Pennsylvania, velvetleaf and giant
    foxtail control in, 257
Perennial weeds
    in conservation tillage systems,
    131-132
    generation time of, 40
    invasion by, 253
    in reduced-tillage sysetms, 243
    species shifts of, 243
Pesticides, environmental hazard
    evaluation of, 235
*Phaseolus vulgaris* (dry bean), 62
Phenolic acid, 81
*Phomopsis convolvulus*, 128
Phosphatase, 132
Phosphinothricin acetyltransferase, 85
Phosphorus
    accumulation by weeds, 104-105
    weed species' requirements for,
    105
Photoperiodism, neutral, 142
Photoperiod sensitivity, of crops, 68

Phytochemicals. *See also*
    Allelochemicals
    metabolic pathways of, 85
*Phytophythora palmivora*, 127
Pigweed (*Amaranthus*), biological
    control of, 128
Pigweed, redroot (*Amaranthus
    retroflexus*)
    bioeconomic model-based control
    of, 214
    herbicide control of, 169-171
    mechanical control of, 170-171
        emergence time-related, 196,199
    threshold levels of, 16
    yield loss caused by, 42
Plant breeding
    as integrated weed management
    component, 10
    for weed competitiveness, 59-76
    biotechnololgy use in, 70-71
    genetic variability in, 67-68
    standard methods for, 69-70
Planting patterns, as weed
    management technique, 247
Planting time, 60
Plant morphology, of weed
    competitive crops,
    61-62,63,65-66,68
Plant residue management, 68,78. *See
    also* Cover crops; Green
    manures
Poison ivy (*Toxicodendron*), 243
Poisonous weeds, 151
*Polygonum. See* Smartweed
Polyploidy, 142
Population decrease, weed-arthropod
    comparison of, 39,40
Population dynamics, 96-97
    of annual weeds, 97-100
    interspecific interference and,
    98,100
    seedbank persistence and, 97-99
    seedling establishment and,
    98,99-100
Population ecology, weed-arthropod
    comparison of, 38-44

Population shifts. *See also* Species shifts
  in weed seed banks, 148-149,152-153
*Portulaca oleracea* (purslane), 154
Potassium, accumulation by weeds, 104-105
Potato (*Solanum tuberosum*)
  green manuring of, 107
  soil management systems study of, 111-113,114
  weed threshold density in, 13
Predation
  macrofaunal, 129-130
  as seed bank loss cause, 98,99
Predator-prey relationship, 96-97
Prediction
  of weed seed bank behavior, 149,150-151,158
  of weed populations, 45
Preventive approach, in weed seed bank management, 150
Profit, relationship to field working days, 179-180
Progeny testing, 83
Purslane (*Portulaca oleracea*), 154

Quizalofop, 202

Ragweed
  common (*Ambrosia artemisiifolia*), 16
  giant (*Ambrosia trifida*), 243,259
Rapeseed (*Brassica rapus*), 107
Rectangular hyperbolic yield loss equation, 11-12,15-16,18-19
Reduced-tillage cropping systems
  perennial weeds in, 243
  weed seed distribution in, 153
Regional habitat management, 245,255-258
Regression equations, 15,16-17
Regression modeling, of weed thresholds, 44-45

Remote sensing, use in site-specific weed management, 228-229,230
*Rhinocyllus conicus*, 248
Rhizobacteria, 100,127,128
Rhizosphere, 125-126
Rice (*Oryza sativa*)
  biocontrol agents for, 127
  germplasm of, allelopathic potential of, 82-83
  weed competitiveness of, 64
    genetic assessment of, 67-68
Risk
  communication of, in decision-making systems, 231-232,236
  farmer's perception of, 168-169
Risk management, in integrated weed management, 167-224
  assumed risk, 173
  economic returns and, 172-176
  efficiency frontiers in, 174-175,177
  implication for reduced herbicide use, 169-178
  risk efficiency, 174-176
  timeliness in, 176-183
Root galls, 127
Root rot, 126,127
Roots, exudates of, effect on microflora, 126
Rots, 126,127
Rotary hoeing
  comparison with herbicide treatments, 170-171
  cost of, 173,174
  field work time requierd for, 179
  risk efficiency of, 174,176
  weed emergence-based timing of, 189-205
    economics of, 195-196
    emergence percentage in, 191,192
    effect on net economic returns, 195-196,199-204
    relationship to herbicide use, 195,196,199-203

sensitivity analysis of, 202-204
effect on weed densities,
196-199,202
Round leaf mallow (*Malva pusilla*),
127
Row crops, weed management of, 246
Row spacing, narrow, effect on weed
competitive ability, 60,62,68
*Rubus* (bramble), 243
*Ruhus* (sumac), 243
*Rumex acetosella* (red sorrell), 259
Rye (*Secale cereale*)
allelopathic effects of, 85
as cover crop, 80-81
germplasm of, allelopathic
potential of, 82
as smother crop, 78
Ryegrass, annual/Italian (*Lolium*),
259-260

Safe sites, for weed seeds,
98,99-100,102-103
St. Johnswort, common (*Hypericum
perforatum*), 229-230
"Sanding," 247
Saprophytes, 126
Scentless plant bugs (*Niesthrea
louisianica*), 128
*Sclerotinia sclerotiorum*, 128
Sclerotization, of seeds, 126
*Secale cereale. See* Rye
Seed(s)
abortion of, 126
age of, 129-130
animal manure content of, 105-106
biology of, lack of information
about, 158
discoloration of, 126
of herbicide-resistant crops,
world-wide trade in, 260
mechanical damage to, 126
moisture content of, effect on
fungal colonization, 126
mortality of, 131
mortality/survival rate, 254-255

necrosis of, 126
sclerotization of, 126
stromatization of, 126
of weeds. *See also* Seed banks, of
weeds
colonization by, 142
decay of, 40,42,142,147,148,
150-151
dispersal of, 141-142,147-148,
253,259-260
embryo food supply of, 141,142
fecundity of, 42
genetic diversity of, 141,143
longevity of, relationship to
economic thresholds, 175
microbial biocontrol of, 101
in no-till cropping systems,
101-102
persistence of, 142
predators of, 102
size of, relationship to
allelopathic chemical
response, 109,111
spatial patterns of, 145-146
in species multiplication,
141,143
viability of, 129-130,156-157
wind-borne, 41,51,131-132
Seed bank model, of weed
management, 247
Seed banks, of weeds,
129-131,139-166
active, 147,148
shift of seeds within, 153
aggregate behavior in, 144
annual life cycle of, 146-148
biodiversity of, 143-144,154,261
bioeconomic model analysis of,
184
definition of, 140
depth of, 152,153
dormant, 147,148
shift of seeds within, 153
germination microsites of, 157
heterogeneity of, 143-144
matrix of, 141,144-149

seed sampling from, 158
multiple-year changes in, 148-149
niches in, 141
research issues regarding, 155-159
seed dispersal and colonization in,
    141-142
seed embryo food reserves in, 142
seed inputs and losses of,
    147-148,156
seed longevity in, 149,152
seed persistence in, 97-99,141,142
seed viability in, 156-157
size of, 149
effect of soil microorganisms on,
    129-131
spatial distribution of, estimation.
    of, 157
species shifts in, 251
temporal changes of, 146-149
weed management of,
    149-152,247,248
    lack of research information
        about, 157-158
    model-based vs.
        standard-herbicide
        treatments, 214-215,220
Seed coat
    mechanical damage to, 126
    protective function of, 131
Seed embryo, food reserves of, 142
Seedlings
    bacterial diseases of, 127
    biological control of, 128
    establishment of, 98,99-100
    fungal diseases of, 127
    growth rate of, 68
    mortality/survival rate of, 254-255
    predators of, in no-till systems, 102
    safe sites for, 98,99-100
Seed production, by weeds, 254
    effect of crop interference on, 62
    crop suppression of, 61
    economic threshold densities of,
        47-50
    prevention of. *See* Thresholds,
        no-seed

in sub-threshold populations, 17-18
weed density relationship of,
    221-222
Segregant analysis, of allelopathic
    potential, 84
Self-pollination, 142
SELOMA (weed management model),
    208
*Senecio vulgaris* (groundsel), 41,105
Sensor techniques, use in
        species-specific weed
        management, 228-229,230
*Setaria*, caryopses of, fungal
        colonization of, 101,130-131
*Setaria italica* (foxtail millet), 78
Sethoxydim, 193,199,200,201,202,
    212,216
Shattercane (*Sorghum bicolor*), 99,130
Sickle pod (*Cassia obtusifolia*), 127
Simulation models
    of crop-weed competition, 20-21
    use in weed management decision
        making, 234
*Sinapis arvensis* (wild mustard),
    109,110
Single seed descent method, use in
        weed competitive crop
        breeding, 69
Site-specific weed management,
    71,249
    definition of, 226
    knowledge-based decision support
        strategies (KBDSS) for,
        225-258
    fundamentals of, 227-232
    goal of, 232
    implementation of, 232,233
    knowlege incorporation into,
        233-235
    wisdom incorporation into,
        235-236
Smartweed (*Polygonum*), 128
Smartweed, Pennsylvania (*Polygonum
    pensylvanicum*), 196,199,214
Smother crops, 78
Smuts, 127

Soil
  moisture content of
    cover crop-related depletion of,
      111
    effect on field working days,
      181
    effect on seed mortality, 129
    in surface-managed residue
      systems, 132
    effect on weed seed banks,
      129-131
  pH of, 104
  quality of, 96
  salinity of, 96,104
  structure of, effect on weed seed
    germination, 132
  temperature of, 129,130
  types of, effect on weed community
    distribution, 255-256
Soil erosion, 64,68,78,96,106
Soil management, weed-related,
    10,95-101
  Maine Potato Ecosystem Project
    study of, 111-113,114
  objectives of, 114
  strategies for
    cover crops and green manures,
      106-113,114
    crop rotation, 111-113,114
    fertility management, 104-106
    tillage, 101-104,114
    unamended vs. amended soil
      management systems,
      111-113,114
  weed population dynamics and,
    97-100
Soil microorganisms, in weed
    management, 123-138
  interactions of, 126-129
  location in soil environment,
    125-126
  in no-till cropping systems,
    131-132
  effect on weed seed banks,
    129-131,132

*Solanum ptycanthum* (eastern black
    nightshade), 21-22,127-128
*Solanum tuberosum. See* Potato
Sorghum
  germplasm of, allelopathic
    potential of, 82
  as smother crop, 78
Sorrell, red (*Rumex acetosella*), 259
South Carolina, witchweed
    management in, 256
Soybeans
  glyphosate-tolerant, 260
  integrated weed management for
    effect on economic returns and
      risk, 172-176
    effect on grain yield, 169-172
    risk-efficient, 174-175,176
    timeliness of, 176-178
  mechanical weeding of, weed
      emergence time-based,
      192-204
    economics of, 195-196
    effect on net returns, 199-202
    effect on yield, 199-202
    relationship to herbicide use,
      199,200-202
  mulching of, 102
  mulch-related frost injury to, 79
  seed weight of, 67
  velvetleaf management in, 50
  weed competitiveness of, 62,66,68
  weed economic thresholds for,
    32,34,37
  weed management of
    biocontrol agents for, 127
    bioeconomic models of,
      248-249
    cost of, 173-174
    herbicide vs. mechanical
      cultivation, 169-172
    multiple weed control
      techniques for, 168
  weed threshold levels in, 12,16
  yield
    economic returns and risk of,
      172-176

effect of integrated weed
management on, 169-172
yield loss, weed-related, 18,66
SOYHERB (weed management
model), 208
Spatial statistical analysis, 71
Species shifts, of weeds, 71
in conservatively-tilled land,
131-132
of perennial weeds, 243
in weed seed banks,
148-149;153-154
Spores, mycorrhhizal fungal, 100
Spotted knapweed (*Centauria
maculosas*), 128
Sprayers
patch, 249
"smart" (weed-detecting), 54
Stale seedbed, 247
Standard weed units, 15
*Stellaria media* (chickweed), 14,102-103
Stem rot, 127
Stranglervine (*Morrenia odorata*), 127
Streptomyces, *pat/bas* gene of, 85-86
*Striga asiatica* (witchweed), 256
Stromatization, of seeds, 126
Sugarbeets (*Beta vulgaris*)
barnyardgrass management in,
47,48-50,53-54
weed economic thresholds for,
34,37,47,48-50,53-54
weed threshold density in, 13
yield loss of, prediction of, 21
Sumac (*Ruhus*), 243
Sunflower (*Helianthus annus*), 12,37
Survival, off-season, weed-arthropod
comparison of, 39,42
Survival strategies, of annual weeds,
254


*Taraxacum officinale* (dandelion),
128,214
Temperature sensitivity, of crops, in
multiple cropping systems, 68
Thifensulfuron, 212,216

Thiram, 130
Thistle
Canada (*Cirsium arvense*), 128
musk (*Carduus nutans*), 248
Thresholds, 9-29,31-58
action, 36
competitive, 11,35
contraindications for, 21-22
definitions of, 11,31,35-38
ecological implications of, 31-58
economic, 11,19,32
in cereal crops, 12
comparison with arthropod
thresholds, 32,33,38-44
critique of, 31-58
definitions of, 35-38
models of, 183
population dynamics studies of,
46-50
population ecology factors in,
38-44
relationship to
herbicide-resistant weed
development, 34,45-46
relationship to seed longevity,
175
relationship to weed invasions,
34,50-51
relationship to weed seed
production, 47-50
single-season, 32,34,46,48-49,50
in weed management decision
making, 234
economic injury level,
34-35,40,41,43,44,168
economic optimum, 17-18,36,37,44
guidelines for application of, 21-22
as integrated weed management
system component, 9-10
limitations of, 11
mechanistic models of, 20-21
for multiple weed species, 13-16
no-seed, 36,38,43,44-54
cost of, 44,45,46,47-48,51-52,53
farm examples of, 51-53
predictive, 36,37

regression modeling of, 44-45
safety, 36
for single weed species, 11-12
stability of parameter estimates for,
 18-20
statistical, 35
visual, 36
weed competitive indices for, 14-16
weed emergence times and, 16-17
zero, 38
Tillage, as weed management
 technique, 150-151,235,247.
 *See also* No-till cropping
 systems; Reduced-tillage
 cropping systems
effect on soil properties, 101-104
timing of, 150-151,155
weed control efficacy of, 63
efect on weed-crop competition, 68
effect on weed seed banks, 145,158
for weed seed control, 22,152
Tillering, as weed competitive
 characteristic, 62
Time density equivalents, 182-183
Tolerance, of crops, to weeds, 61,66,
 152,252
Tomato (*Lycopersicon esculentum*)
barnyardgrass weed management
 in, 47,53-54
weed economic thresholds for,
 34,37,47,53-54
weed interference with, 105
Tomato fruitworm, economic
 thresholds for, 41
Total competitive loads, 15
*Toxicodendron* (poison ivy), 243
Triazine, 248
*Trichosirocalus horridus*, 248
Trifluralin, 211,212
*Trifolium alexandrinum* (berseem
 clover), 53
*Trifolium incarnatum* (crimson clover),
 as green manure, 107-109
*Trifolium pratense* (red clover), as
 green manure, 109, 110
*Trifolium subterraneum* (subterranean
 clover), 253

*Triticum aestivum. See* Wheat
Trophic levels, weed-arthropod
 comparison of, 38,39,40

Unit production ratio, 15
University of California at Davis, 48
University of Maine Cooperative
 Extension, 111
University of Minnesota Southwest
 Experimental Station, 180
Urea-ammonium nitrate fertilizer, 112

Variable rate controllers, 226
Velvetleaf (*Abutilon theophrasti*), 257
bioeconomic model-based control
 of, 209
biological control of, 128,130,131
economic thresholds for, 16,37,47,
 50,209
eradication of, 151
herbicide control of, modeling
 approach to, 183
no-seed thresholds for, 47,53-54
population changes in, 259
population dynamics model of, 47
seed survival rate of, 254-255
threshold parameter stability of,
 18-19
*Vicia faba* (faba bean), 14
*Vicia villosa* (hairy vetch), 112-113
Vigor, early, as weed competitive trait,
 69
Vines, growth habits of, 62
Virginia, velvetleaf and giant foxtail
 control in, 257
Viruses
pathogenic, insect-assisted
 transmission of, 128
as seed decay cause, 126

Water management, as weed
 management technique, 247
Water quality, effect of herbicides on,
 64,68,235

Weed(s). *See also* individual weed
    species
  beneficial, in weed seed banks,
      154-155
  density-dependent mortality of, 254
  distribution of, patchiness of, 45
  economic costs of, 2,242
  ecosystem roles of, 258
  environmental interactions of, 241
  surface-residing, in no-till systems,
      132
*WeedCast* software, 195,197-199
Weed competitive indices, 14-16
Weed control
  comparison with integrated weed
      management, 2,168
  conceptual basis of, 244
  effect on ecosystems, 241
  goal/focus of, 244
  as integrated weed management
      system component, 10
  as level I of integrated weed
      management, 242-244,246
  limitations of, 242-243,246
  main objective of, 242
  methods of, 244
  spatial scale of, 242,244
  temporal scale of, 242,244
Weed control science and technology,
    2,3
Weed eradication, 151,260-261
Weed escapes, thresholds for, 17,21
WEEDIR (weed management model),
    208
Weed management, 2
  alternative systems in, 3-4
    development of, 4-5,6
  conceptual basis of, 244
  expanded context of, 1-7
  goals/focus of, 2,3-4,244,246,248,
      260-261
  hierarchy of, 241
  as level II of integrated weed
      management, 244,246-250
  limitations of, 250
  long-term effects on weed
      populations, 259

  long-term strategies of, 4-5
  methods of, 244
  multiple technology approach in,
      246,247,248,249-250
  research questions in, 244
  of seed banks, 149-155
  short-term strategies in, 4,5
  site-specific, 71
  spatial scale of, 244,246,248
  systems approach to, 10
  temporal scale of, 244
  theoretical basis of, 249
  weed control versus, 2
Weed removal, early, 66-67
Weed resistance, of cropping systems,
    252-253
WEEDS (weed management model),
    208
Weed science principles, 2-3
Weed science research, categories of,
    2-3
WEEDSIM (bioeconomic model), 234
Weed suppression, goal of, 261
Weed thresholds. *See* Thresholds
Weight, of weeds, as competitive
    index basis, 15
West Central Experiment Station,
    University of Minnesota,
    193,195
Wheat
  competitive index for, 15
  weed thresholds for, 12
  winter, competition with wild oats,
      21
Whiteflies (*Bemisia*), economic
    thresholds for, 41
Willowherb (*Epilobium*), 41
Wilt diseases, vascular, 126,127
Witchweed (*Striga asiatica*), 256
Woolly cupgrass (*Eriochloa villosa*), 154

*Xanthium strumarium*. *See* Cocklebur,
    common

Yellow starthistle (*Centaurea
    solstitialis*), 229-230

Yield
  of crop plants
    interference and, 66
    tolerance and, 66
    as weed competitive trait, 69
    weed density and, 242
  of weeds, 143
Yield loss
  as function of weed pressure, 10
  multiple weed species-related,
    13-14

weed-arthropod comparison of,
    39,42
  weed density and, 11-12,100
  weed emergence time and, 16-17
  weed interference and, simulation
    models of, 21
  weed thresholds and, 18-20
Yield monitors, 71,226


*Zea mays. See* Corn

T - #0499 - 101024 - C0 - 229/152/16 - PB - 9781560220633 - Gloss Lamination